First Edition

Marine Fire Fighting

Barbara Adams
Project Manager/Editor

Contributing Authors:
John F. Lewis
John Lewis & Associates, Inc.

Donald Merkle
Maritime Institute of Technology and Graduate Studies

Jon Swain
Texas A&M University System

Robert B. Wright
University of Maryland

Validated by the International Fire Service Training Association
Published by Fire Protection Publications, Oklahoma State University

Cover photo courtesy of R. Wright/Maryland Fire and Rescue Institute/United States Coast Guard.

The International Fire Service Training Association

The International Fire Service Training Association (IFSTA) was established in 1934 as a "nonprofit educational association of fire fighting personnel who are dedicated to upgrading fire fighting techniques and safety through training." To carry out the mission of IFSTA, Fire Protection Publications was established as an entity of Oklahoma State University. Fire Protection Publications' primary function is to publish and disseminate training texts as proposed and validated by IFSTA. As a secondary function, Fire Protection Publications researches, acquires, produces, and markets high-quality learning and teaching aids as consistent with IFSTA's mission.

The IFSTA Validation Conference is held the second full week in July. Committees of technical experts meet and work at the conference addressing the current standards of the National Fire Protection Association and other standard-making groups as applicable. The Validation Conference brings together individuals from several related and allied fields, such as:

- Key fire department executives and training officers
- Educators from colleges and universities
- Representatives from governmental agencies
- Delegates of firefighter associations and industrial organizations

Committee members are not paid nor are they reimbursed for their expenses by IFSTA or Fire Protection Publications. They participate because of commitment to the fire service and its future through training. Being on a committee is prestigious in the fire service community, and committee members are acknowledged leaders in their fields. This unique feature provides a close relationship between the International Fire Service Training Association and fire protection agencies which helps to correlate the efforts of all concerned.

IFSTA manuals are now the official teaching texts of most of the states and provinces of North America. Additionally, numerous U.S. and Canadian government agencies as well as other English-speaking countries have officially accepted the IFSTA manuals.

ISBN 0-87939-177-4   *Library of Congress # 00-101804*

First Edition, First Printing, February 2000   *Printed in the United States of America*
10   9   8   7   6   5   4

If you need additional information concerning the International Fire Service Training Association (IFSTA) or Fire Protection Publications, contact:

Customer Service, Fire Protection Publications, Oklahoma State University
930 North Willis, Stillwater, OK 74078-8045
800-654-4055   Fax: 405-744-8204

For assistance with training materials, to recommend material for inclusion in an IFSTA manual, or to ask questions or comment on manual content, contact:

Editorial Department, Fire Protection Publications, Oklahoma State University
930 North Willis, Stillwater, OK 74078-8045
405-744-4111   Fax: 405-744-4112   E-mail: editors@ifstafpp.okstate.edu

Table of Contents

Preface

It was evident to many individuals in the marine community that written training materials for shipboard fire fighting were lacking and that existing materials were sorely out of date. In 1995, the Technical Committee on Fire Service Training of the National Fire Protection Association (NFPA) responded to this need by preparing NFPA 1405 *Guide for Land-Based Fire Fighters Who Respond to Marine Vessel Fires*. As a result of the efforts of NFPA 1405 Committee Chair and International Fire Service Training Association (IFSTA) Executive Board member John W. Hoglund and others from the Maryland Fire and Rescue Institute, the Maritime Institute of Technology and Graduate Studies, and the U.S. Coast Guard, the IFSTA Executive Board recommended development of a marine fire fighting training manual. Early in 1995, Gene Carlson, IFSTA representative on the NFPA 1405 Committee and Fire Protections Publications (FPP) staff member, presented the proposed scope and purpose statement for the **Marine Fire Fighting** manual to the IFSTA Executive Board. A validation committee was formed, and the first committee meeting was held at the July 1995 IFSTA Validation Conference.

The proposed manual was to be written for mariners having commercial shipboard duties that included fire fighting and other emergency operations. The manual would cover all subjects pertaining to shipboard fire fighting and the requirements of maritime regulatory organizations such as the U.S. Coast Guard, Canadian Coast Guard, and International Maritime Organization (IMO). The manual was to improve the basic and advanced knowledge and skills of shipboard crew members and also provide information to land-based fire fighting personnel.

At its first meeting, the IFSTA Marine Fire Fighting committee recognized a need for two manuals: one for shipboard fire fighting and a second for land-based firefighters who respond to shipboard fires. This first edition of **Marine Fire Fighting** is the first book resulting from these endeavors, and it addresses shipboard fire fighting.

A dedicated and hard-working IFSTA committee compiled and validated this **Marine Fire Fighting** manual. FPP gratefully acknowledges and thanks these committee members for their continuing input and diligent work.

IFSTA Marine Fire Fighting Validation Committee

Chair
John F. Lewis
John Lewis & Associates, Inc.
Surrey, British Columbia, Canada

Vice Chair
Jon Swain
Texas A&M University System
College Station, TX

Secretary
Les Omans
California Maritime Academy
San Jose, CA

Committee Members

J. David Badgett
Los Angeles City Fire Department
Los Angeles, CA

V. Frank Bateman
Boots & Coots
Martinez, CA

S. W. (Stan) Bowles
BowTech Maritime Consultancy, Inc.
North Vancouver, British Columbia, Canada

Gaylen Brevik
Alaska Department of Public Safety
Juneau, AK

William J. (Bill) Guido
New York City Fire Department Marine Division
Staten Island, NY

Leto L. Lanius III
International Training Consultants
Oroville, WA

Donald Merkle
Maritime Institute of Technology and Graduate
 Studies
Linthicum, MD

Robert E. "Smokey" Rumens
Virginia Beach Fire Department (retired)
Ocala, FL

Steven Stokely
Lamar University Institute of Technology
Beaumont, TX

Mark Turner
Marine Institute
St. John's, Newfoundland, Canada

Robert B. Wright
Maryland Fire and Rescue Institute
College Park, MD

Special recognition is given to John F. Lewis, Donald Merkle, Jon Swain, and Robert B. Wright — the Marine Fire Fighting Committee Subgroup — who contributed their time and writing talents to provide Fire Protection Publications with all the finishing details that were needed to complete this manual. The sponsoring organizations of the Marine Subgroup are commended for their contributions. Special credit goes to John W. Hoglund, who was instrumental in creating the subgroup and encouraging the manual's completion.

During the completion phase of this manual, several persons provided valuable technical assistance to the Marine Subgroup members and to Fire Protection Publications. Their technical expertise was invaluable to this project and very much appreciated. Special thanks also go to the military and civilian employees located at the U.S. Coast Guard YARD (including the YARD Fire Department, YARD Safety Office, and Activities Baltimore) in Baltimore, MD, for their continuing support of IFSTA Marine Fire Fighting Committee members.

Richard E. Bordner
WorldWide Consulting Service, Inc.
Boutte, LA

Howard Chatterton
Maryland Fire and Rescue Institute
College Park, MD

Colm Currivan
Transport Canada
Surrey, British Columbia, Canada

Matthew T. Gustafson
Marioff, Inc.
Baltimore, MD

Jim Henry
U.S. Coast Guard
Baltimore, MD

William M. Riley
American Admiralty Bureau, Ltd.
Bowie, MD

Albert G. Kirchner, Jr.
U.S. Coast Guard
Arlington, VA

Robert J. Schappert, III
Maryland Fire and Rescue Institute
College Park, MD

Robert D. McMenamin
U.S. Coast Guard
Baltimore, MD

Michael Stalzer
Young & Stalzer Consulting Engineers
Stillwater, OK

Brooks A. Minnick
U.S. Coast Guard
Baltimore, MD

Barry F. VanVechten
Calhoon M.E.B.A. Engineering School
Easton, MD

Many thanks go to the Marine Subgroup and their sponsoring organizations for providing most of the photographs and ideas for illustrations for this manual. Committee members Robert E. "Smokey" Rumens and Mark Turner also contributed photographs along with U.S. Coast Guard Engineer, Thomas F. Kierman. Fire service contributors were Luke Carpenter and Aaron Hedrick of the Seattle (WA) Fire Department and former FPP staff member David Ward. We are also grateful for permissions for illustrations from the Fire Service College, International Maritime Organization, Fire Protection Association Australia, Alpha Omega Instruments Corporation, and Marioff, Inc., Finland. Special recognition goes to Robert B. Wright for responding quickly to every photo request and content question and for reviewing every illustration before publication.

Many people and organizations contributed during the early development stage of this manual, and we acknowledge their many contributions.

Tom S. Anderson
American Steamship Company
Williamsville, NY

Bob Fenner
Fire Service College
Morton-In-Marsh, England

Jamie Ballester
Maritima Jovellanos
Gijon/Asturias, Spain

Rich E. Fredericks
Smit Americas, Inc.
Baltimore, MD

Matthias Borchert
Hannover, Germany

Tom Henning
Emergency Response Group
Brea, CA

Joseph A. Cafasso, Jr.
North American Group
Carteret, NJ

Forest Herndon, Jr.
Military Sealift Command
Freehold, NJ

W. D. (Dave) Cochran
Boots & Coots
Houston, TX

Janice A. Kenefick
Marine Fire & Safety Training
Surrey, British Columbia, Canada

Steven T. Edwards
Maryland Fire and Rescue Institute
College Park, MD

Eric Lavergne
Williams Fire & Hazard Control
Mauriceville, TX

Michael J. Romstadt
Great Lakes Fire Training Center
Swanton, OH

Joseph J. Leonard, Jr.
U.S. Coast Guard
Galena Park, TX

Paul Smith
Justice Institute
New Westminster, British Columbia, Canada

Klaus Maurer
Koeln, Germany

Ted Thompson
Washington, DC

Steven News
Donjon Marine Company, Inc.
Hillside, NJ

David W. Ward
Insurance Services Office
Springtown, TX

Dan Norton
U.S. Coast Guard
Corpus Christi, TX

Jim Walsh
Carnival Corporation
Miami, FL

Last, but certainly not least, gratitude is also extended to the members of the Fire Protection Publications Marine Fire Fighting Project Team whose contributions made the final publication of this manual possible.

Marine Fire Fighting Project Team

Project Manager/Editor
Barbara Adams, Associate Editor

Technical Reviewer
Michael D. Finney, Senior Fire Protection
  Publications Editor

Production Coordinator
Don Davis, Coordinator, Publications
  Production

Proofreader
Marsha Sneed, Associate Editor

Illustrators and Layout Designers
Ann Moffat, Graphic Design Analyst
Desa Porter, Senior Graphic Designer

Draft Editor
Carol M. Smith, Senior Editor

Graphics Assistants
Ben Brock
Tara Carman

Research Technicians
Bob Crowe
Brian Charlesworth
Jack Krill
Lee Noll
Mike Krebs

Freelance Artists
Matt Gambrell
Tony Storm

Introduction

Of all the perils at sea, one of the most frightening is fire. Difficult to deal with and devastating in its effects, fire at sea leaves the mariner caught between two unforgiving elements. Preventing and extinguishing fire have been concerns of sailors since they first put to sea in wooden or hide-covered boats. **Marine Fire Fighting,** first edition, is for mariners having fire prevention and tactical fire fighting duties, but all who work aboard vessels may benefit from this text. The contents are directed to mariners from entry level to the highest level of management (commander, captain, or master). Seafarers require training and certification in shipboard fire fighting and fire prevention endeavors.

Through the ages, man has used the waterways of the world for transportation. Rivers, lakes, seas, and oceans have been used for exploration and cargo and passenger movement. Seafarers faced many dangers and storms as they traveled the world. Even though cargo and passenger movement are still the major reasons for sea travel, many modern-day passengers travel for pleasure and personal exploration. Shipping is vital to the world's commerce today. Modern seagoing vessels are more sophisticated and the science of meteorology is advancing, but dangers are still present.

No other transportation mode has used every type of propulsion possible. The seafarer has used manpower rowing, sails, and mechanical types of power. Every type of fuel has been used: wood, coal, hydrocarbon, and nuclear energy. Navigation was accomplished by simple guesswork in the beginning. The Polynesians in the South Pacific used wave charts made of sticks laid in patterns of wave direction with shells representing the islands to which they wanted to travel. Today, sophisticated electronic charts, integrated radar, and computerized satellite systems are available.

During World War II, the United States (U.S.) Merchant Marines (civilians transporting supplies and personnel to the war zones) lost move lives per capita than any other group involved in the war except the U.S. Marines. Merchant mariners traveled in unarmed vessels in many cases. When attacked by enemy forces, fires were inevitable. Fires onboard these vessels were fought with little more than guts and ingenuity and very minimal training and equipment (Figures I.1 a and b).

Protective clothing for the mariner from the 1940s and even on some vessels today consisted of rain gear and a hard hat. Breathing apparatus consisted of the oxygen re-breathing type known as oxygen breathing apparatus (OBA) or Chemox. The quantities of this equipment ranged from two to four sets per vessel. The situation today on some vessels is better. Some mariners have National Fire Protection Association (NFPA) approved personal protective equipment, positive-pressure self-contained breathing apparatus (SCBA), air cascade systems, and more in-depth training.

There are now international regulations that require all mariners to receive a minimum of two days of fire fighting training. This training must include hands-on live fire exercises. All seagoing vessels are required to hold regular drills for training and exercising emergency response skills. Some countries and shipping companies have come a long way to improve a vessel's fire fighting capabilities. In the United States, some companies, trade unions such as the International Organization of Masters, Mates, and Pilots International Union, Marine Engineers Benevolent Association (District One), the Seamen's International Union (SIU), and others have provided better training and equipment consistent with NFPA standards.

 ## Scope

This book provides training and skills development for all who work on the waters and seas of the world. Whether tug or barge, fishing vessel or cruise ship, oil rig or deep-sea vessel, the need to prevent or contain and extinguish fires is essential. Every year lives are lost at sea from sinking, collision, fire, or other causes. When on board, vessel personnel must be their own fire department, police, and ambulance. Covered within these pages are the requirements of the International Maritime Organization (IMO), Standards of Training, Certification, and Watchkeeping for Seafarers (known as STCW 95), U.S. Coast Guard (Basic and Advanced Fire Fighting), Title 46 U.S. Code of Federal Regulations (CFR), Marine Emergency Duties (MED) – Fire Fighting Canada Shipping Act, and other port state control organizations.

 ## Purpose

The best fire training uses a combination of well-written materials, shoreside live fire training, and meaningful drills conducted while working in the maritime environment.

Mariners need up-to-date training for shipboard fire fighting and fire prevention endeavors. **Marine Fire Fighting** covers all aspects of fire prevention and suppression performed by vessel personnel aboard, whether sailing deep-sea, coastal, or inland waters. It covers areas from the basics to command. Examples of various aspects of marine fire fighting may be found in historically significant cases given in Appendix A, Case Histories. References for more detailed information on related subjects are given in Appendix B, References and Supplemental Readings. Some of the major topics covered in this manual include the following:

- Chemistry of fire
- Fire prevention
- Extinguishing agents and fire fighting equipment
- Fire detection and suppression systems
- Shipboard fire fighting evolutions and training
- Fire fighting strategies for various types of vessels and incidents
- Tactical applications during emergencies
- Guidance for those directing and commanding shipboard damage control teams
- Interface with land-based fire fighting emergency service personnel and port state control authorities

Figures I.1 a and b Equipment and procedures are different, but the courage is unchanged (a) *SS Pennsylvania Sun*, torpedoed on July 16, 1942. *Courtesy of Maritime Institute of Technology and Graduate Studies.* (b) Modern mariner firefighter dressed in personal protective equipment carrying a low-velocity fog applicator. *Courtesy of Maritime Institute of Technology and Graduate Studies.*

Chapter 1

Fire Science and Chemistry

We first learn about fire as children. Fire heats our homes, cooks our food, and generates power, but in its hostile mode, fire is dangerous. We know that fire consumes fuel, needs air, and gives off heat and light. Normally, that degree of understanding is all that one needs. Vessel crew members are responsible for fire fighting in addition to regular daily assignments, so they must have a more in-depth understanding of fire in order to carry out the duties of fire prevention and fire suppression. In particular, they must know more about the chemical process, the methods of heat transfer, the composition and nature of the fuels (materials that burn), the environment a fire needs, and the characteristics of fire development. It is this knowledge that enables a crew member to prevent and fight fires effectively.

Fire is actually a by-product of a larger process called combustion. Fire and combustion are two words used interchangeably by most people; however, crew members should understand the difference. *Combustion* is the self-sustaining process of rapid oxidation (chemical reaction) of a fuel, which produces heat and light. *Fire* is the result of this rapid oxidation reaction. The oxygen in air most commonly oxidizes fuels. The normal oxygen content of air is 21 percent, 78 percent of air is nitrogen, and the remaining 1 percent is composed of trace amounts of other elements. A fire must be oxidized either by oxygen or by some other chemical compound or mixture that contains oxygen. Some fuels do not need oxygen because they have their own oxidizers bound in their chemical formulas. The root letters *oxy* or *ox* in their names (for example, *organic peroxide*) can sometimes identify these fuels.

Combustion was defined as the process of rapid oxidation (resulting in fire), but oxidation is not always rapid. It may be very slow or it may be instantaneous. Neither of these extremes produces fire (combustion) as we know it, but they are common occurrences in themselves. Very slow oxidation is more commonly known as *rusting* or *decomposition*. A light film of oil placed on metal prevents rusting by keeping air and its oxygen from the metal so that it cannot react and oxidize. Instantaneous oxidation is an *explosion* such as what occurs inside the casing of a bullet cartridge when the primer is ignited. The speed of the oxidation process determines the rate of released heat and the violence of the reaction. The presence of an oxidizer can accelerate a fuel that would normally burn slowly into an explosion.

This chapter discusses fire terminology and the causes and behaviors of fire so that a crew member can better understand how to prevent and suppress fire. Sources of heat energy, heat transfer, and principles of fire behavior and fire development help identify fire prevention methods. Knowledge of fire suppression considerations, fire extinguishment theory, and classification of fires also aid in fire suppression techniques.

◆ Terminology

In order to understand the material in this chapter and other parts of this manual, it is important that the reader know some basic terms and measurements that pertain to fire science and chemistry.

Liquids and Gases

◆ **Combustible liquid** — Any liquid that gives off flammable vapors at or above 80°F (26.7°C) according to the U.S. Coast Guard rating on engine fuel. However, the definition by the National Fire Protection Association (NFPA) uses a temperature of 100°F (37.8°C) and above for shoreside situations, while the U.S. Department of Transportation (DOT) and the International Maritime Organization (IMO) use the temperature of 143°F (61.7°C) and below when rating cargo. The maritime definition varies from how it is defined in shoreside industries and fire fighting, but the important issues are that combustible liquids are hazardous and volatile.

◆ **Flammable liquid** — Any liquid that gives off flammable vapors at or below 80°F (26.7°C) according to the U.S. Coast Guard rating on engine fuel. However, the definition by NFPA uses a temperature of 100°F (37.8°C) and below for shoreside situations, while the U.S. DOT and IMO use the temperature of 143°F (61.7°C) and below when rating cargo. The maritime definition varies from how it is defined in shoreside industries and fire fighting, but the important issues are that flammable liquids are hazardous and volatile.

◆ **Fluid** — Any substance that can flow; a substance that has definite mass and volume at a constant temperature and pressure but no definite shape. A fluid is unable to sustain shear stresses.

◆ **Gas** — Compressible substance with no specific volume that tends to assume the shape of its container; a fluid (such as air) that has neither independent shape nor volume but tends to expand indefinitely. The term gas is most accurately used to describe the state of a pure gaseous substance (for example, propane), rather than a fume, vapor, or mixture of gases.

◆ **Liquid** — Incompressible substance that assumes the shape of its container. Molecules flow freely, but substantial cohesion prevents them from dispersing from each other as a gas would.

◆ **Vapor** — Any substance in the gaseous state as opposed to the liquid or solid state. Vapors result from the evaporation of a liquid such as gasoline or water.

 Temperature and Heat

◆ **Autoignition temperature** — Lowest temperature at which a combustible material ignites in air without a spark, flame, or other source of ignition.

◆ **Boiling point** — Temperature of a substance when the vapor pressure (measure of a substance's tendency to evaporate) exceeds atmospheric pressure. At this temperature, the rate of evaporation exceeds the rate of condensation. At this point, more liquid is converting into vapor than vapor is converting back into a liquid.

◆ **Fire point** — Temperature at which liquid fuels produce vapors sufficient to support continuous combustion once ignited by an outside ignition source. The fire point is usually a few degrees above the flash point.

◆ **Flammable or explosive limits (flammable ranges)** — Minimum and maximum percentages of a substance (vapor) in air that burns once it is ignited. Most substances have an upper (too rich) flammable limit (UFL) and a lower (too lean) flammable limit (LFL).

◆ **Flash point** — Minimum temperature at which a liquid fuel gives off sufficient vapors to form an ignitable mixture with the air near the surface. At this temperature the ignited vapors flash but do not continue to burn. The flash point of a fuel is usually a few degrees below the fire point, but the flash point and fire point of some fuels are almost indistinguishable.

◆ **Heat** — Form of energy that raises temperature. Heat can be measured by the amount of work it does; for example, the amount of heat needed to make a column of mercury expand inside a glass thermometer.

◆ **Ignition temperature** — Minimum temperature to which a fuel in air must be heated to start self-sustained combustion independent of an outside ignition source.

Pressure and Weight

◆ **Atmospheric pressure** — Force exerted by the weight of the atmosphere at the surface of the earth. See sidebar for more information.

◆ **Density** — Weight per unit volume of a substance. The density of any substance is obtained by dividing the weight by the volume.

◆ **Specific gravity** — Weight of a substance compared to the weight of an equal volume of water at a given temperature. A specific gravity of less than 1 indicates a substance lighter than water; a specific gravity greater than 1 indicates a substance heavier than water.

◆ **Specific weight** — Established weight per unit volume of a substance. For example, the specific weight of pure water is generally accepted to be 62.4 pounds per cubic foot (9.81 kilonewtons per cubic meter). The specific weight of seawater is given as 64 pounds per cubic foot (10.1 kilonewtons per cubic meter).

◆ **Vapor density** — Weight of a given volume of pure vapor or gas compared to the weight of an equal volume of dry air at the same temperature and pressure. A vapor density of less than 1 indicates a vapor lighter than air; a vapor density greater than 1 indicates a vapor heavier than air.

◆ **Vapor pressure** — Measure of the tendency of a substance to evaporate; a higher value means it is more likely to evaporate, and a lower one means it is less likely.

Fuels and Reactions

◆ **Boiling liquid expanding vapor explosion (BLEVE)** — Situation in which a liquid in a container has been heated to above the point of boiling, resulting in rupture of the container and violent release of the vapor and contents; results in a mushroom-type fireball if the liquid is flammable or combustible.

◆ **Flame spread** — Progression of flame across a fuel surface away from the ignition source.

◆ **Fuel load** — Type (class) and amount of fuels in a given space.

◆ **Organic materials** — Substances containing carbon such as plant and animal materials and hydrocarbon fuels.

◆ **Oxidation** — Complex chemical reaction of organic materials with oxygen or other oxidizing agents resulting in the formation of more stable compounds. More stable compounds are simply those with less closely associated chemical energy. They become more stable by releasing some of their energy as heat and light during combustion. Examples are fire, explosions, and rusting (decomposition).

Measurements

◆ **British thermal unit (Btu)** — Amount of heat required to raise the temperature of one pound (lb) of water one degree Fahrenheit (F). 1 Btu = 1,055 joules (J).

◆ **Calorie (Cal)** — Amount of heat required to raise the temperature of one gram (g) of water one degree Celsius (C). 1 Cal = 4.187 joules (J).

◆ **Celsius (C) or centigrade** — Unit of temperature measurement in the International System of Units (SI). On the Celsius scale, 0 degrees is the melting point of ice; 100 degrees is the boiling point of water.

◆ **Fahrenheit (F)** — Unit of temperature measurement in the English or Customary System (primarily used in the United States). On the Fahrenheit scale, 32 degrees is the melting point of ice; 212 degrees is the boiling point of water.

◆ **Joule (J)** — Unit of work or energy in the International System of Units; the energy (or work) when unit force (1 newton) moves a body through a unit distance (1 meter).

- ◆ **Kilowatt (kW)** — Unit of power in the International System of Units. 1 kW = 1,000 watts (W).

- ◆ **Newton (N)** — Unit of force in the International System of Units. 1 newton = 1 kilogram per meter per second squared.

- ◆ **Pounds per square inch (psi)** — Unit for measuring pressure in the English or Customary System. The International System of Units equivalent is kilopascal (kPa). Another equivalent unit of pressure is bar. 1 bar = 14.5038 psi. 1 bar = 100 kPa. This manual uses the units of psi as the primary measurement of pressure. Metric equivalent pressures expressed in bar and kPa are also given. See sidebar for more information.

 Psi Gauge or Psi Absolute

Unless stated otherwise, pressures given throughout this manual are gauge pressure — not absolute pressure. Therefore, units shown simply as psi indicate psig (pounds per square inch gauge). Engineers make the distinction between a gauge reading and actual atmospheric pressure. The notation for actual atmospheric pressure is psia (pounds per square inch absolute). Absolute zero pressure is a perfect vacuum. Any pressure less than atmospheric pressure is simply a vacuum. When a gauge reads -5 psig (-34.5 kPa) {-0.35 bar}, it is actually reading 5 psi (34.5 kPa) {0.35 bar} less than the existing atmospheric pressure (at sea level, 14.7 - 5, or 9.7 psia [101 - 34.5 or 66.5 kPa]) {1.01 - 0.345 or 0.665 bar}).

 Atmospheric Pressure

Atmospheric pressure is greatest at low altitudes; consequently, its pressure at sea level is used as a standard. At sea level, the atmosphere exerts a pressure of 14.7 psi (101 kPa) {1.01 bar}. A common method of measuring atmospheric pressure is to compare the weight of the atmosphere with the weight of a column of mercury: the greater the atmospheric pressure, the taller the column of mercury. A pressure of 1 psi (7 kPa) {0.069 bar} makes the column of mercury about 2.04 inches (52 mm) tall. At sea level, then, the column of mercury is 2.04 × 14.7, or 29.9 inches tall (759 mm). See the following diagram (Figure 1.1).

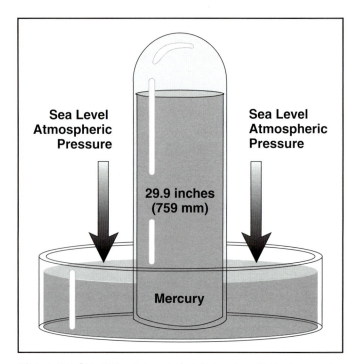

Sea Level Atmospheric Pressure

Sea Level Atmospheric Pressure

29.9 inches (759 mm)

 Mercury

Figure 1.1 Simple barometer showing atmospheric pressure.

◆ Sources of Heat Energy

Besides being a form of energy that raises temperature, *heat* may also be described as a condition of *matter in motion* caused by the movement of molecules. All matter contains some heat, regardless of how low the temperature, because molecules are constantly moving. When a body of matter is heated, the speed of the molecules increases; therefore, the temperature increases. Anything that sets the molecules of a material into faster motion produces heat in that material. It is important to understand how fuels become heated in order to reduce the possibilities of unwanted ignition and uncontrolled fire. The five general categories of heat energy are given in the following list and described in the sections that follow. Knowledge of these forms of energy is important in both fire prevention (how to prevent these energies from causing fires) and fire suppression (how best to extinguish particular fires).

* Chemical
* Electrical
* Mechanical
* Nuclear
* Solar

Chemical Heat Energy

Chemical heat energy is generated as the result of some type of chemical reaction. Four types of chemical reactions result in heat production: (1) heat of combustion, (2) spontaneous heating, (3) heat of decomposition, and (4) heat of solution. The following sections describe these chemical reactions.

Heat of Combustion

Heat of combustion is the amount of heat generated by the combustion or oxidation reaction. The amount of heat generated by burning materials varies, depending on the material. This phenomenon is why some materials (such as some cargoes or paneling in cabins) are said to burn "hotter" than others. For example, fiberglass burns hotter than wood.

Spontaneous Heating

Spontaneous heating is the heating of an organic substance without the addition of external heat. Some hazardous materials (marked as *pyrophoric, spontaneous combustible,* or *dangerous when wet*) undergo spontaneous heating when they contact air or water. Spontaneous heating occurs most frequently where sufficient air is not present and heat produced cannot dissipate — heat that is produced by a low-grade chemical breakdown process. An example would be oil-soaked rags that are rolled into a ball and carelessly thrown into a corner. If ventilation (air circulation) is not adequate for the heat to escape, eventually it becomes sufficient to cause ignition of the rags. Good housekeeping routines are important, especially in engine spaces.

A common example of heat production in confined spaces is found in cargo vessels loaded with iron filings. Oxidization of the filings confined within the hold of the vessel generates heat that cannot be dissipated because of its location. This heat is conducted to the hull and subsequently to the water outside the vessel. When the vessel is in motion, the heat is transferred to the water and goes unnoticed. When the vessel is stationary, however, the fact that heat is being conducted to the surrounding water becomes apparent when the water near the vessel begins to boil. While the temperature does not usually increase to the point that fire occurs, the condition can be quite dramatic.

Heat of Decomposition

Heat of decomposition is the release of heat from decomposing (decaying) compounds, usually due to bacterial action. In some cases, these compounds may be unstable and release their heat very quickly; they may even detonate. In other cases, the reaction and resulting release of heat is much slower. This reaction is commonly seen when viewing a compost pile. The decomposition of organic materials creates heated vapors and steam that can be seen on a cold day by jabbing holes into the pile. Wet brown (soft) coal loaded into a cargo hold undergoes a decomposition process that can result in a cargo-hold fire. A similar situation occurs when flooding in a lower peak hold soaks bales of cotton rags. After the water is pumped out of the hold, the bales begin to dry. Three months later, the bales are examined and found too hot to touch in the center. If left in place longer, they would ignite.

Heat of Solution

Heat of solution is the heat released by the solution (dissolving) of matter in a liquid. Some acids can produce violent reactions, spewing hot water and acid with explosive force when they are dissolved in water.

Electrical Heat Energy

It is not uncommon to see "electrical" listed as a cause of a fire. Electricity has the ability to generate temperatures capable of igniting combustible materials near the heated area. Electrical heating can occur in a variety of ways. The following sections highlight some of the more common methods: resistance heating, dielectric heating, leakage current heating, heat from arcing, and static electricity.

Resistance Heating

Resistance heating refers to the heat generated by passing an electrical current through a conductor such as a wire or an appliance. Resistance heating is increased if the conductor is not large enough in diameter for the amount of current flow. Fires may occur when several pieces of electrical equipment are connected by a single extension cord, causing an overloaded circuit.

Dielectric Heating

Dielectric heating occurs as a result of the action of pulsating either direct current (DC) or alternating current (AC) at high frequency on a nonconductive material. The material is not heated by the dielectric heating but is heated by being in constant contact with electricity. This is somewhat similar to bombarding an object with many small lightning bolts. Dielectric heating is used in microwave ovens.

Leakage Current Heating

Leakage current heating occurs when a wire is not insulated well enough to contain all the current, and some leaks out into the surrounding material such as behind the wall paneling of a compartment (enclosed space) or room. This current causes heat that may result in a fire.

Heat from Arcing

Heat from arcing is a type of electrical heating that occurs when the current flow is interrupted, and electricity "jumps" across an opening or gap in a circuit. Arc temperatures are extremely high and may even melt the conductor in an uncontrolled situation such as at the site of a loose connection or defective switch. A common arc used in industrial applications is the arc welder. In this case, the welding rod (conductor) melts as metals are joined together.

Static Electricity

Static electricity is the buildup of positive charge on one surface and negative charge on another. The charges are naturally attracted to each other and seek to become evenly charged again. This condition is shown when the two surfaces come close to each other — such as when a person's finger touches a metal doorknob — and an arc occurs, producing a spark.

Static electricity may cause fires when flammable liquids are being transferred between containers that are not properly electrically isolated. For this reason,

cargo hose strings and metal arms used for the shipboard transfer of flammable liquids (whether to or from shore or barge) are fitted with an insulating flange or a single length of nonconducting hose to ensure electrical discontinuity between the vessel and shore.

Heat generated by lightning is static electricity on a very large scale. The heat generated by the discharge of billions of volts from earth-to-cloud, cloud-to-cloud, or cloud-to-ground can be in excess of 60,000°F (33 316°C).

Mechanical Heat Energy

Mechanical heat is generated in two ways: friction and compression. *Heat of friction* is created by the movement of two surfaces against each other, which generates heat and/or sparks. On board a vessel, friction from the rubbing of adjacent parts of the vessel structure or objects (such as cargo) against the vessel or against each other may create enough heat for ignition. *Heat of compression* is generated when a gas is compressed. This principle explains how diesel engines ignite fuel vapor without a spark plug. It is also the reason that self-contained breathing apparatus (SCBA) cylinders feel warm to the touch after they are filled.

Nuclear Heat Energy

Nuclear heat energy is generated when atoms are either split apart (fission) or combined (fusion). In a controlled setting, fission is used to heat water to drive steam turbines and produce electricity. Currently, fusion cannot be controlled and has no commercial use.

Solar Heat Energy

Solar heat energy is the energy transmitted from the sun in the form of electromagnetic radiation. Typically, solar energy is distributed fairly evenly over the face of the earth and in itself is not capable of starting a fire. However, when solar energy is concentrated on a particular point such as through a lens, it may ignite combustible materials.

 ## Heat Transfer

A number of the natural laws of physics are involved in the transmission of heat. For example, the *Law of*

Heat Flow specifies that heat tends to flow from a hot substance to a cold substance. The colder of two bodies in contact absorbs heat until both objects are at the same temperature. Heat can travel throughout a vessel by one or more of four methods: conduction, convection, radiation, and direct flame contact (impingement). The following sections describe how each heat transfer process takes place.

Conduction

Conduction is the transfer of heat energy from one body to another by direct contact of the two bodies or by an intervening heat-conducting medium. An example of conduction is when a compartment fire heats pipes hot enough to ignite material inside adjacent spaces. The amount of heat that is transferred and its rate of travel depend upon the conductivity of the material through which the heat is passing. Not all materials have the same heat conductivity. Aluminum, copper, and iron are good conductors. Fibrous materials such as felt, cloth, and paper are poor conductors, thus they are heat insulators (materials that slow heat transfer by separating conducting bodies). Air is a relatively poor conductor, so sandwiched airspaces between bulkheads or portholes/windows provide additional insulation from outside air temperatures. Certain solid materials, such as fiberglass, that are shredded into fibers and packed into batts make good insulation because the material itself is a poor conductor and air pockets exist inside the batting (Figure 1.2).

Convection

Convection is the transfer of heat by the movement of heated air or liquid. When water is heated in a glass container, the movement within the vessel is observed through the glass. If sawdust is added to the water, the movement is more apparent. As the water is heated, it expands and grows lighter; hence, the upward movement. In the same manner, as air near a steam radiator becomes heated by conduction, it expands, becomes lighter, and moves upward. As the heated air moves upward, cooler air takes its place at the lower levels. When liquids and gases are heated, they begin to move within themselves. This movement is responsible for heat transfer by convection. Convection is the reason a hand held over a flame feels heat even though the hand is not in direct contact with the flame (Figure 1.3).

Because heated air expands and rises, fire spread by convection is mostly in an upward direction; however, air currents can carry heat in any direction. Convection currents are generally the cause of heat movement from deck to deck, compartment to compartment, and area to area. The spread of fire through passageways, through corridors, up stairwells, up pipe chases, up elevator shafts, through ventilation ducts, between bulkheads, and through void spaces/hidden

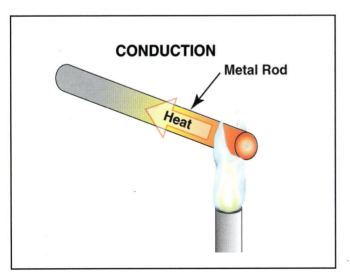

Figure 1.2 An example of conduction: The temperature along the rod rises because of the increased movement of molecules from the heat of the flame.

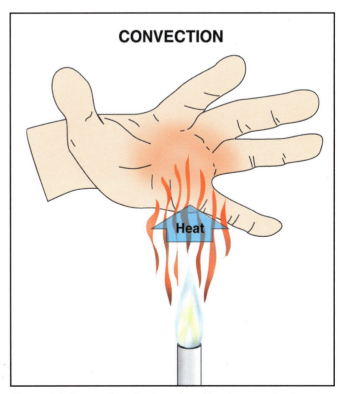

Figure 1.3 Convection: the transfer of heat energy by the movement of heated liquids or gases.

areas is caused mostly by the convection of heat currents caused by air and gases taking the path of least resistance.

If the convecting heat encounters a *deckhead* (also known as *overhead* or *ceiling*) or barrier that keeps it from rising, it spreads out laterally (sideways) along the deckhead or barrier. This phenomenon is commonly referred to as *mushrooming*. When the heat reaches the bulkheads, it is pushed by more heated air rising behind it and travels down the bulkheads toward the deck. Convection has more influence upon the positions for fire attack and ventilation than either radiation (see following section) or conduction.

Radiation

Although air is a poor conductor, it is obvious that heat can travel where matter does not exist. *Radiation* is the transmission of energy as an electromagnetic wave without an intervening medium. A hand held a few inches (millimeters) to the side of a flame feels heat by radiation (Figure 1.4). The heat of the sun reaches the earth even though it is not in direct contact with the earth (conduction), nor is it heating gases that travel to the earth (convection). The

sun's heat transmits by the radiation of heat waves. Heat and light waves are similar in nature, but they differ in length per cycle. *Heat waves* (sometimes called *infrared rays)* are longer than light waves. Radiated heat travels through space until it reaches an opaque object. As the object is exposed to heat radiation, it in return radiates heat from its surface. Radiated heat is one of the major sources of fire spread to exposures (fuel separate from an original fire to which fire could spread). Radiation is important as a source of fire spread and demands immediate attention at locations where radiation exposure is severe.

Direct Flame Contact (Impingement)

Direct flame contact (impingement) is a combination of heat transfer by convection and radiation at close range. When a substance is heated to the point where flammable vapors are released, these vapors may ignite, creating a flame. As other flammable materials come into contact with the burning vapors, or flame, they may be heated to a temperature where they, too, ignite and burn.

◆ Principles of Fire Behavior

Fuel may be found in any of three states: *solid, liquid,* or *gas.* Only vapors or gases burn. The initiation of combustion of a liquid or solid fuel requires its conversion into a gaseous state by heating. Fuel gases evolve from solid fuels by a *pyrolysis process (sublimation)* — the chemical decomposition of a substance through the action of heat. Fuel vapors evolve from liquids by *vaporization.* This process is the same for either water evaporating by boiling or water evaporating in sunlight. In both cases, heat causes the liquid to vaporize.

Generally, the vaporization process of liquid fuels requires less heat input than does the pyrolysis process for solid fuels. Control and extinguishment of gas fuel fires is difficult because reignition is much more likely. Gaseous fuels can be the most dangerous because they are in the natural state required for ignition. No pyrolysis or vaporization is needed to ready the fuel. These fuels are also the most difficult to contain.

Knowledge of fire behavior, fuels, and how fuel affects fire behavior can make the difference between success or failure in fire suppression. The following sections identify and explain fuel charac-

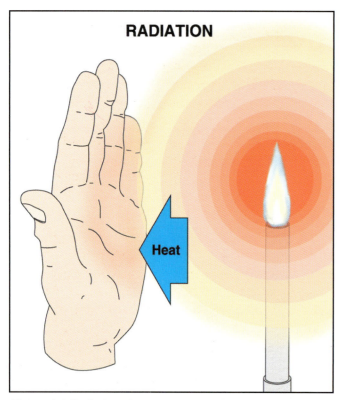

Figure 1.4 Radiation: the transmission of energy as an electromagnetic wave without an intervening medium.

teristics, explain fuel vapor-to-air mixtures, and discuss the burning process.

Fuel Characteristics

The degree of fuel flammability is determined by fuel characteristics that vary from space to space. Seafarers need an awareness of the hazards posed by different states and distributions of fuels. The shape, size, position, density, physical properties, water solubility, and volatility of fuels are important factors in the combustion process.

Shape and Size

Solid fuels have definite shapes and sizes that significantly affect their ignitability. A primary consideration is the *surface-to-mass ratio:* the ratio of the surface area of the fuel to the mass of the fuel. As the fuel particles become smaller and more finely divided (for example, sawdust as opposed to logs), this ratio increases; ignitability also increases tremendously. As the surface area increases, heat transfer is easier and the material heats more rapidly, thus speeding pyrolysis.

Position

The physical position of a solid fuel is also a great concern to fire fighting crew members. Fire spread is more rapid when a solid fuel is in a vertical position rather than a horizontal position. The rapidity of fire spread is due to increased heat transfer through convection as well as conduction and radiation.

Physical Properties

Liquid fuels have physical properties that increase the difficulty of extinguishment and the hazards to crew members. A liquid assumes the shape of its container. When a spill occurs on a vessel at sea, the liquid assumes the shape of the deck or platform (flat or cambered), and it flows and accumulates in low areas. Movement of the vessel causes the liquid to flow in different directions.

Density

The concept of density is an important fuel characteristic to understand. As mentioned earlier, *density* is a measure of how tightly the molecules of a substance are packed together (see Terminology section). Dense materials are heavy (for example, saltwater has a density of 1.025, whereas freshwater has a density of 1).

As also learned earlier, *specific gravity* is the density of liquids in relation to water (see Terminology section). Water is given a value of 1. Liquids with a specific gravity less than 1 are lighter than water, while those with a specific gravity greater than 1 are heavier than water. If the other liquid also has a density of 1, it mixes evenly with water (which is not the same as dissolving in water). Most flammable liquids have a specific gravity of less than 1. This means that if a crew member is confronted with a flammable liquid fire and flows water on it, the water may vaporize explosively (thus spreading the fire), or the whole fire can just float away on the flow of water and ignite everything in its path. Dry chemicals or foam is generally the extinguishing agent chosen for liquid fuel fires (see Fire Extinguishment Theory section).

Vapor density (the density of gas or vapor in relation to air) is a concern with volatile liquids and gaseous fuels (see Terminology section). Gases tend to assume the shape of their containers but have no specific volume. If a vapor is less dense than air (air is given a value of 1), it rises and tends to dissipate. If a gas or vapor is heavier than air, which is a more common situation, it tends to stay at the lowest level (decks) and travel as directed by sloping surfaces and the wind.

It is important for crew members to know that every base hydrocarbon except the lightest one, methane, has a vapor density greater than 1. Common gases such as ethane, propane, and butane are examples of hydrocarbon gases that are heavier than air. Natural gas (composed largely of methane) is lighter than air. A cryogenic liquid (for example, liquefied natural gas [LNG]) that vaporizes may produce a vapor that is initially heavier than air but that becomes lighter than air as it warms.

Water Solubility

The solubility of a liquid fuel in water is also an important factor in combustion. Alcohols and other polar solvents (such as paint thinners and acetone) dissolve in water (soluble). If large volumes of water are used, alcohols and other polar solvents may dilute to the point where they will not burn. As a rule, hydrocarbon liquids (nonpolar solvents) do not dissolve in water (insoluble), which is why water alone cannot wash oil from the hands. Soap must be used with water to dissolve the oil. Consideration must be given to which extinguishing agents are effective on fires involving hydrocarbons (insoluble) and which ones are effective on polar solvents and alcohols (soluble). Multipurpose foams are currently available that can control both types of liquid fuel fires (see Fire Extinguishment Theory section).

Volatility

The volatility or ease with which a liquid gives off vapors influences fire control. All liquids give off vapors in the form of simple evaporation. Liquids that give off large quantities of flammable or combustible vapors can be dangerous because they are easily ignited at almost any temperature.

Fuel Vapor-to-Air Mixture

For combustion to occur after a fuel has been converted into a gaseous state, it must be mixed with the oxygen (oxidizer) in air in the proper ratio — that is, within the flammable limits (see Terminology section). The flammable (explosive) range of a fuel is reported using the percent by volume of gas or vapor in air for the *lower flammable limit (LFL)* (the minimum concentration of fuel vapor and air that supports combustion) and for the *upper flammable limit (UFL)* (concentration above which combustion cannot take place). Concentrations that are below the LFL are *too lean* (not enough fuel vapor) to burn. Concentrations that are above the UFL are *too rich* (too much fuel vapor) to burn.

Table 1.1 presents the flammable ranges for some common materials. The flammable limits for combustible gases are presented in chemical handbooks and documents such as NFPA's *Fire Protection Guide to Hazardous Materials* and *Fire Protection Handbook*. The limits are normally reported at ambient (surrounding air) temperatures (generally 70°F or 21°C) and atmospheric pressures. Variations in temperature and pressure can cause the flammable range to vary considerably.

The Burning Process

In order for burning to occur, not only must the proper fuel vapor-to-air mixture be present, but also the material must be at its ignition temperature or the point where self-sustained combustion continues. However, an accumulation of vapors can occur even below fire or flash points in an enclosed container such as a barge.

Fire burns in two basic combustion modes: *flaming* and *smoldering*. Educators previously used the fire triangle to represent the smoldering mode of combustion, until it was proven that another factor was involved besides the presence of fuel, heat, and an oxidizer (oxygen). Scientific experimentation showed the presence of an uninhibited chemical chain reaction (flaming mode). To illustrate this more accurate description of the flaming mode of the burning process, the fire tetrahedron was developed. The *fire tetrahedron* includes the three parts of the old fire triangle (fuel, heat, and an oxidizer), but it adds the fourth dimension of an uninhibited chemical chain reaction (Figure 1.5). Remove any one of the four components, and combustion will not occur. The four components are briefly discussed in the sections that follow.

Fuel

The fuel segment of the diagram is any solid, liquid, or gas that can combine with oxygen in the oxidation chemical reaction (see Terminology section). A fuel with a sufficiently high temperature ignites if an oxidizing agent is liberated. Combustion continues as long as enough energy or heat is present. Under most conditions, the oxidizing agent is the oxygen in air. Some materials, such as sodium nitrate and potassium chlorate, release their own oxygen during combustion and can cause fuels to burn in an oxygen-free atmosphere.

Table 1.1
Flammable Ranges for Selected Materials

Material	Lower Flammable Limit (LFL)	Upper Flammable Limit (UFL)
Acetylene	2.5	100.0
Carbon Monoxide	12.5	74.0
Ethyl Alcohol	3.3	19.0
Fuel Oil No. 1	0.7	5.0
Gasoline	1.4	7.6
Hydrogen	4.0	75.0
Methane	5.0	15.0
Propane	2.1	9.5

Source: *Fire Protection Guide to Hazardous Materials*, 12th edition, 1997, by National Fire Protection Association.

Heat

A self-sustaining combustion reaction of solids and liquids depends on *radiative feedback:* radiant heat providing energy for continued vapor production. When sufficient heat is present to maintain or increase this feedback, the fire either remains constant or grows, depending on the heat produced. A *positive heat balance* occurs when heat is fed back to the fuel. A positive heat balance is required to maintain combustion. If heat is dissipated faster than it is generated, a *negative heat balance* is created.

Oxygen

The amount of oxygen available to support combustion is important. As stated previously, air contains about 21 percent oxygen under normal circumstances. However, crew members may frequently encounter situations where less than 21 percent oxygen is present (mostly because of iron oxidation — rusting). Oxygen-deficient conditions are most commonly found in empty ballast or cargo tanks, double bottoms or void spaces, inerted spaces, and compartments involved in fire. Often, an oxygen-deficient atmosphere is created when tanks are purged and residual gases or vapors remain. A worker or rescuer entering a tank without breathing equipment may soon become unconscious and die. The oxygen-deficient atmosphere found in an enclosed compartment fire is caused by the fire using the available oxygen in order to continue the burning process.

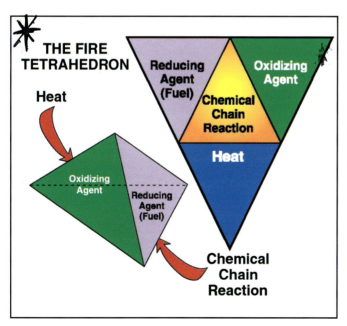

Figure 1.5 The components of the fire tetrahedron: fuel, heat, oxygen, and chemical chain reaction.

Oxygen concentrations below the normal 21 percent adversely affect both fire production and life safety. Table 1.2 lists the symptoms associated with operating in an oxygen-deficient atmosphere from a life-safety standpoint. From a fire standpoint, the intensity of a fire begins to decrease below an 18 percent oxygen concentration. Oxygen concentrations below 15 percent do not support combustion.

Chemical Chain Reaction

A *chain reaction* is a series of reactions that occur in sequence with the results of each individual reaction being added to the rest. While scientists only partially understand what happens in the combustion chemical chain reaction, they do know that heating a fuel can produce vapors that contain substances that combine with oxygen and burn. Once flaming combustion or fire occurs, it can only continue when enough heat energy is produced to cause the continued development of fuel vapors. The self-sustained chemical reaction and the related rapid growth are the factors that separate fire from slower oxidation reactions.

Table 1.2 Physiological Effects of Reduced Oxygen (Hypoxia)	
Oxygen in Air (Percent)	**Symptoms**
21	None — normal conditions
17	Reduced motor functions and impaired coordination
14 to 10	Faulty judgment and quick fatigue
10 to 6	Unconsciousness; death can occur within a few minutes

Source: *Fire Protection Handbook,* 18th edition, 1997, by National Fire Protection Association.

Note: These data cannot be considered absolute because they do not account for difference in breathing rate or length of time exposed.

These symptoms occur only from reduced oxygen. If the atmosphere is contaminated with toxic gases, other symptoms may develop.

◆ Principles of Fire Development

When the four components of the fire tetrahedron come together, ignition occurs. For a fire to grow beyond the first material ignited, heat must be transmitted beyond that first material to additional fuel groups. Early in the development of a fire, heat rises and forms a plume of hot gas. If the fire is in the open (on the main deck or in a large space, for example), the plume rises unobstructed; air is drawn (entrained) into it as it rises. Because the air being pulled into the plume is cooler than the fire gases, this action has a cooling effect on the gases above the fire. The spread of fire in an open area is primarily due to heat energy transmitted from the plume to nearby fuels. Fire spread in outside fires is increased by wind and a sloping surface, which allows exposed fuels to preheat.

The development of fire in a compartment is more complex than one in the open. The growth and development of fires is usually controlled by the availability of fuel and oxygen. When the amount of fuel available is limited, a fire is *fuel controlled*. When the amount of available oxygen is limited, the condition is *ventilation controlled*. Researchers have attempted to describe compartment fires in terms of stages or phases that occur as the fire develops. These stages include the following:

- Ignition
- Growth
- Flashover
- Fully developed fire
- Decay

Figure 1.6 shows the development of a compartment fire in terms of time and temperature rise. Describing the stages as a fire develops in a space where no suppression action is taken helps to explain the reaction that occurs. The development of fire is very complex and is influenced by many variables. As a result, all fires may not develop through each of the stages described. Fire is a dynamic event that depends on many factors for its growth and development.

Understanding into which stage a fire has developed determines, in part, the tactics used to extinguish it. A small fire in the ignition or growth stages (also called the *incipient phase*) might easily be extinguished by a portable extinguisher, while a fully developed fire may require using fire hoses or fixed suppression systems. It is essential that crew members understand the time-versus-temperature-rise concept and develop a sense of urgency in responding to fire because every minute is critical. Unchecked, even the smallest fire may cause the loss of a vessel and all aboard. The factors that impact fire development are also important considerations in effective fire suppression techniques.

Ignition

The *ignition stage* is the period when the four elements of the fire tetrahedron come together and flaming combustion begins. The physical act of ignition can be *piloted* (caused by a spark or flame) or *spontaneous* (caused when a material reaches its

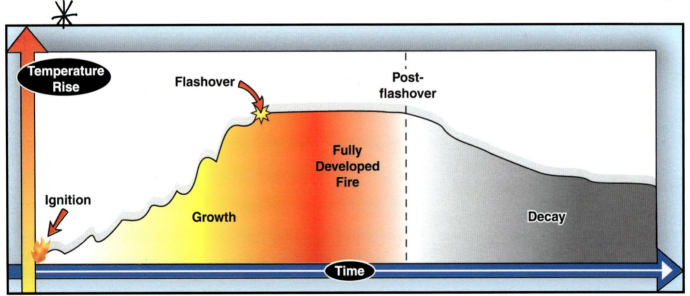

Figure 1.6 Stages of fire development in a compartment: ignition, growth, flashover, fully developed fire, and decay.

autoignition temperature as a result of self-heating). At this point, the fire is small and generally confined to the material (fuel) first ignited. All fires — whether in an open area or within a compartment — occur as a result of some type of ignition.

Growth

Shortly after ignition, a fire gas plume begins to form above the burning fuel. Air is drawn into the plume, and convection causes the heated gases to rise. The initial growth is similar to that of an outside unconfined fire. Unlike an unconfined fire, the plume in a compartment is affected by the bulkheads and deckhead of the space. The hot gases rise until they reach the deckhead, and then they begin to spread outward. When the gases reach the deckhead, the deckhead covering materials absorb heat energy by conduction. The absorption of heat and the cooling effect of the entrained air cause the temperature at the highest level to decrease as the distance from the centerline of the plume increases. Figure 1.7 shows plume development, air entrainment, and fire gas temperatures. Fire also spreads because radiated heat from the flame increases the temperature of surrounding fuels to their fire or flash points.

The growth stage continues if enough fuel and oxygen are available. Compartment fires in the growth stage are generally fuel controlled. As the fire grows, the temperature in the compartment increases, as does the temperature of the gas layer at the upper level (Figure 1.8).

Flashover

Flashover is the transition between the growth and the fully developed fire stages but is not a specific event. During flashover, conditions in the compartment change very rapidly: Temperatures are rapidly increasing, and additional fuel groups are becoming involved. The hot gas layer that develops at the

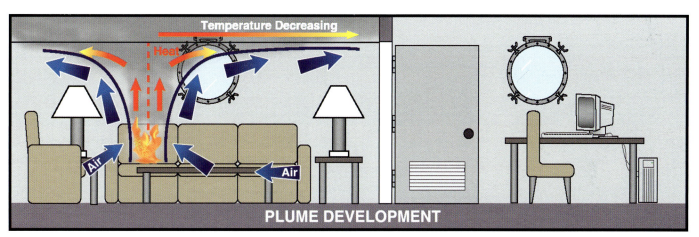

Figure 1.7 Plume development: Initially, the temperature of the fire gases decreases as they move away from the centerline of the plume.

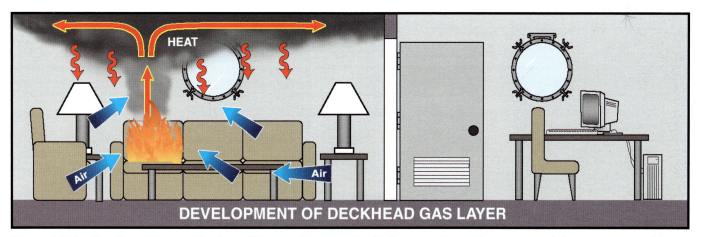

Figure 1.8 Continuation of the growth stage: The overall temperature in the compartment increases, and the temperature of the gas layer at the deckhead level increases as the fire grows.

deckhead level during the growth stage causes radiant heating of combustible materials remote from the origin of the fire. This heating causes pyrolysis to take place in these materials. The gases generated during this time are heated to their ignition temperature by the radiant energy from the gas layer at the deckhead. The combustible materials and gases in the compartment ignite as flashover occurs — the compartment is fully involved in fire (Figure 1.9).

Scientists define *flashover* in many ways, but most base their definition on the temperature in a compartment that results in the simultaneous ignition of all combustible contents in the space. While there is no exact temperature associated with this occurrence, a range from approximately 900°F to 1,200°F (483°C to 649°C) is widely used. This range correlates with the ignition temperature of carbon monoxide (1,128°F or 609°C), one of the most common gases given off by pyrolysis. Persons who have not escaped from a compartment before flashover are not likely to survive.

Even in protective equipment, crew members in a compartment at flashover are at extreme risk.

Fully Developed Fire

After the flashover stage, all combustible materials in the compartment are involved in fire (Figure 1.10). During this period, the burning fuels are releasing the maximum amount of heat possible and producing large volumes of unburned fire gases. The fire can frequently become ventilation controlled, depending on the number and size of ventilation openings in the compartment. If these fire gases flow from the compartment into adjacent spaces, they often ignite when mixed with abundant air.

A condition known as *rollover* (also called *flameover*) occurs when flames move through or across unburned gases during a fire's progression (Figure 1.11). This condition may occur during the growth stage as the hot-gas layer forms at the deckhead or when unburned fire gases discharge from a compartment during the growth and fully developed stages. As these

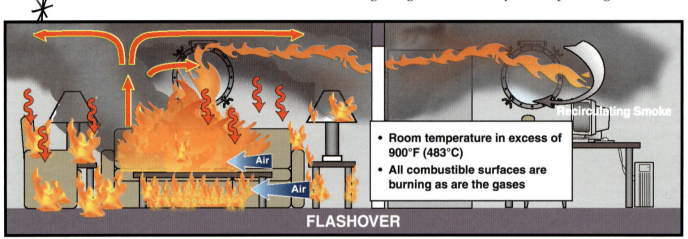

- **Room temperature in excess of 900°F (483°C)**
- **All combustible surfaces are burning as are the gases**

FLASHOVER

Recirculating Smoke

Figure 1.9 Flashover: simultaneous ignition of all combustible contents in a compartment.

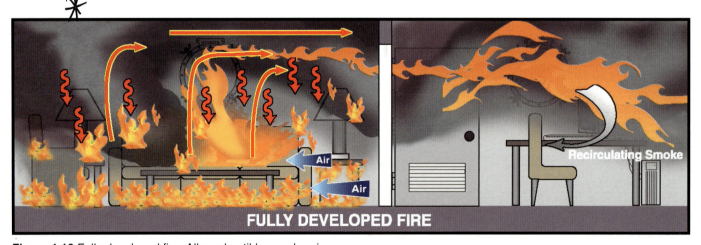

FULLY DEVELOPED FIRE

Recirculating Smoke

Figure 1.10 Fully developed fire: All combustibles are burning.

hot gases flow from the burning compartment into the adjacent space, they mix with oxygen, and flames often become visible in the layer. Rollover is distinguished from flashover because only the gases are burning in rollover, not the entire contents of the compartment.

Decay

As the available fuel in the compartment is consumed by fire, the amount of heat energy released begins to decline. Once again, the fire becomes fuel controlled, the amount of fire diminishes, and the temperatures within the compartment begin to decline. The remaining mass of glowing embers can, however, result in moderately high temperatures in the compartment for some time. For example, a steel structure retains heat for a long time.

Time-Versus-Temperature-Rise Concept

The fact that fire follows a time-versus-temperature-rise curve during development is an important concept. Understanding this concept makes it clear to crew members just how quickly a fire can develop, spread, and grow out of control. Following a fire through each of the previously described stages, Figures 1.12 and 1.13 show two scenarios in which fires develop at different rates. The objective of fire fighting is to interrupt a fire's progress as early as possible during its development. The more time delays there are in attacking a fire, the further it develops—more damage is sustained—and it becomes more difficult to control the fire and save property or even the entire vessel.

Fires clearly do not develop at the same rate, given differing factors. But fires can, and often do, develop

with amazing speed. Fires have been known to reach the flashover stage in less than 2 minutes. In other situations, the factors affecting fire development may

Figure 1.12 Fire developing at various rates: A wastebasket fire in a crew member's room develops slowly.

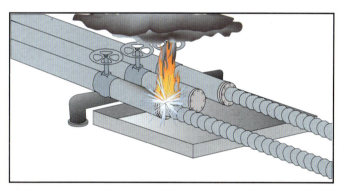

Figure 1.13 Fire developing at various rates: A leak from a pipe flange under pressure at a manifold can cause a sudden, more developed fire that can spread rapidly.

Figure 1.11 Rollover: unburned gases igniting and rolling over the deckhead.

result in flashover being reached much later, after 8 minutes or so. In still other situations such as a small fire with limited fuel in a large, otherwise empty, closed hold, fire may never reach flashover. In the latter case, the fuel would be consumed before the dynamics of the fire carried it through all of the stages. The chances of such a fire occurring are unlikely, and one must not assume the best-case scenario.

Crew members must always prepare for the worst-case scenario and respond to a fire and attack it with a sense of urgency. The traditional marine response of "closing up" the compartment is a good initial action that may actually achieve extinguishment in a steel vessel. This response method would not likely contain the fire in other types of structures such as a house. The crew must understand that the primary aim of this initial action is to acquire time — time to don personal protective equipment (PPE), time to prepare extinguishing equipment, and time to implement a plan of attack. As already stated, fires may reach flashover in as little as 2 minutes, so any time gained is valuable. Crew members must not lose the sense of urgency that is required to quickly and effectively extinguish a fire. The first few minutes of a fire emergency present the *golden window of opportunity*. Once that window is closed, there is no way to reopen it.

 ## Factors Affecting Fire Development

Factors that impact the development of a fire in a compartment are as follows:

- Size and number of ventilation openings
- Volume of the compartment
- Height of the compartment
- Size and location of the first fuel group ignited and total fuel load in the compartment

For a fire to develop, sufficient air must be available to support burning beyond the ignition stage. The size and number of ventilation openings determines how the fire develops within the space. The compartment's size, shape, and height determine whether a significant hot-gas layer forms. The size and location of the initial fuel group is also very important in the development of the hot-gas layer. The plumes of burning fuel in the center of a compartment entrain more air and thus are cooler than those against bulkheads or in corners of the compartment.

The temperatures that develop in a burning compartment are the direct result of the energy released as the fuels burn. In a fire, the resulting energy is in the form of heat and light. The amount of heat energy released in a fire (heat release rate) is measured in Btu per second (Btu/s) or kilowatts (kW). This heat energy is directly related to the amount of fuel consumed over time and to that fuel's heat of combustion (amount of heat a specific mass of a substance gives off when burned). Materials with high heat release rates (foam-padded furniture or polyurethane foam mattresses, for example) would burn rapidly once ignition occurs. Fires in materials with lower heat release rates are expected to take longer to develop (see Table 1.3 for heat release rates for some common materials).

The heat generated in a compartment fire is transmitted from the initial fuel group to other fuels in the space by all modes of heat transfer. The heat rising in the initial fire plume is transported by convection and ventilation. As the hot gases travel over surfaces of other fuels in the compartment, heat is transferred to them by conduction. Radiation and direct flame contact play significant roles in the transition from a growing fire to a fully developed fire in a space. As the hot-gas layer forms at the deckhead, hot particles in the smoke begin to radiate energy to the other fuel groups in the compartment. As the radiant energy increases, the other fuel groups pyrolyze and give off ignitable gases. When the temperature in the compartment reaches the ignition temperature of these gases, the entire space becomes involved in fire (flashover).

 ## Special Considerations for Fire Suppression

Crew members need an understanding of several conditions or situations that occur during a fire's growth and development that impact fire suppression efforts. The following sections provide an overview of thermal layering of gases, backdraft, and products of combustion along with the potential safety concerns for each.

Thermal Layering of Gases

Thermal layering of gases is the tendency of gases to form layers according to temperature. Other terms sometimes used to describe this layering of gases by heat are *heat stratification* and *thermal balance*. The hottest gases tend to be in the top layer, while the cooler ones form the bottom layer. Smoke is a heated mixture of air, gases, and particles, and it rises. If a hole is made in a deckhead, smoke will rise from the space or compartment to the next deck or to the outside.

Thermal layering is critical to fire fighting activities. As long as the hottest air and gases are allowed to rise, the lower levels will be safe for fire crews (Figure 1.14). If water is applied continuously to the upper level of the layer where the temperatures are highest, the rapid conversion to steam can cause the gases to mix rapidly. This swirling of smoke and steam disrupts normal thermal layering and causes hot gases to mix throughout the compartment. This mixing *disrupts the thermal balance* or *creates a thermal imbalance* that can burn crew members. Once the normal layering is disrupted, forced ventilation procedures (such as using fans) must be used to clear the area. Other-

wise fire team members will be forced to retreat, and their attack becomes ineffective.

The proper attack procedure is to ventilate the compartment (if possible), allow the hot gases to escape, and direct the fire stream to the deckhead in short bursts to maintain the thermal balance (Figure 1.15). Use a straight fire stream or narrow power cone (20- to 30-degree) fog pattern. After the initial application of water, direct the fire stream at the base of the fire, keeping it out of the upper layers of gases. Using this same technique in an unventilated space causes the least amount of thermal disruption.

Backdraft

Crew members operating at fires in compartments or other confined areas must use care when opening doors, portholes, or other potential ventilation openings. During the growth, fully developed, and decay phases of fire development, large volumes of hot, unburned fire gases can collect in unventilated spaces. These gases are at or above their ignition

Table 1.3 Heat Release Rates for Common Materials		
Material	**Maximum Heat Release Rate**	
	kW	**Btu/s**
Wastebasket (0.53 kg or 1.2 lb) with milk cartons (0.40 kg or 0.9 lb)	15	14.2
Upholstered chair (cotton padded) (31.9 kg or 70.3 lb)	370	350.7
4 Stacking chairs (metal frame, polyurethane foam padding) (7.5 kg or 16.5 lb each)	160	151.7
Upholstered chair (polyurethane foam) (28.3 kg or 62.4 lb)	2,100	1,990.0
Mattress (cotton and jute) (25 kg or 55 lb)	40	37.9
Mattress (polyurethane foam) (14 kg or 30.9 lb)	2,630	2,492.9
Mattress and box spring (cotton and polyurethane foam) (62.4 kg or 137.6 lb)	660	626.0
Upholstered sofa (polyurethane foam) (51.5 kg or 113.5 lb)	3,200	3,033.0
Gasoline/kerosene (2 ft² or 0.19 m² pool)	400	379.0
Christmas tree (dry) (7.4 kg or 16.3 lb)	500 (minimum)	474.0

Source: NFPA 921, *Guide for Fire and Explosion Investigations*; NBSIR 85-3223 *Data Sources for Parameters Used in Predictive Modeling of Fire Growth and Smoke Spread;* and NBS Monograph 173, *Fire Behavior of Upholstered Furniture.*

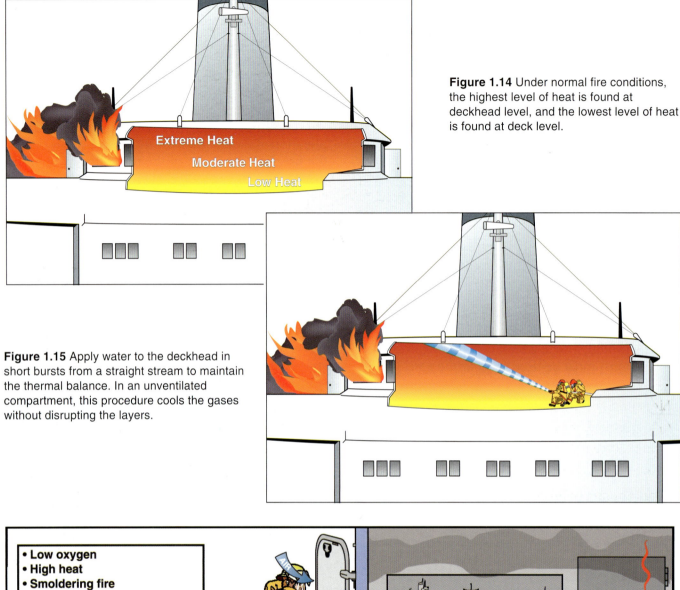

Figure 1.14 Under normal fire conditions, the highest level of heat is found at deckhead level, and the lowest level of heat is found at deck level.

Extreme Heat

Moderate Heat

Low Heat

Figure 1.15 Apply water to the deckhead in short bursts from a straight stream to maintain the thermal balance. In an unventilated compartment, this procedure cools the gases without disrupting the layers.

- **Low oxygen**
- **High heat**
- **Smoldering fire**
- **High fuel vapor concentrations**

PREBACKDRAFT

Figure 1.16 Prebackdraft conditions: low oxygen, high heat, smoldering fire, and high fuel vapor concentrations. Attempting to open this door is dangerous.

temperatures, but insufficient oxygen is available for them to ignite. Any action during fire fighting operations that allows air to enter this space and mix with these hot gases can result in *backdraft* — an explosive ignition of the gases (Figures 1.16 and 1.17). Backdrafts can injure or kill crew members attempting to enter these spaces. The potential for backdraft is reduced with vertical ventilation (opening at the highest point) because the unburned gases rise. Opening the space at the highest possible point (such as a ventilation

• Introduction of oxygen causes fire of explosive force

BACKDRAFT

Figure 1.17 Backdraft: Introducing oxygen into a fire space can cause a fire explosion.

duct or escape scuttle) allows gases to escape before crew members enter. The following conditions may indicate the potential for backdraft:

- Pressurized smoke exiting small openings
- Black smoke becoming dense gray yellow
- Confinement and excessive heat
- Little or no visible flame
- Smoke leaving the space in puffs or at intervals (appearance of breathing)
- Smoke-stained porthole/window glass
- Large amount of blistering or burning paint on the outside of the fire space on the upper portion of a bulkhead

Products of Combustion

As a fuel burns, the chemical composition of the material changes. This change results in the production of new substances (products of combustion: heat, light, smoke, fire gases, and flame) and the generation of energy (Figure 1.18). As a fuel burns, some of it is actually consumed. As mentioned earlier, the *Law of Conservation of Mass* explains that any mass lost converts to energy. This energy is in the form of *heat* and *light* in the case of fire. Burning also results in the generation of *smoke* — airborne *fire gases,* particles, and liquids. Heat is responsible for the spread of fire. It also causes burns, dehydration, heat exhaustion, and injury to the respiratory tract.

Flame is the visible, luminous body of a burning gas. When a burning gas is mixed with the proper amount of oxygen, the flame becomes hotter and less luminous. The loss of luminosity is caused by a more

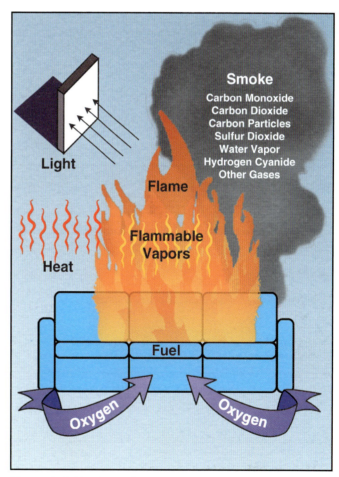

Smoke
Carbon Monoxide
Carbon Dioxide
Carbon Particles
Sulfur Dioxide
Water Vapor
Hydrogen Cyanide
Other Gases

Light

Flame

Flammable Vapors

Heat

Fuel

Oxygen **Oxygen**

Figure 1.18 Products of combustion: heat, light, smoke, fire gases, and flame.

complete combustion of carbon. For these reasons, flame is considered to be a product of combustion. Of course, it is not present in those types of combustion that do not produce a flame such as smoldering fires.

While the heat energy from a fire is a danger to anyone directly exposed to it, smoke causes most deaths in fires. Smoke is composed of materials that vary from fuel to fuel, but generally smoke is considered toxic. Some fuels produce more smoke than others. Smoke contains asphyxiants, irritants, flammable gases and vapors, and toxins. Carbon monoxide is the most common of the hazardous materials contained in smoke. While it is not the most dangerous of the materials found in smoke, it is almost always present when combustion occurs. Carbon monoxide is easily detected in the blood of fire victims. As a result of the variety of dangerous products contained in smoke and the fact that either alone or in combination they are deadly, crew members operating in smoke must use SCBA for protection.

◆ Fire Extinguishment Theory

Limiting or interrupting one or more of the essential elements in the combustion process (fire tetrahedron) extinguishes fire. Thus, a fire may be extinguished by four methods: (1) reducing its temperature, (2) eliminating available fuel, (3) excluding oxygen, or (4) stopping the uninhibited chemical chain reaction (Figure 1.19).

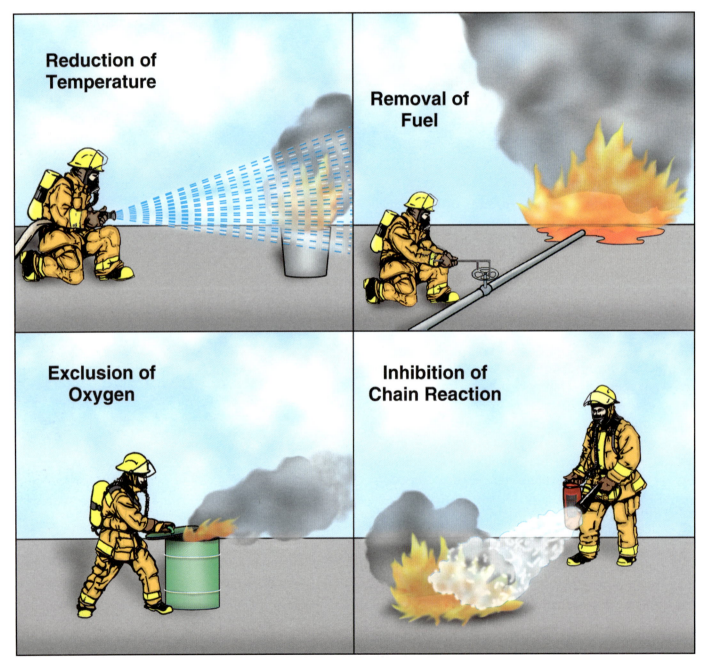

Figure 1.19 Four methods of fire extinguishment: temperature reduction, fuel removal, oxygen exclusion, and chemical flame inhibition.

Temperature Reduction

One of the most common methods of extinguishment is cooling with water. Extinguishment depends on reducing the temperature of a fuel to the point where it does not produce sufficient vapor to burn. Solid fuels and liquid fuels with high flash points can be extinguished by cooling. Fires involving low flash point liquids and flammable gases cannot be extinguished by cooling with water because vapor production cannot be sufficiently reduced. The reduction of temperature depends on the application of an adequate flow of water to establish a negative heat balance. Water best absorbs heat when it is in droplet form because it covers more surface area.

Fuel Removal

In some cases, a fire is effectively extinguished by removing the fuel source. For example, stopping the flow of liquid or gaseous fuel and removing solid fuel in the path of a fire are ways of removing the fuel source. Another method of fuel removal is to allow a fire to burn until all fuel is consumed.

Oxygen Exclusion

Reducing the oxygen available to the combustion process reduces a fire's growth and may totally extinguish it over time. In its simplest form, this method is used to extinguish cooking stove fires when a cover is placed over a pan of burning food. Flooding an area with an inert gas such as carbon dioxide, which displaces the oxygen and disrupts the combustion process, can reduce oxygen content. Blanketing a fuel with fire fighting foam also separates oxygen from the fuel. Neither of these latter methods works on those rare fuels that are self-oxidizing.

Chemical Flame Inhibition

Extinguishing agents such as some dry chemicals and halogenated agents (halons) interrupt the flame-producing chemical reaction and stop flaming combustion. This method of extinguishment is effective on gas and liquid

fuels because they must flame to burn. These agents do not easily extinguish smoldering fires. Very high agent concentrations and extended periods of application are necessary to extinguish smoldering fires, making these agents impractical. Thus, cooling is the only practical way to extinguish a smoldering fire.

◆ Classification of Fires

A description of how fires are classified is important when discussing extinguishment. Each class has its own requirements for extinguishment. Different continents and individual countries have their own systems for classifying fires. However, a manual of this size prohibits an in-depth discussion of each system. Therefore, for the purposes of this manual, the North American classification system is described. The sidebar contains a brief summary of how fires are currently classified in most European countries. See Chapter 5, Portable and Semiportable Fire Extinguishers, for information on fire extinguisher rating systems based on the classes of fires.

Class A

Class A fires involve ordinary combustible materials such as wood, cloth, paper, rubber, and many plastics (Figure 1.20). Water is used to cool or quench the burning material below its ignition temperature. The addition of *Class A fire fighting foams* (sometimes referred to as *wet water*) may enhance water's ability to extinguish Class A fires, particularly those that are deep-seated in bulk materials (such as rolls of paper, coal, iron filings, wood chips, cargo containers of clothing, etc.). The Class A foam agent reduces the water's surface tension, allowing it to penetrate more easily into piles of the material. Class A fires are difficult to extinguish using oxygen-exclusion methods

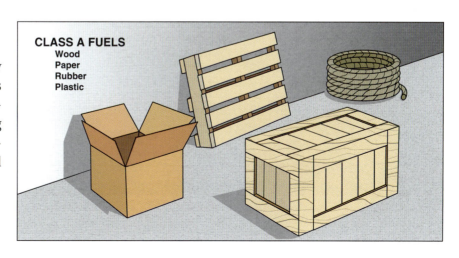

CLASS A FUELS
Wood
Paper
Rubber
Plastic

Figure 1.20 Class A fuels: ordinary combustibles (wood, cloth, paper, plastics, etc.).

because they do not give the cooling needed for total extinguishment.

Class B

Class B fires involve flammable and combustible liquids and gases such as gasoline, oil, lacquers, paints, mineral spirits, and alcohols (Figure 1.21). The smothering or blanketing effect of oxygen exclusion is most effective for extinguishment and also helps reduce the production of additional vapors. Other extinguishing methods include removal of fuel, temperature reduction when possible, and interruption of the chain reaction with dry chemical agents.

Class C

Fires involving energized electrical equipment are Class C fires (Figure 1.22). Electrical appliances, transformers, engineering department switchgears, and computerized equipment are examples. A non-conducting extinguishing agent such as halon, dry chemical, or carbon dioxide can sometimes control class C fires. The fastest extinguishment procedure is to first de-energize the electrical device or circuit and then treat the fire appropriately, depending upon the fuel involved.

Class D

Class D fires involve combustible metals such as aluminum, magnesium, titanium, zirconium, sodium, and potassium (Figure 1.23). These materials are particularly hazardous in their powdered forms. Given a suitable ignition source, airborne concentrations of metal dusts can cause powerful explosions. The extremely high temperatures of some burning metals make water and other common extinguishing agents ineffective. In fact, the use of these agents on combustible metal fires could result in violent reactions. No single agent is available that effectively controls fires in all combustible metals. Special extinguishing agents are available for control of fire in each of the metals, and they are marked specifically for that metal. These agents cover the burning material and smother the fire.

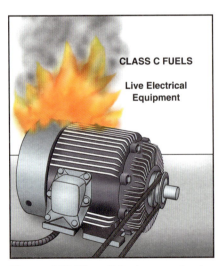

Figure 1.21 Class B fuels: flammable and combustible liquids (gasoline, kerosene, oil, etc.).

Figure 1.22 Class C fuels: energized electrical equipment (electrical appliances, computers, etc.).

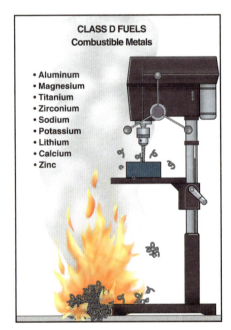

Figure 1.23 Class D fuels: combustible metals (aluminum, magnesium, titanium, etc.).

Chapter 2

Onboard Fire Causes and Prevention

The old saying *"An ounce of prevention is worth a pound of cure"* is never more true than when it is referring to destructive fires. Depending on the magnitude of a fire, an ounce may be worth a ton. Thus, the philosophy behind this chapter is that onboard *fire prevention* is the key to "curing" the onboard fire problem. If a fire occurs, crew members may rely on an early detection system to minimize losses, and portable fire extinguishers may be able to extinguish the fire in its beginning stage. Fixed automatic fire suppression systems may also hold a fire in check or extinguish it. If these extinguishing efforts fail and a fire is not halted, a well-trained, properly equipped, and organized fire team will need to attack the fire. While all these resources and programs are quite necessary, the negative side is that fire damage has already occurred at the time they go into operation; that is, they are reactionary remedies. Quick, effective fire suppression may be the difference between minimal loss and total loss.

Fire prevention, on the other hand, is a proactive approach and makes its impact before any damage occurs. No fire prevention program can provide 100 percent assurance that fire will not occur. However, an effective prevention program is the least expensive approach to fire protection because every fire has the potential to cause a total loss. If a prevention program is effective, it may prevent a fire that would destroy an entire vessel and claim the lives of everyone aboard.

Before developing an effective fire prevention program, one must first understand the major causes of shipboard fires. This chapter addresses common fire causes and how to prevent fires from occurring through the use of good housekeeping routines, preventative maintenance/repair procedures, safety inspections, and onboard educational/training sessions. A fire prevention program and a model fire prevention/inspection checklist are included.

Common Causes of Onboard Fires

The major causes of onboard fires are discussed in the following sections, along with actions that crew members can take to reduce the possibility of fire. These causes of fire — these situations and actions — are common to all vessels and are the responsibility of all crew members. Some fires may be purely accidental, and others may be caused by circumstances beyond a person's control. But many fires result from the acts or omissions of crew members. Carelessness and irresponsible or ill-advised actions can cause disastrous fires. Omissions — not taking the proper preventive measures when hazardous situations are discovered — can "allow" many fires to occur. No matter how a shipboard fire starts, it can result in the loss of the vessel and perhaps lives. It is therefore extremely important that crew members are constantly alert for situations that could cause fire on board the vessel.

Virtually all shipboard fires and explosions can be identified as having originated from one of the following causes. Each of these categories are addressed in the following sections.

- Smoking
- Chemical reactions
- Electrical equipment
- Stowage

- Galley operations
- Fuel oil systems and transfer of fuel (bunkering)
- Hot-work operations
- Shoreside workers aboard performing cargo movement, repair, and maintenance
- Shipyard operations
- Tanker loading, unloading, and cleaning operations
- Collisions
- Incendiary/arson fires

Smoking

The most common cause of shipboard fires is careless smoking and the improper discarding of smoking materials and matches. A glowing cigarette contains enough heat to ignite materials such as paper products, clothing, bedding, some plastics, rope, and most wood-based products. No-smoking regulations must be enforced in hazardous parts of the vessel: cargo holds, weather deck, engine and boiler rooms for example. Smoking while in bed is particularly dangerous and can be deadly. A smoldering cigarette touching bedding material can start a smoldering fire. The resulting smoke can cause drowsiness and possible asphyxiation before the fire is discovered. Following one simple but important rule can prevent such fires: *Don't smoke in bed under any circumstances* (Figure 2.1).

Disposing of Smoking Materials

Glowing ashes/tobacco contain enough heat to start a fire in materials such as dunnage, paper, cardboard, excelsior, rope, and similar materials. Deposit smoking materials in noncombustible receptacles — a requirement in every designated smoking area. Place these receptacles wherever smoking is permitted. It is also a good idea to soak smoking materials with water before discarding them. The soaking gives added protection against fire. Empty ashtrays only when they contain no glowing embers — soaking under a faucet ensures cooling. Then, deposit ashes into covered, noncombustible containers (Figure 2.2).

Smoking and Alcohol

A person who has been drinking alcohol tends to become careless. A person who is also smoking can be extremely careless and dangerous. To a person who

Figure 2.1 Smoking while in bed is a hazardous situation.

Figure 2.2 Smoking materials that are discarded correctly cannot become sources of ignition.

has had one or two drinks, a few glowing embers dropped from a pipe (or smoking materials that are not quite extinguished or a lit cigarette that someone has left in an ashtray) may not seem important. But all these situations are actually small shipboard fires. If they come into contact with a nearby flammable material, these fires will not stay small very long. For example, curtains or coveralls on hooks swing with vessel movement and may contact glowing cigarette ends. Observe a smoker who is "under the influence" very carefully. Everyone is responsible for seeing that the actions of this person do not jeopardize the safety of the vessel and its crew/passengers (Figure 2.3).

No-Smoking Areas

Certain parts of a vessel are designated as no-smoking areas because of the hazards they contain. Inform everyone aboard a vessel of where smoking is permitted and where it is not. Inform any noncrew member who comes aboard of the smoking regulations of the vessel. Several no-smoking areas are described in the following paragraphs.

Cargo holds and weather deck. Smoking is forbidden in the holds of cargo vessels and on the weather deck when the hatches are open. Designating cargo holds as no-smoking areas prevents fires. Break bulk cargo vessels are especially vulnerable to cargo-hold fires during loading/unloading. Monitor cargo holds closely during loading/unloading operations to ensure compliance with this regulation (Figure 2.4). Some cargoes such as grain or coal produce dust that may explode if ignited. A fire in a cargo hold may not be

discovered until the vessel is out to sea — sometimes days later. By that time, much of the cargo may be involved in fire, and the fire may be difficult to extinguish or control.

Engine and boiler rooms. Engine rooms and boiler rooms contain relatively large amounts of petroleum products such as fuel oil, lubricating oil, and grease. Even the thickest of these products tends to vaporize and mix with the warm air of the engine room or boiler room. A lighted match or glowing tobacco can ignite this flammable air-vapor mixture, resulting in a volatile fire. Engine room fires are difficult to extinguish and very hazardous for the engine room crew. If a fire is serious enough, it may mean loss of propulsion and control of the vessel — an extremely dangerous situation. For these reasons, strictly enforce no-smoking regulations in engine rooms and boiler rooms.

Figure 2.4 Smoking and careless disposal of smoking materials have caused many serious fires in cargo holds. Prohibit smoking in cargo spaces, and monitor these spaces during cargo handling.

Figure 2.3 Smoking and drinking alcohol — a dangerous combination.

Storage and work spaces. Prohibit smoking in storage rooms and workrooms. These spaces (paint and rope lockers and carpenter shops) contain large amounts of flammable materials. In some cases, the mere presence of an ignition source is all it takes to start a destructive blaze.

Chemical Reactions

Many materials carried as cargo or used to operate a vessel are capable of chemical reactions. A chemical reaction occurs through the interaction of two or more substances, one of which is often air or water. Precautions for stowing many hazardous substances are required by both federal/national and international cargo regulations. Carefully screen cargo manifests (invoices) to identify the presence of dangerous materials. For example, chlorine produces a violent reaction when it combines with certain organic materials. Ammonium nitrate, a common fertilizer, is a very strong solid oxidizer and becomes an explosive when contaminated with petroleum liquids or other organic substances. Take necessary precautions to prevent a reaction that could result in fire, explosion, or the release of toxic substances.

Certain cargoes have been known to mix in transit and/or react with vessel construction materials to cause explosive chemical reactions. Many incompatible products such as oxidizing agents, inhibited polymers (acrylonitrile), and explosives are carried in vessels. Check all hazardous cargoes for compatibility with other cargoes, and stow in accordance with the International Maritime Organization's (IMO) *International Maritime Dangerous Goods (IMDG) Code.*

Spontaneous ignition is a common chemical reaction encountered in marine situations and is often overlooked as a cause of fire on board a vessel. Yet many common materials are subject to this dangerous chemical phenomenon. They include materials that are carried as cargo and materials that are used in operating the vessel. An example given earlier of spontaneous ignition that could easily occur on board a vessel is when a rag soaked with vegetable oil or paint is discarded in the corner of a workshop, storage area, or engine room. The area is warm, and often there is inadequate ventilation. The oil on the rag begins to oxidize — react chemically with the oxygen in the warm air around it. After some time, the rag gets hot enough to burst into flames. It then can ignite any

nearby flammable substances, perhaps other rags or stored materials. Another example that can cause spontaneous ignition is the practice of putting flattened cardboard boxes on the deck of a paint locker to absorb spills. In time, the cardboard becomes impregnated with paint, solvent, and other flammable liquids. Under the right conditions, the paint-soaked cardboard may ignite and cause a fire (Figure 2.5). Even baled organic products such as rags, jute, cotton, straw, and brown coal can spontaneously ignite days, weeks, or even months after having been soaked with water.

Although most vessels do not appear susceptible to fire because they are constructed of steel, an enormous fuel load — engine fuel, accommodation fittings, cargoes, and paint — exists on board. Often modifications from the original vessel design plans reduce the fire prevention effectiveness of a vessel. Awareness of the possibilities of chemical reactions in both the vessel's materials and cargoes is important for every crew member.

Vessel Materials

As noted earlier, oily and paint-soaked rags are subject to spontaneous ignition. In this case, fire prevention is simply a matter of good housekeeping routines. However, some materials that are not usually subject to spontaneous ignition can ignite on their own under certain conditions. While some materials can produce their own heat and begin to burn, others can begin to burn when they reach their ignition temperature when heated by an outside source.

Figure 2.5 Careless storage or disposal of materials can lead to spontaneous ignition.

For example, wood, like every other substance, must be heated to a certain temperature before it ignites and burns. Most steam pipes do not get hot enough to ignite wood. Yet, if a piece of wood is in constant contact with a steam pipe or a similar "low-temperature" heat source, it can ignite spontaneously. What happens is that wood is first changed to charcoal by the heat. Then the steam pipe ignites the charcoal, which burns at a lower temperature than wood (Figure 2.6). Even though the change from wood to charcoal may take several days to occur, it could easily go unnoticed. The first sign of a problem would be smoke or flames issuing from the wood. Another example involves refrigerated holds lined with cork insulation held in place by wood and covered with paneling. During heavy weather, the vessel's frames move back and forth against the wood, causing heat from friction and ultimately ignition. It is difficult to find, access, and extinguish these types of fires. To prevent such fires, keep combustible materials away from any heat source. If they cannot be moved, protect them with heat-insulating materials.

Cargo

Many materials that are carried as cargo are subject to spontaneous ignition. A good general principle to follow in preventing the spontaneous ignition of cargo is to separate fibrous materials from oils. Other methods of preventing cargo fires are discussed in the Stowage section later in this chapter. Some examples follow.

Chlorine. As mentioned earlier, chlorine is a strong oxidizer and produces a violent reaction when it combines with finely divided metals or certain organic materials, particularly acetylene, turpentine, and gaseous ammonia. Stow chlorine in well-ventilated holds, and do not stow in the same holds as organic materials.

Metals. The metals sodium and potassium react violently with water and are required to be labeled *dangerous when wet.* Metal powders, such as magnesium, titanium, calcium, and zirconium, oxidize rapidly (and produce heat) in the presence of air and moisture. Under certain conditions, they can produce sufficient heat to ignite. Moisture accelerates the oxidation rate of most metal powders, so they must be kept dry. Metal turnings do not tend to ignite spontaneously when kept dry. However, piles of oily metal borings, shavings, turnings, and cuttings have

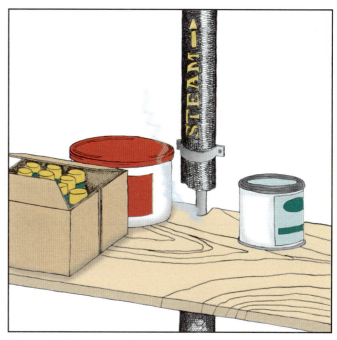

Figure 2.6 A hot steam pipe can change wood to charcoal, ignite the charcoal, and cause a fire.

caused fires by igniting spontaneously. As in the case of oily rags, heat is produced by oxidation of the oil within the pile of shavings. Eventually, enough heat is produced and held in the pile to ignite the most finely divided metal. Then, the coarser shavings and other combustible materials, if present within the pile, ignite and compound the fire problem. Metal shavings are termed *regulated* (hazardous) cargoes and are handled, stowed, and shipped in compliance with special regulations (see Stowage section).

Coal. Coal is also regulated cargo subject to the same regulations as metal shavings. Soft coal may heat spontaneously, depending on several of the following factors:

- Geographic origin
- Moisture content
- Fineness of particles and ratio of fine particles to lump coal
- Chemical makeup, including impurities
- Length of time since coal was crushed

Other products. Other products that present a danger of fire through spontaneous heating are alfalfa meal, charcoal, cod-liver oil, colors in oil, cornmeal feeds, fish meal, fish oil, fish scrap, linseed oil, oiled and varnished fabrics of all kinds, redskin peanuts, and tung-nut meals. Again, screen the manifest for regulated materials, and handle appropriately.

Electrical Equipment

There is no standard power system on board vessels. Voltages may be extremely high (380, 440, 4,100, or even 6,600 volts or more), and the current may be either alternating current (AC) or direct current (DC). For properly insulated and wired equipment, electricity is a safe and convenient source of power. However, when electrical equipment wears out, is misused, or is poorly wired, it can convert electrical energy to heat. Then the equipment becomes a source of ignition and thus a fire hazard. For this reason, electrical equipment must be installed, maintained, tested, and repaired by qualified personnel in accordance with existing regulations.

Another general concern with electricity is that cords and cables are ignored or not recognized for their potential hazards. Extension cords to portable lights or equipment and connections to shore power sources and/or generators may be hazardous due to fraying, water contact, and poor connections. These potential hazards are often overlooked due to the temporary nature of the circuits.

Other potential electrical hazard situations involve makeshift connections, exposed lightbulbs, worn vaportight fixtures/devices, faulty electric motors, and charging storage batteries. Engine rooms have many electrical sources and fuel-line hazards.

Replacement Parts and Equipment

Standard residential or industrial electrical equipment does not last very long at sea. As a result, the equipment or its wiring may overheat or arc, causing a fire if flammable materials are located nearby. Some of the reasons for equipment deterioration are as follows:

- Salt air corrosion
- Vibration of the vessel
- Erratic operation or short-circuiting caused by a steel hull

Electrical equipment approved by either the classification society governing the construction of vessels or the flag state is specially designed and constructed for shipboard use. Given reasonable maintenance, it withstands the strenuous conditions at sea. Thus, only approved replacement parts and equipment are on board a vessel — and only for the use for which they have been approved. Consult the chief engineer if there are any doubts concerning the installation, repair, use, or maintenance of this equipment.

Wiring and Fuses

The insulation on electrical wiring, particularly the type used for appliances, electric hand tools, and cargo and drop lights, does not last forever. With age and use, it can become brittle and crack. Vibration of the vessel may rub the insulation causing it to deteriorate, or careless handling and abuse may cause breaks. No matter how it happens, once the insulation is broken, the bare wire is dangerous. A single exposed wire can arc to any metal object. If both wires are exposed, they can touch and cause a short circuit. Either situation could produce enough heat to ignite the insulation on the wiring or some other nearby flammable material. Replace wires that have inadequate or worn insulation to prevent fires.

If the fuse or circuit breaker in a particular circuit is too large, it will not break the circuit. Instead, an increased current flows, and the entire circuit overheats. Eventually, the insulation begins to burn and ignites combustible material in its vicinity. Prevent this type of fire by installing only fuses and circuit breakers of the proper size for their circuits.

Makeshift Connections

The rigging of improvised electrical outlets to serve additional appliances, particularly common in crew's quarters and galleys, is a dangerous practice (Figure 2.7). The wiring in every electrical circuit is designed to carry a certain maximum load. When this wiring is

Figure 2.7 Overloading circuits is a dangerous practice. Connect only one appliance to each outlet in an electrical circuit.

overloaded with too many operating appliances, it can overheat and burn its insulation. The hot wiring can also ignite flammable materials in the area. Cabins have burned completely by such fires. Avoid the need for makeshift connections by planning appliance use. Regular inspections are important to ensure compliance with this requirement.

Exposed Lightbulbs

An exposed, lighted electric bulb can ignite combustible material by direct contact. A number of shipboard fires have started when crew members left lamps turned on in unoccupied quarters. As the vessel rolled, curtains or other flammable materials ignited when they came in direct contact with the hot bulbs. The result in most cases was destruction of a crew member's quarters.

On weather decks, high-intensity floodlights are usually protected from the elements by canvas or plastic covers. The covers are desirable when the lights are not in use. However, if a cover is left in place while the light is on, the heat of the lamp can ignite the material.

Improperly protected droplights or cargo lightbulbs could similarly ignite flammable materials by contact or by breaking and arcing. Never leave these lights turned on without someone present. What appears to be a safe situation in a calm sea could quickly become dangerous in a rough sea (Figure 2.8).

Vaportight Fixtures/Devices

Vaportight fixtures/devices are designed to protect against the effects of sea air. The vapor protection keeps out moisture, but it also holds heat. This situa-

Figure 2.8 A properly protected lightbulb. *Courtesy of Maritime Institute of Technology and Graduate Studies.*

tion causes the insulation of the fixture/device to dry and crack more rapidly than it does in standard fixtures/devices. Examine vaportight fixtures/devices frequently, and replace them as required to prevent short circuits and possible ignition of insulation.

The vapor seal also prevents an explosive atmosphere from entering into the electrical contacts of the fixture/device where an ignition spark could be generated. Use these types of fixtures/devices (also known as *explosionproof* or *intrinsically safe equipment*) when working in atmospheres containing flammable vapors. These fixtures/devices (such as lights, radios, flashlights, detectors, and fans) are designed so that they are incapable of creating ignition sparks. See Chapter 3, Vessel Types, Construction, and Arrangement, for more information about explosionproof and intrinsically safe equipment.

Electric Motors

Faulty electric motors are prime causes of fire. Problems may result when a motor is not properly maintained or when it exceeds its useful life. Motors require regular inspections, testing, lubrication, and cleaning. Sparks and arcing may result if a motor's winding becomes short-circuited or grounded or if the brushes do not operate smoothly. If a spark or an arc is strong enough, it can ignite nearby combustible material. Lack of lubrication may cause the motor bearings to overheat, which could also ignite other materials.

Engine Room Machinery

Engine rooms are particularly vulnerable to electrical hazards due to the large number of electrical machinery and power sources. Water dripping from ruptured seawater lines can cause severe short-circuiting and arcing in electric motors, switchboards, and other exposed electrical equipment. This, in turn, can ignite insulation and nearby combustible materials. Ruptured fuel and lubrication lines above and near electrical equipment are even more serious threats. An alert engineering staff constantly monitors oil lines for leaks.

Storage Batteries

Storage batteries are potential sources of ignition during charging because they emit hydrogen, a highly flammable gas. A mixture of air and 4.1 percent to 74.2 percent hydrogen by volume is potentially explosive. Hydrogen is lighter than air and consequently rises as it is produced. Charge batteries in a well-ventilated area to prevent hydrogen fires. Hydrogen collects at the deckhead if ventilation is not provided at the highest point in the battery-charging room. Any source of ignition could then cause an explosion and fire. Prohibit smoking and other sources of ignition such as machinery that might produce sparks.

Stowage

Space for stowage is always at a premium on board a vessel. *"A place for everything and everything in its place"* is an important concept. In itself, it is a fire prevention measure, provided the stowage is done safely in the beginning. Fires can result when stowed materials come loose and fall or slide across a deck in rough weather. Loose equipment can rupture fuel lines, damage essential machinery, and smash electrical equipment causing short circuits. In addition, controlling heavy equipment that has come loose during heavy seas is difficult and dangerous. It is important that there be properly constructed storage facilities for all types of equipment and materials.

Even the most dangerous cargo can be transported safely if it is properly packaged and stowed. On the other hand, supposedly "safe" cargo can cause a fire if it is stowed carelessly. Many stowage concerns are discussed in the following sections. For example, some types of cargo (such as regulated/unregulated or bulk) require different stowage procedures. Some cargo is packaged in containers. Proper shoring (support) is essential in many cases to prevent cargo from shifting. Loading and unloading procedures are especially critical to fire prevention.

Makeshift Construction

When unskilled personnel attempt to construct stowage facilities, the results are usually less than satisfactory. In fact, makeshift stowage racks are extremely dangerous. Generally, these racks are too weak to support the material, or they are so poorly designed that they allow material to fall or slide (Figure 2.9). The locations of such construction projects are often chosen without regard for safety. Records show that materials falling where makeshift stowage racks had been constructed caused serious fires. For example, one of the worst places to stow angle iron is directly above a large item of electrical machinery such as a generator.

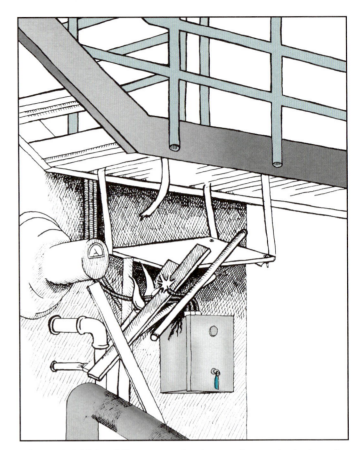

Figure 2.9 Makeshift construction is usually poorly designed and engineered. Materials falling from a makeshift stowage rack can damage equipment or cause a fire.

Regulated/Nonregulated Cargo

Materials carried as cargo on board vessels are divided into two general classifications: regulated and nonregulated. *Regulated cargo* is more generally referred to as *dangerous goods*. Rules governing the classification, description, packaging, marking, labeling, handling, and transporting of regulated cargo may vary from one jurisdiction to another, but they do have one purpose in common: Safeguard the carrier and its personnel. The *IMDG Code* defines very clearly where dangerous goods may be stowed on both passenger and cargo vessels. The regulations include details concerning segregation from other cargo and the proper humidity, temperature, and ventilation requirements. Ultimately, the merchant vessel *master* (also called *captain* or *commander* depending on vessel type) is responsible for compliance with these regulations; however, every member of the crew should be aware of the purpose of the regulations and the consequences of noncompliance.

Cargo that is not specifically covered by regulations is referred to as *nonregulated cargo*. Nonregulated cargo can present a fire hazard if it or its packing is combustible. It may be subject to spontaneous ignition, and careless smoking or faulty electrical equipment may ignite it. It could then have an effect on hazardous cargo that is stowed nearby.

Loading/Unloading Operations

A vessel's deck officers closely supervise loading and unloading operations. The master or a representative always monitors loading—even when port personnel prepare stowage plans in advance of a vessel's arrival. Carefully observe loading and unloading; vigilance and close monitoring are especially important.

Reject leaking cargo immediately during loading. Remove or otherwise render harmless any liquid that leaks into the hold. For example, a vegetable oil that leaks onto baled cotton, rags, or other fibrous material could cause spontaneous ignition. Packaging can be damaged if cargo is allowed to bump either hatch coamings (frames) or other cargo during handling. Packaging can also be damaged during loading if cargo is roughly placed or dropped into the hold. Such damage could go undetected and cause serious leaking problems after the vessel leaves port. Shoreside personnel leave the vessel after loading or unloading the holds; therefore, they have no vested interest in fire prevention on board. A careless act by a shoreside worker can result in a fire starting hours or days later.

Shoring

A vessel can move in many different directions at sea. Proper shoring of cargo to keep it from shifting in rough seas is, of course, important for vessel stability. It is also important from a fire-safety standpoint. If stowed cargo is allowed to shift, hazardous materials that are incompatible can mix and ignite spontaneously or release flammable fumes. Further, metal bands on baled goods can produce sparks as they rub against each other — one spark is enough to ignite some fumes. If not properly shored, heavy machinery can also produce sparks or damage other packaging and thus release hazardous materials. As a precaution, hazardous materials containers must be inspected frequently during a voyage for shifting, leakage, and possible intermixing with other materials. These requirements are included in applicable regulations such as Title 49 U.S. Code of Federal Regulations (CFR), *IMDG Code,* and Transport Dangerous Goods, Canada (TDG).

Bulk Cargo

Bulk cargoes (loose materials stored in a hold), whether solids (grain, ore, or coal) or liquids (oil, molasses, or chemicals), have specific hazards that must be considered (Figure 2.10). The danger is obvious with a chemical such as toluene di-isocyanate (TDI), but it is less apparent with cargoes such as tallow or grain. The hazards of bulk cargoes are as follows:

- *Liquidity* — Any cargo that flows may affect the stability of a vessel if it is improperly stowed. Even ore and other solids may flow if the moisture content is high.

- *Volatility* — Flammable or toxic vapors are emitted by many products. The ambient temperature in relation to the boiling point is a factor in vapor pressure.

- *Static electricity* — Many dust-producing cargoes such as grain may explode due to generation of static electricity. Liquids moving through pipes also generate static electricity that can provide an ignition spark.

- *Chemical instability* — Some chemicals are inherently unstable. Examples are monomers such as acrylonitrile. Although an inhibiting agent is added before transport, heating of the cargo may initiate a rapid polymerization reaction, causing fire and explosion.

- *Reactivity* — Contact with other substances may cause a dangerous reaction, even if only residues are involved. For example, traces of bulk fertilizer under a coaming may combine with a later cargo of diesel fuel to form an explosive mixture.

Combustible bulk cargo such as grain can be extremely hazardous if required precautions are not

Figure 2.10 Closely supervise bulk cargo loading/unloading. *Courtesy of R. Wright/Maryland Fire and Rescue Institute/ United States Coast Guard.*

followed. Before loading, the lighting circuits in the cargo compartments are de-energized at the distribution panel or panel board. A sign warning against energizing these circuits is posted at the panel. In addition, periodic inspections are made to guard against reenergizing the circuits.

The explosive potential of some fuels is significant due to the release of gases from a fuel cargo. For example, methane is released from coal when it is heated. Monitoring of methane and oxygen levels may be necessary when carrying coal cargoes. Proper ventilation is necessary. Take precautions against inhaling any released fumes or vapors.

Cargo Containers

Previously the loading/unloading of intermodal containers (suitable for multiple carriers) or other types of cargo containers did not receive much attention, but it is scrutinized more carefully today (Figure 2.11). Vessel personnel have little control over the contents because containers are usually packed many miles (kilometers) from the point where they are finally loaded on board. This lack of control makes cargo-container safety a matter of great concern. Take the following precautions to reduce the chance of fire involving cargo containers and their contents:

- Stow cargo containers with hazardous contents in accordance with applicable hazardous materials regulations (*IMDG Code*).

- Do not allow a container on board a vessel if it shows any sign of leakage or shifting of cargo.

Figure 2.11 Intermodal containers require proper stowing when loading to prevent hazardous conditions. *Courtesy of R. Wright/Maryland Fire and Rescue Institute/United States Coast Guard.*

- Use extreme caution if a container must be opened for any reason in case a potentially dangerous fire, explosion, or toxic-fume condition has developed inside.

- Consider damaged containers a cause for concern.

Galley Operations

A vessel's galley is a busy place, regardless of whether it is on a small harbor tug or a large passenger liner — and it has a great potential for fire. The intense activity, the many people, the long hours of operation, and the basic hazards — open flames, fuel lines, food and packaging rubbish, and grease accumulations — all add to the danger of fire in galley operations. For these reasons, it is extremely important to never leave a galley unattended when it is in use. Also, leave galley spaces unlocked after normal operating hours so that fire/security rounds can be made through these areas. Extra care is particularly important around energy sources, ranges, and deep fryers. Good housekeeping practices are necessary to prevent fires. Awareness of fire hazards and fire prevention techniques in these areas are discussed in the following sections.

Energy Sources

The most common energy source for cooking is electricity. Diesel oil is used to a lesser degree, and liquefied petroleum gas (LPG) is used on some smaller vessels. Electric ranges are subject to the same hazards as other electrical equipment: short circuits, brittle and cracked insulation on wiring, overloaded circuits, and improper repairs. When liquid fuels are used for cooking, take extreme care to avoid accidental damage to fuel lines. All galley personnel should be alert to leaks in fuel lines and fittings and know the locations of fuel-line shutoff valves. Emergency shutoff valves must be readily accessible. If a leak occurs, close the proper valves at once. Ensure that competent, qualified personnel make repairs.

Ranges

Ranges present several fire dangers. First, the heat of an operating range can cause a galley fire unless galley personnel exercise extreme caution. Clothing, towels, rags, and other fabrics or papers used around a range can easily ignite if they are handled carelessly. Do not stow any materials above a range, and use range battens (support bars) at all times when a vessel is at sea (Figure 2.12). Second, a gas range's fuel can also become involved in a fire. The main burners must light

Figure 2.12 When underway, range battens keep cooking pots from sliding off the cooking surface. *Courtesy of R. Wright/Maryland Fire and Rescue Institute/United States Coast Guard.*

promptly when they are turned on. If pilot lights do not operate properly, gas fumes could leak into the galley. Any source of ignition could then cause an explosion and fire. Extinguish all burners, pilot lights, and other sources of ignition if a gas leak is discovered. Then close the emergency shutoff valves.

Deep Fryers

Deep fryers (also known as *deep-fat fryers*) are a source of both heat and fuel for a galley fire. Use deep fryers with caution, and monitor them carefully during operation. Keep a deep fryer stationary so that it cannot shift with vessel movement. Do not stow anything above a deep fryer. When cooking, do not place food that is excessively wet into a deep fryer because moisture causes hot grease to splatter. Grease also splatters or overflows if the basket is overfilled. Once ignited, grease burns rapidly. Most importantly, do not leave a deep fryer unattended while it is operating.

Housekeeping Practices

Most galley fires are caused by poor housekeeping practices and carelessness. Galley activities generate abundant heat and fuel for fires. Thus, good housekeeping practices are of the utmost importance, particularly concerning grease accumulations, cleanliness, and trash disposal. Place used boxes, bags, paper, and even leftover food in covered, noncombustible refuse cans where they cannot be ignited by carelessly discarded matches or other smoking materials. Segregate plastics from other refuse, and dispose of all trash properly.

Grease accumulations in and around the range (particularly in the hoods, filters, and ductwork) can fuel a galley fire. If the ductwork becomes involved because of heavy grease accumulations, a fire can extend to other areas and decks. Therefore, thoroughly clean hoods, filters, and ductwork periodically. Fixed, automatic extinguishing systems for ductwork are most efficient in extinguishing grease fires. One type of system that prevents fires from occurring by performing an automatic cleaning operation and assists in controlling rangetop fires is discussed in Chapter 8, Fixed Fire Suppression Systems.

Fuel Oil Systems

Fuel oil for the ship's propulsion is stored in double-bottom tanks, deep tanks, cross bunkers, and day or service tanks in the vicinity of the engine room. The capacity of these tanks, using diesel oil as an example, can be as high as 24,000 barrels (2,900 metric tons), depending on the size of the vessel. The types of fuels most commonly used are No. 6 fuel oil, Bunker C, and diesel oil. Bunker C and No. 6 fuel oil are both heavy, tarry substances that require preheating before they can be transferred or burned. Both have flash points of approximately 150°F (66°C) and ignition temperatures of 695°F to 765°F (368°C to 407°C). Double-bottom tanks and deep tanks are fitted with steam pipe grids and coils near the suction pipe to preheat the oil. Diesel oil does not require heating to be transferred and burned. Its flash point is 110°F (43°C), and its ignition temperature is 500°F (260°C). Fuel, diesel, and lubricating oils must be purified or separated by a centrifuge or separator before use. The purifier (or separator) may be located in a room or on a flat (a partial deck in an engine room). This area has extremely high fire potential.

The large quantities of fuel oil on board require proper transfer and maintenance procedures by crew members to prevent fires. Oil burner maintenance and diesel engine operations also require hazard awareness and fire prevention measures. These concerns are discussed in the following sections along with the need to continuously monitor bilge areas for oil leaks.

Transfer of Fuel (Bunkering)

Refueling or bunkering is a frequent operation on any vessel whether conducted between shore and vessel or between vessels (Figure 2.13). Strict regulations apply in most ports. Inspect the transfer system to ensure that strainers are in place and flanged joints are properly tightened before fuel is transferred. Fuel is transferred under pressure, but the liquid fuel itself is not a fire hazard if no mistakes occur. Watch the transfer hose carefully to avoid fire, leakage, or pollution. The fuel vapors that may be released are very hazardous. After fuel is stored and heated if necessary, it is then pumped to the service tanks or settling tanks. From there, it moves to a gravity or day tank (or to a fuel oil service pump), from which it is pumped to the fuel oil burners or diesel engines. Check the system continuously for leaks during the pumping process. Both the overfilling of fuel tanks and leaks in the transfer system increase the danger of fire.

Overfilling. When a tank is overfilled, the fuel rises through the overflow pipe and eventually through a vent pipe that terminates topside. Besides creating a fire hazard, this overflow may also create a marine pollution problem. The engine room crew monitors the transfer process carefully and constantly to prevent overfilling. However, strict control of flames, sparks, and smoking is enforced in the event that overfilling does occur.

Transfer system leaks. If there is a leak in the transfer piping, the pressurized fuel will spray out through the break. Spraying tends to atomize the fuel, and the smaller particles are easily ignited. Thus, line breaks are very hazardous if steam pipes, electric motors, electric panel boards, and so forth are in the area (Figure 2.14). This is also true of lubricating oil leaks near steam pipes. For example, a diesel oil line break resulted in a serious engine room fire in the passenger liner *SS Hanseatic* in New York Harbor in 1966. Fire spread upward from the engine room and involved every deck of the vessel.

Oil Burner Maintenance

Oil burner tips require regular cleaning and maintenance for proper atomization and operation. An improperly operating oil burner tip can cause incomplete burning of the fuel and a buildup of unburned fuel in the windbox of the boiler. This fuel will eventually ignite. If sufficient fuel is present, the flames can spread away from the boiler and involve other materials and equipment. Install oil burner tips with care because improper installation can also cause fuel buildup and ignition.

Diesel Engine Operations

Modern marine diesel engines are equipped with blast/explosion doors to prevent mechanical damage to the engine in the event of crankcase overpressure. Fire may occur in the engine room if these doors

Figure 2.13 An example of bunkering (transfer of fuel) between two vessels. The container vessel is anchored, and the tanker is the bunker barge. Note the derrick with hose attached through which the fuel flows. *Courtesy of Captain John F. Lewis.*

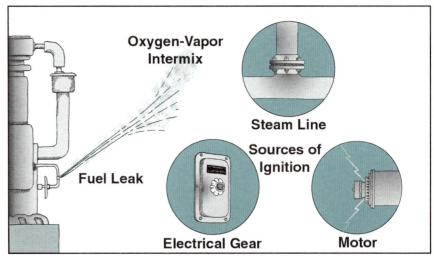

Figure 2.14 Fuel line leaks can spray atomized fuel far enough to be ignited by steam lines or electrical equipment.

malfunction. Leaks in a diesel engine fuel system are also common causes of engine room fires on board diesel-propelled vessels.

Bilge Area Inspections

Fires occur in bilge areas because of accumulations of oil. The oil vaporizes, and flammable vapors accumulate in and around the bilge area. Once these vapors are mixed with air in the right proportion, a spark can ignite the vapors and cause a fire. Bilge fires can move very quickly around machinery and piping, and for this reason, they are not easily controlled. They are more difficult to extinguish than most engine and boiler room fires (Figure 2.15). Most often, the oil leaks into the bilge from an undetected break in a fuel or lubricating oil line. Inspect oil lines until the leak is found. Inspect oil/water separators frequently to prevent overflow, which can also be a source of large accumulations of oil in the bilge area. Monitor the bilge area closely for oil accumulations.

Hot-Work Operations

Hot work refers to any operation that requires heat to do a job or generates heat during the job, most commonly welding or cutting, but it also includes burning, brazing, and grinding. Welding is accomplished either by burning a mixture of gas and oxygen or by electricity. Temperatures produced are extremely high. An oxyacetylene torch can produce a flame that has a temperature of 6,000°F (3 316°C). Sparks and hot debris can travel up to 35 feet (10.7 m) horizontally. The following conditions can contribute to onboard fires during hot-work operations:

- Conduction to adjacent areas
- Ignition of materials within the work area or adjoining areas (Figure 2.16)
- Insufficient elimination of flammable vapors, gases, or dust
- Malfunction or misuse of welding or cutting equipment

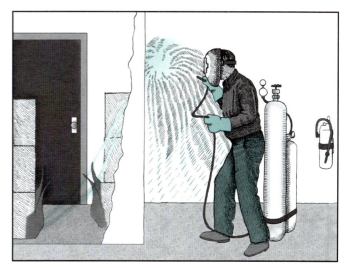

Figure 2.16 Failure to remove combustible materials or to establish a competent fire watch are major causes of fires during burning and welding operations.

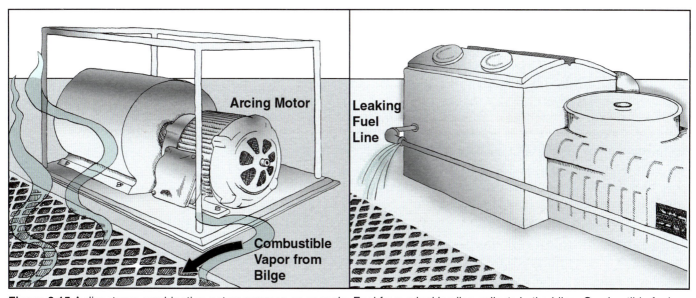

Figure 2.15 A disastrous combination not uncommon on vessels: Fuel from a leaking line collects in the bilge. Combustible fuel vapors from the bilge mix with air as they move toward the arcing motor. Ignition of the fumes by the motor can cause an explosion and fire.

Before beginning hot-work operations, take the following precautions:

- Remove combustible materials or protect materials from ignition by covering or isolating them.

- Test the air in the working area as well as adjacent spaces for flammable vapors.

- Remove any flammable vapors found in the working area or adjacent spaces.

- Ensure the space is kept free of flammable vapors throughout the operation.

- Have the proper type of portable fire extinguisher available and a charged fire hose ready for use.

- Provide a fire watch (person trained to closely observe for the outbreak of fire) around the entire hot-work area (including the opposite side of the bulkhead) until all materials have cooled (at least 30 minutes).

Allow only qualified individuals to perform welding and burning operations. Refer to welding permit laws and local regulations for standards. Many ports have strict regulations governing welding and burning operations. Some port requirements are stricter than federal/national regulations. Workers must pass qualifying examinations before they are licensed to operate burning equipment in some ports. Check local regulations before permitting shoreside workers aboard to perform welding or burning operations.

In some jurisdictions, using an oxyacetylene torch or performing similar welding or burning is prohibited on waterfront facilities or on vessels moored to them during the handling, storage, stowing, loading, discharging, or transporting of explosives. Hot work, electric welding, and the operation of equipment are also prohibited during these handling operations. This restriction applies regardless of whether it is the personnel of the facility or those of the vessel handling the cargo. An exception can only be made by the authority having jurisdiction such as the captain of the port (person having authority over vessels in a port area in the United States) or harbourmaster (equivalent person in the United Kingdom) on a case-by-case basis.

The issuance of a hot-work permit means only that the work may be performed, not that it will be done safely. For safe performance, the operator and assistants must comply with all the safety requirements that are part of the permit. Fire prevention procedures and common sense are integral parts of every welding or burning operation. No welding or burning operations are performed by shoreside workers or crew members without the knowledge and approval of the master or a representative who ensures that safety regulations are followed.

The high temperatures, molten metal, and sparks produced in welding and burning operations are extremely serious fire hazards. Table 2.1 gives recommended practices for preventing and controlling fires during burning and welding practices.

Shoreside Personnel Aboard

Generally, shoreside personnel do not have as much concern for or interest in the vessel as do members of the crew. This situation is perhaps understandable because many shoreside workers do not fully realize the dangers involved in a vessel fire. Indifferent attitudes and lack of interest in fire prevention measures can result in vessel fires. Extremely close supervision and extraordinary alertness on the part of the crew is necessary when shoreside personnel are aboard, particularly during operations involving cargo movement, repairs, and maintenance.

Cargo Movement

Because of the frequency with which they come aboard, the nature of their duties, their access to the vessel's holds, and the materials they handle, shoreside personnel require close supervision. The hazards involved in cargo handling have already been discussed, but they are important enough to summarize as follows:

- Illegal smoking in the hold or on deck during loading and unloading or careless smoking in areas where it is allowed

- Careless discarding of smoking materials and matches

- Careless handling of cargo and the loading of damaged cargo

- Improper stowage of cargo so that it could shift under rough sea conditions — particularly dangerous if two types of cargo are incompatible because they can ignite spontaneously when mixed

Repairs and Maintenance

Contractors who come aboard to do repair or maintenance work, particularly welding, burning, or other hot work, also require close and intensive supervision. Follow the safety measures discussed earlier,

Table 2.1
Recommended Hot-Work Procedures

Factors Contributing to Fire	Recommended Practices
Failure to provide a competent fire watch	• Post a fire watch on continuous duty for each welder or torch operator, and post an additional fire watch opposite the bulkhead or deck upon which the welding or burning operation is being conducted. • Ensure that fire watches have no other duties. • Instruct fire watches to inspect one-half hour after completion of the welding/torch operations and follow by a final inspection one-half hour later (hot metal and slag can retain heat for a long time).
Failure to remove or otherwise protect combustible materials from ignition ***Case Study Reference:*** *CV Manulani Seattle,* August 30–September 7, 1997 (see Appendix A, Case Histories)	• Remove materials in the work area and the areas belowdeck and on the opposite sides of bulkheads if they are within 35 feet (10.7 m) of the welding or burning operation. • Protect materials with flameproof covers or metal guards or curtains if they or the hot-work operation cannot be moved. • Cover openings in the bulkheads, decks, or ducts if they are within 35 feet (10.7 m) of the hot-work operation.
Burning or welding near a concentration of dust or combustible vapors ***Case Study Reference:*** *M/V Carnival Ecstasy, June 20, 1998* (see Appendix A, Case Histories)	• Use appropriate monitoring equipment to determine if the atmosphere is safe to begin hot-work operations and continue to monitor the atmosphere for unsafe concentrations as the work proceeds and concludes. • Ensure that all vapors and dusts are removed from the work area and adjacent areas before beginning hot-work operations. • Ensure that flammable vapors do not return to the area to create a combustible atmosphere during hot-work operations.
Failure to provide an effective means of extinguishing fires at the work site	• Provide a fire extinguisher of the appropriate type and size. • Provide a charged fire hose ready for immediate use.
Failure to properly maintain, secure, and protect burning and welding equipment from physical damage	• Place compressed gas cylinders in an upright position and support them securely. Place cylinders a safe distance from the hot-work operation. • Protect gas and oxygen hoses from mechanical damage, flying sparks, slag, and hot metals resulting from the work operation. • Remove gas and oxygen hoses from the hot-work area when torches are disconnected or not in use. • Use welding and burning equipment approved by either the flag state or classification society. Only use approved equipment that is in good repair.

Continued on next page

Table 2.1 *(Continued)*
Recommended Hot-Work Procedures

Factors Contributing to Fire	Recommended Practices
	• Handle oxygen and gas cylinders equipped with regulators (devices that prevent excess pressures and provide for proper mixing of oxygen and gas) with extreme care.
	• Use only standard gas and oxygen hoses (oxygen is green and acetylene is red) and fittings.
	• Do not improvise repairs such as using tape to seal a leak in a line.
	• Make sure all gas line connections are tight.
Inexperienced or unqualified workers	• Permit only well-trained, qualified operators to handle the equipment.
	• Ensure that the person using the equipment (either crew member or shoreside worker) has the proper knowledge and experience before allowing the hot-work operation to begin.
	• Use welding permit laws and other local regulations to aid in determining whether an operator is qualified.
Failure to provide a means of turning off welding and burning gases and other flammable and oxidizing agents piped into the work area from an outside location	• Provide remote shutoffs for all gases or other fuels piped or otherwise supplied into the work area before beginning hot-work operations.
Failure to provide adequate ventilation to confined spaces	• Provide and maintain adequate ventilation of ALL spaces throughout hot-work operations. An otherwise safe atmosphere in a confined space can become hazardous (either from a fire or a life-safety standpoint) just from the products of combustion created by the operation and the consumption of oxygen.

and assign crew members to watch for any unsafe practices and report them promptly. Assign a crew member to accompany every work party, and never leave shoreside personnel who are working aboard unsupervised. Because the ship's crew members are the ones endangered by fire at sea, they must assume the responsibility for seeing that fire prevention procedures are followed by shoreside workers. Crew members should take the following general safety precautions:

• Ensure that the work area is free of combustible rubbish and waste material.

• Enforce no-smoking regulations.

• Test any machinery or equipment that has been repaired by a contractor. Improperly repaired equipment, particularly electrical equipment, can be a source of problems later.

• Inspect handheld power tools for the proper type of grounding plug and for frayed wiring.

• Inspect and test any fixed fire detection or extinguishing system that has been repaired to ensure that the repairs were done properly.

Shipyard Operations

The hazards of shipyard operations are closely related to the hazards of repair/maintenance operations performed by shoreside workers aboard but on a much larger scale (Figure 2.17). A vessel is normally placed in a shipyard for major repairs, refitting, or conversion — operations beyond the capability of the crew. Thus, the vessel may be swarmed by shoreside workers whose poor housekeeping habits and indifferent attitudes can add to the fire hazards normally encountered. The following situations may be present during shipyard operations:

- Welding and burning operations throughout the vessel

- Temporary shutdown of fire detection and extinguishing systems for modification or to allow repair operations

- Few crew members aboard to monitor work procedures and safety precautions by workers

- Limited or nonoperating systems for power, ventilation, and lighting on board (such a condition is referred to as a *dead ship*)

The mere existence of regulations on shipyard operations does not ensure that they are followed. Hazardous conditions are created at shipyards through carelessness, indifference, lack of knowledge, oversight, or deliberate violation of the regulations. A crew's defense against such practices is vigilance while the work is being done, immediately after it is completed, and for as long as the vessel is in service. Shipyard practices that can lead to hazardous conditions include the following:

- Dry-docking a vessel or making major alterations without prior approval from the appropriate agencies

- Failing to request an inspection following repairs or alterations

- Installing unapproved or substandard equipment not designed for use on board a vessel

- Performing improper or poor workmanship on bulkheads and decking that destroys their resistance to fire

- Concealing poor repairs to tanks, bulkheads, and so forth by conducting inadequate pressure tests

- Failing to complete repairs on fire detection or suppression systems and equipment before the vessel leaves the shipyard

Figure 2.17 A ferry in a floating dry dock: a dock that is partly submerged to permit entry of the vessel and then raised to keep the vessel dry while major repairs are made. *Courtesy of Captain John F. Lewis.*

- Leaving the vessel unprotected by removing all fire fighting equipment for testing at the same time

- Failing to establish a water supply, thus leaving the vessel's fire main dry

- Cutting holes in hull and bulkheads, thus rendering fixed extinguishing systems ineffective

- Failing to replace watertight doors following repairs

- Making openings in bulkheads in violation of fire-safety standards

- Making fire doors inoperable by leaving hoses in the openings

- Failing to use fire curtains if doors are open

- Failing to gas free tanks, piping, and adjacent tanks for bulkhead welding before beginning welding or burning operations

- Dismantling fuel pipes that are under pressure

- Using improper electrical wiring practices such as the following:
  - Using wire of a gauge insufficient to carry the intended load
  - Bypassing overload protection
  - Extending wires through bilges or other areas in the vicinity of water piping

Tanker Operations

Tanker operations involve the movement of combustible or flammable liquid cargo from vessel to shore, shore to vessel, or vessel to vessel. Carelessness, neglect, inattention to duties, poor equipment, or violation of the regulations can have dire consequences. Tanker accidents lead to the destruction of vessels, the loss of lives, and environmental pollution. Resulting fires and explosions can be so severe that shoreside installations can also be seriously affected. No transfer operation can begin without a person in charge at each end of the operation. The licensed officer or tanker crew member in charge must know and perform the required duties and responsibilities according to the regulations of Standards of Training, Certification, and Watchkeeping for Seafarers (known as STCW 95).

Under many flags and in many nationally controlled waters, strict rules and regulations (such as those from STCW 95, *International Convention for the Safety of Life at Sea [SOLAS]*, U.S. Coast Guard, and Titles 33 and 46 U.S. CFR, *Shipping* and *Navigation and Navigable Waters*) govern the operation of tank vessels. Many codes require that the persons in charge jointly and independently inspect both the vessel and the shoreside facility before any flammable liquid or other hazardous products are transferred. These are very formal inspections during which a listing of the inspected items is completed and signed by both parties. A sampling of the items on this checklist is as follows:

- Mooring lines are adequate for anticipated conditions.

- Cargo hoses and/or loading arms are long enough for intended use.

- Cargo hoses are adequately supported to prevent strain on couplings.

- Transfer system is properly aligned for discharging or receiving liquid.

- Adequate spill containments are provided for couplings.

- Scuppers or other overboard drains are closed.

- Communications system is provided.

- Emergency shutdown system is available and operable.

- Adequate lighting of the work areas and manifold areas is provided.

- Persons in charge have met to ensure mutual understanding of the transfer operation.

Procedures such as proper fendering (cushioning), coordination and communication during cargo transfers, use of cargo transfer hose, and vessel-to-vessel transfers are critical cargo transfer operations. Cargo expansion, static electricity, and sources of open flames or sparks are situations that also require cautions during transfer operations. Crew members also need awareness of hazards that are present in pump rooms, with cargo heating systems, and during tank cleaning. These tanker operations are discussed in the sections that follow.

Cargo Transfers

Thoroughly plan every cargo transfer with close coordination throughout the operation. The person in charge on the vessel is responsible for ensuring the safety of the transfer operation, which extends to areas around the vessel, as well as to shore installations and other vessels. The entire crew is required to cooperate with the person in charge during transfer procedures. Test emergency shutdown procedures and the means of communication between the persons in charge before a transfer is started. The person in charge on a vessel must be able to shut down the transfer flow or request shutdown through a communication system that is used for no other purpose. Emergency shutdown systems must be provided on vessels and on shore.

Transfer systems are only as effective as the people who are charged with the responsibility for using them. Even a momentary lapse can permit an overflow resulting in a spill on the vessel or vessels, at the terminal, in the water, or at all locations. Errors or omissions can result in fire or explosion. Make sure fire-fighting equipment is ready for use before any transfer operation is conducted. Take the following steps:

- Flake fire hoses on deck and connect them to the fire main.

- Place fire extinguishers near the deck side fuel transfer manifold.

- Unlock and prepare foam monitors for operation.

- Pressurize the fire main by running the main fire pumps. See sidebar.

Once shoreside facilities and the vessel (or vessels) are inspected, communications are tested, emergency shutdown procedures are tested, and fire fighting equipment is positioned, cargo transfer can begin.

Fire Main

Because an outlet for the water is required and the deck scuppers are sealed to avoid pollution, the easiest way to achieve a discharge is through the anchor washes. Thus, the fire main retains pressure without the pump overheating. If a fire occurs, opening the hydrants gives water instantly, and closing the anchor washes gives full pressure. This method also prevents the fire main from freezing in cold weather.

Awareness of hazards involving fendering, cargo transfer hose operations, cargo expansion, static electricity or open flames/sparks, and pump rooms continues throughout the transfer process. Some additional precautions are needed for vessel-to-vessel transfers.

Fendering. Improper or inadequate fendering between the two objects involved in the transfer can generate sparks, particularly during vessel-to-vessel operations. Because the vapors given off by petroleum products are heavier than air, they tend to drift down to the water where they can be ignited by sparks from metal-to-metal contact caused by inadequate cushioning.

Cargo transfer hose. Either the vessel or the shoreside facility (or either vessel in the case of vessel-to-vessel transfer) may supply the cargo transfer hose. However, both persons in charge must inspect it to ensure its quality and stability. The following precautions are taken to prevent or contain leaks during a transfer:

- Position the cargo transfer hose so it cannot be pinched between the vessel and the dock or between vessels.

- Allow sufficient slack in the cargo transfer hose to allow for changing tide conditions and lightening of the load.

- Do not place the cargo transfer hose near a hot surface.

- Support the cargo transfer hose by derrick or crane to prevent chafing.

- Inspect the cargo transfer hose frequently for leaks during the transfer; prepare to shut down the transfer operation if necessary.

- Use proper containment devices such as spill trays (permanent devices are under the manifold); place temporary spill trays under flanges in the cargo transfer hose.

Static electricity. Precautions must be taken to prevent the generation of static sparks during cargo transfers. Clothing and footwear worn by persons involved in transfers must be types that do not produce static electricity. The best way to prevent sparks during transfers is to ensure electrical isolation. Cargo hose strings and metal arms fitted with an insulating flange provide electrical isolation between the vessel and the shoreside facility and protect against arcing. A ship/shore bonding cable is **not** effective as a safety device and may be dangerous. The persons in charge are responsible for ensuring that isolation is done properly. See the *International Safety Guide for Oil Tankers and Terminals* for more information (Appendix B, References and Supplemental Readings).

Certain cargoes such as kerosene jet fuels and distillate oils can generate static electricity as they are moved. Water suspended in these cargoes increases the possibility of static spark generation. Start the operation with a low loading rate to permit the water to settle to the bottom of the tank easily. Avoid the use of synthetic line, steel ullage tapes, metal sampling cans, and metal sounding rods when these static-producing cargoes are loaded and for at least 30 minutes after loading. The waiting time permits suspended water to settle and static electricity to dissipate.

Oil that is splashed or sprayed may become electrostatically charged. For this reason, never load oil into a tank through an open cargo transfer hose. Oil splashing around the opening can generate static electricity.

Open flames or sparks. Ignition of flammable vapors by an open flame or a spark is the most obvious fire hazard during transfer operations. Safety measures include posting signs indicating when radio equipment and boiler and galley areas may operate, controlling hot-work and other repair operations, securing ventilation and air-conditioning intakes, and securing doors and ports on or facing cargo tank areas. Sources of sparks and flames include the following:

- Smoking materials

- Boiler and galley operations

- Radio equipment

- Welding and burning operations

- Machinery operation

- Electrical equipment in living quarters
- Unapproved flashlights or portable electrical equipment
- Chipping and grinding of metalwork prior to painting
- Ignition sources on nearby vessels or on shore
- Tools that produce sparks
- Clothing or footwear that produces static electricity

Cargo expansion. One cause of cargo overflow is failure to allow for expansion of the product caused by temperature increases. This problem is prevented by following proper loading procedures that allow room for expansion. Most cargoes are not loaded greater than 98 percent of tank capacity and in some cases may be loaded to a lesser percent such as when loading in winter in North America or Europe and unloading or discharging in the Tropics.

Pump room hazards. Because it is subject to vapor accumulation, the cargo pump room is the most hazardous area on a tank vessel. To ensure that vapors are removed during cargo transfers, operate the ventilation systems in pump rooms continuously. As a safety precaution, be sure the ventilation system is working and has operated for at least 15 minutes before entering the pump room. No repair work should be permitted in the pump room unless absolutely necessary. In fact, proper maintenance helps avoid both repairs and leaks in piping and pump seals. Any piece of equipment that might cause a spark is prohibited because of the possibility of igniting vapor accumulations.

Vessel-to-vessel transfers. When vessel-to-vessel transfers are underway, take the following additional precautions:

- Provide adequate fendering.
- Anticipate changes in weather, sea, and current conditions.
- Understand clearly who on each vessel is in charge of the operation.
- Consider the effect of drifting vapors on both vessels.

Cargo Heating System
High-viscosity cargoes become so thick at low temperatures that heating is required before they can be pumped. Steam pipes or coils that run through the bottoms of the tanks heat the liquids. Regulation of the temperature to which these liquids are heated is critical. Overheating is hazardous because dangerous flammable gases can be generated and released.

The tank heating system must be well maintained. A steam leak at the tank bottom can lead to the same problem as overheating—chemical reactions and the production of dangerous flammable gases. The cargo could also leak into the steam coils with equally dangerous results. Many SOLAS or IMO regulations limit the heating of fuel oil in storage tanks to a maximum of 120°F (49°C).

Tank-Cleaning Operations
A vessel is very vulnerable to fire during tank-cleaning operations because the tank atmosphere passes through the flammable range. After a tank has contained a volatile cargo (for example, gasoline), its atmosphere is *rich.* During cleaning, it changes to *lean* and then *clean* (no gas present). If any ignition source or electrostatic probe is introduced during the time the tank's atmosphere is explosive, it will explode. Using an inert gas system (IGS) ensures safe operation during tank cleaning. The IGS produces exhaust gas (either from a gas generator or from the main engine exhaust) that is introduced into the tanks until the atmosphere is less than 8 percent oxygen (usually less than 3 percent oxygen), which makes ignition impossible. See Inert Gas Systems on Tank Vessels section in Chapter 8, Fixed Fire Suppression Systems.

Collisions
Fires caused by collisions, particularly when tankers are involved, have resulted in serious damage and great losses of property and life. An example is the *Sea Witch/Esso Brussels* incident (see Appendix A, Case Histories). The container vessel *Sea Witch* lost her steering in New York Harbor on June 2, 1973, and struck the anchored tanker *Esso Brussels*. The resultant fire in the blazing waters as both vessels (locked together) drifted towards the Verrazano Narrows Bridge involved the response of many agencies, tugs, and other vessels. The cause of the loss of steering was the failure of a small pin, which certainly raises concerns about steering systems, independent backup systems, testing of emergency systems, and emergency procedures.

Some collision incidents in the past were beyond the capacity of the crew to prevent. However, many important lessons have been learned. Regulations on allowable quantities of flammable or reactive substances, enforcing no-smoking requirements, and other controls do work in preventing accidents. As a result, the number of these casualties and losses is decreasing.

Incendiary/Arson Fires

In most onboard fires, a human factor is involved. In certain instances, this factor is minimal, but in some cases it is the single, greatest contributing factor. *Incendiary or arson fires* are purposeful destruction of property with a willful disregard for property and safety. The reasons for a person setting an intentional fire result from a number of factors, including the following:

- Pyromania (a psychological disorder)
- Attempted suicide
- Revenge
- Profit (fraud)
- Terrorism

Any of these reasons can be used as an excuse for arson, but only firesetters can explain their behaviors. An appropriate regulatory agency is responsible for making a fire cause determination after conducting an investigation. This section only introduces the mariner to the subject of intentionally set fires. See IFSTA's **Introduction to Fire Origin and Cause** manual for more details.

 ## Fire Prevention Program

If most onboard fires are preventable, then who is responsible for preventing them? Fire prevention is the shared duty of each and every member of the crew — not just the master or the chief engineer or any particular individual or group of individuals. No fire prevention effort or fire and life safety program can be successful unless it involves everyone aboard the vessel. Fire prevention is not easily defined, perhaps because it is primarily a matter of attitude, and its benefits are not easy to measure until after they are lost. For these reasons, it is difficult to persuade people to actively practice fire prevention. Acceptance requires continuing effort and strong guidance and leadership.

Every seafarer probably fears the consequences of a serious fire at sea. Unfortunately, awareness of the possibility of fire does not always lead to the attitudes and actions necessary to prevent it. Some individuals are sensitive to the hazards of fire and the means of preventing it. Others are completely irresponsible, perhaps because of indifference. Only good luck keeps some of these people from becoming victims of their own carelessness. The majority of people are somewhere between these extremes: those who are very careful in some respects and foolishly careless in others. A lack of knowledge is a factor in carelessness.

Crew members should analyze their own attitudes toward fire and life safety and toward fire prevention in particular. This analysis may only require the answers to two simple questions: *"Do I know the causes of onboard fires?"* and *"Have I considered the damage and loss of life that can result from a fire?"* A carefully planned and conducted fire prevention program can ensure that both questions are answered with a strong, unconditional *"Yes!"*

Every person aboard has responsibilities in a fire prevention program: master, engineering and deck officers and other supervisory personnel, and crew members. Elements of an effective program include formal/informal fire and life safety education/training sessions, good housekeeping practices, and preventive maintenance and repair procedures. It is important that new crew members also receive a thorough orientation to the vessel, their fire prevention responsibilities, and their emergency duties. The examples shown by vessel owners and officers and the recognition of crew members' efforts are motivational factors that cannot be overlooked.

Crew Member Responsibilities

Because every crew member is responsible for the prevention of fire on board, every person has a role in the vessel's fire prevention program. A fire and life safety committee guides development of the program. The committee is composed of representatives from all departments on the vessel. Committee members are appointed by their department heads or elected by coworkers. The master appoints some members. The fire and life safety committee reviews and inspects any incidents or accidents and remedies dangerous situations before an accident occurs. Committee members observe the workplace and act on items brought to their attention.

Because a person's attitude is a part of fire and life safety, it is also an important part of a fire prevention program. To a great extent, the attitude of the crew reflects that of the master. Concisely stated, the following lists give the fire and life safety responsibilities at each personnel level.

Program Elements

The *International Safety Management (ISM) Code* is the current benchmark for safety programs, including fire prevention programs. A fire prevention program must be carefully planned and structured to be successful. The details of the program are tailored to the vessel for which it is developed. Thus, a fire prevention program for a tug is much less formal than one for a tank vessel; however, each program reflects the master's concern for fire and life safety. A fire and life safety committee develops the program, and the master and deck and engineering officers (who give it the high priority that it merits) implement it. On any vessel, a fire prevention program includes the following elements:

- Fire and life safety education/training (both formal sessions and informal discussions)
- Preventive maintenance and repair procedures
- Good housekeeping practices
- Orientation for new crew members, including the following:
  — Fire prevention measures, including good housekeeping practices
  — Fire reporting procedures

 The Master

- *Develops and maintains positive and cooperative attitudes among crew members.*
- *Exhibits a concern for fire prevention.*
- *Emphasizes that every crew member is expected to assist in fire and life safety goals.*
- *Participates in managing, contributing to, and approving the programs of a fire and life safety committee.*
- *Exhibits continued interest in fire prevention efforts.*
- *Supervises all coordinating fire fighting and lifesaving efforts as designated on station bill.*

 Deck and Engineering Officers

- *Take an active role in the functions of a fire and life safety committee.*
- *Evaluate subordinates for fire and life safety attitudes and levels of training.*
- *Provide daily, on-the-job supervision that develops good fire prevention attitudes and habits (Figure 2.18).*
- *Provide formal and informal fire and life safety instructional sessions to subordinates, correcting unsafe actions immediately.*

- *Provide day-to-day training to crew members about the development of fire prevention standards, emergency duties, and the operation of equipment.*
- *Instill a sense of pride in crew members for earning and maintaining a fire-safe record.*
- *Keep up to date on the causes of fires on board vessels.*
- *Schedule inspections, maintenance, and repairs.*
- *Coordinate fire fighting and lifesaving efforts at the emergency scene as designated on station bill.*

 Crew Members

- *Report and eliminate hazardous conditions on board, wherever they are found.*
- *Operate the vessel's equipment safely. Ask for instruction when unsure of how to use unfamiliar machinery or equipment.*
- *Develop and maintain a positive fire and life safety attitude.*
- *Perform emergency duties as designated on station bill.*

- Crew members' duties and assigned locations as listed and described on the muster list and station bill
- Use of portable fire extinguishers, fire fighting equipment, self-contained breathing apparatus (SCBA), and personal protective equipment (PPE)

• Recognition programs

These elements are discussed in the sections that follow. It is important to emphasize that the fire prevention program itself is the subject of continual review by both the master and the fire and life safety committee. Both the scope and content of the program are modified as necessary to improve fire and life safety. Fire and life safety committee members cannot relax once they have developed and implemented a fire prevention program. Examine the program every time an unsafe situation is discovered, and expand it to ensure that such a situation does not occur again. To wait until a fire occurs is to await possible disaster.

Figure 2.18 On-the-job training gives deck and engineering officers the opportunity to teach safe practices. It ensures that crew members get correct information, and it establishes an avenue of communication.

Fire Prevention Education/Training

The fire and life safety education/training of crew members may be difficult and, at times, frustrating, but it is an important factor in any fire prevention program. It is a continuing process that includes both formal educational/training sessions and informal discussions. Do not miss any opportunity to develop an awareness of fire safety. The objective of this education/training process is to teach every crew member to think fire prevention — before, during, and after every action. Each crew member must ask: *"Is it safe?" "Could it cause a fire?"* This attitude toward fire prevention might be called *"taking one second for safety."*

Formal educational/training sessions. The fire and life safety committee plans and schedules the formal educational/training sessions. In addition to the required fire and lifesaving drills, conduct sessions on a regular basis during each voyage, and start them as soon as possible (Figure 2.19). Convey the master's attitude toward fire prevention and life safety to all crew members before sessions begin. Hold sessions at different times of the day (for example, morning and afternoon) so that all watches are accommodated. Post schedules in advance. Total participation is just as important to these educational/training sessions as it is to the overall fire prevention program.

Training aids such as films, transparencies/slides, compact discs (CDs) and videotapes add interest to educational/training sessions. While repetition is necessary to achieve learning, it should be done in such a way as to not become mundane. People lose interest and pay little attention to material that is presented over and over again in the same way. Vary the topics,

Figure 2.19 Formal educational/training sessions are the foundations of fire prevention programs.

presentation, and approach as much as possible. Along with lectures, use several instructional delivery methods such as discussion, illustration, and demonstration. Role-play and brainstorming activities add interest, and case study reviews encourage discussions. Schedule practice training sessions (for example, performing equipment maintenance or inspections) so crew members can apply their learning.

This book could serve as a basis for a fire prevention and fire-fighting curriculum. In addition, each vessel should have its own fire library. Encourage crew members to use it. Up-to-date materials promote active library use. The reference materials and publications named in this book would make excellent additions to such a library. For example, IFSTA's **Fire and Life Safety Educator** manual and **Fire and Emergency Services Instructor** manual give many resources for formal educational/training sessions and library materials. See Appendix B, References and Supplemental Readings, for more sources.

Informal discussions. Informal discussions are also effective teaching tools. When crew members hold discussions in a relaxed atmosphere, everyone gets a chance to speak and to listen. Discussions give a free exchange of information and ideas, which leads to a better understanding of the responsibilities of crew members relative to their specific skills, the general safety of the crew and vessel, and fire prevention in particular (Figure 2.20).

Visual reminders, posters, warning signs, and personal messages to the crew are also effective informal education media. Here again, it is important to vary the message and the instructional medium. The same posters or messages left in the same locations week after week indicate a lack of interest on the part of the program's planners. Unfortunately, this lack of interest can easily become contagious.

Good Housekeeping Practices

Basically good *housekeeping* means the care and management of property. However, from the fire prevention standpoint, it means the elimination of sources of fuel for fires, that is, the elimination of fire "breeding grounds." Include good housekeeping practices in every five-prevention/training program. Housekeeping problem areas are given in the fire prevention/inspection checklist given at the end of the chapter. Samplings of some critical practices are as follows:

- Store cleaning rags and waste in covered metal containers.
- Place oily rags in covered metal containers and dispose of as soon as possible (Figure 2.21).
- Store paints, varnish, etc., in paint locker when not in use (Figure 2.22).
- Clean up spills or leaks immediately.
- Clean grease filters and galley hoods regularly.
- Avoid accumulations of dust in holds and on lightbulbs.

Figure 2.20 Informal discussions provide opportunities for crew members to learn from each other, stimulate interest in fire prevention, and establish the proper attitude toward safety.

Figure 2.21 Place oily rags in covered metal containers to prevent fires by spontaneous ignition.

Figure 2.22 These paints, varnish, and others items are properly stored in this paint locker. *Courtesy of R. Wright/ Maryland Fire and Rescue Institute/United States Coast Guard.*

Preventive Maintenance/Repair Procedures

After training is complete, establish schedules for periodic maintenance procedures. Strong leadership and the backing of management are necessary ingredients of preventive maintenance programs. Information should be channeled from the master through deck and engineering officers to crew members. Standardized maintenance schedules are absolutely necessary, but they are effective only when they are implemented. The basic elements of a preventive maintenance/repair program include the following:

- Lubrication and care
- Testing and inspection
- Repair or replacement
- Record keeping

Perform equipment lubrication tasks, testing and inspections, and repair or replacements according to definite schedules that depend on the equipment in question. For example, some equipment needs servicing at various intervals during each watch. Other equipment might require maintenance once each watch, daily, weekly, annually, or at even longer intervals. The equipment manufacturer's manual is the best guide for establishing periodic maintenance schedules. Keep appropriate records of maintenance/repairs on each piece of equipment.

It would be beyond the scope of this book to outline complete programs for preventive maintenance on typical vessels. Such programs already exist on most vessels. Instead, the following sections discuss the basic elements of a preventive maintenance program for the care of machinery and equipment (including boilers, piping, fittings, and bearings) and the program's relation to fire prevention. The elements of testing, inspection, repair or replacement, and record keeping are also discussed. A well-run maintenance program is an important defense against fire.

Machinery and equipment. Probably the most basic element in a preventive maintenance program for machinery and equipment is regular and proper lubrication. Scheduling alone is not sufficient to ensure that lubrication will be performed: Personnel may tend to neglect machinery that is difficult to reach. Institute controls to ensure that manufacturers' lubrication recommendations are followed and lubrication schedules are met.

Boilers and appurtenances. Inspect and test boilers, pressure vessels, piping, and other machinery at regular intervals. Boilers and appurtenances require special care because they involve heat and high pressures. Perhaps no other type of equipment pays higher dividends for proper preventive maintenance. On the other hand, neglect of this equipment can result in poor operation, explosion, and fire.

Piping and fittings. Repair leaks in piping and fittings that carry fuel, chemicals, flammable products, water, or steam immediately. In some cases, it is only necessary to replace a gasket and tighten a fitting. In others, a section of piping may have to be replaced. Whatever the repair, take care to ensure that it is done properly. The repair should leave the piping properly aligned and supported. Do not use piping and fittings for handholds or footholds or for securing chain falls. The results of such misuse may not be evident immediately, but continued misuse can only weaken the equipment and lead to a slow leak or sudden rupture.

Bearings. Lubricate bearings properly. Do not operate any piece of machinery until its bearings are lubricated with the proper lubricant. Do not use any piece of machinery when its bearings are in poor condition except in a dire emergency. A number of onboard fires have been caused by overheated bearings. Ensure that the machine operator knows the approximate normal operating temperature of the bearings so checks can be made during operation to determine whether they are overheating.

Testing and inspection. Inspect and test every piece of equipment on board before the vessel leaves port. The status of the equipment on a vessel before its sailing is just as critical as that on an aircraft. Before takeoff, for example, an aircraft pilot runs through an extensive checklist. Vessel management personnel (mainly the master, chief officer, chief engineer, and second engineer) also require that careful tests and inspections are performed. A formal report of the results of these tests and inspections is given to management personnel or entered in the official, deck, or engine logbooks.

Repair or replacement. Regulatory agencies require notification when equipment is repaired or replaced. Only approved types of machinery and equipment are acceptable replacements. Approval from the appropriate agency is based on past performance, and safety is an important criterion. It is important that competent and knowledgeable people perform repairs. Establish controls to ensure that repairs are made properly regardless of whether workers are vessel personnel, shoreside contractors aboard, or workers in a shipyard. An improper repair to an electric range in the galley, a leaky joint in a fuel line, or a defective boiler can have the same results — fire on board.

Record keeping. Keep a log of machinery and equipment maintenance as it is performed. Include a record of all tests, inspections, malfunctions, repairs, adjustments, readings, and injuries/casualties. Such a file provides the history of each piece of equipment from the day it was installed. Records are a great help in diagnosing problems and in deciding when to replace machinery.

New Crew Member Orientation

Hold new crew member orientation sessions promptly. It is essential to correctly instruct new crew members joining a vessel on their emergency-related duties on board. Encourage new crew members to ask questions, report any unusual circumstances, and take nothing for granted. Show them the SOLAS emergency instruction books, and explain where they are stored so that they are easily accessed for study. Brief descriptions of the areas covered in orienting new crew members follow. See Chapter 14, Fire and Emergency Training and Drills, for more information on new crew member training.

- *Emergency-related duties* — Explain that all crew members shall acquaint themselves with their positions and assigned locations given on the station bill and muster list. Identify muster stations and emergency escape routes. The station bill also specifies the duties to be performed in the event of an emergency (such as fire, man overboard, or abandon ship). Some vessels, depending on their types, may list other emergency-duty situations such as fuel spills, piracy, collision, and grounding.

- *Alarm signals, safety information symbols, and safety signs* — Explain the meaning of alarm signals, safety information symbols, and safety signs used on the vessel. Explain how to sound an alarm. Show that exit signs lead to the main deck, and explain how to plan alternative routes in case a primary route is blocked. Alarm signals are described on the vessel's station bill. Some of the more common emergency signals are as follows:

  — *Fire and emergency:* Continuous blast of the whistle for at least 10 seconds followed by continuous ringing of the general alarm for at least 10 seconds

  — *Man overboard:* Letter O (three long blasts about 5 seconds each) sounded at least four times on the whistle followed by the same signal sounded on the general alarm

  — *Abandon ship:* At least seven short blasts followed by the same signal sounded on the general alarm

  — *Boat handling:* One short blast on the whistle means to lower lifeboats; two short blasts means stop lowering lifeboats

- *Fire and lifesaving equipment* — Point out the locations of fire and lifesaving equipment, especially in each crew member's area of work (such as galley, engine room, etc.). Have everyone find the portable fire extinguisher nearest to their cabin and work area and the nearest exit. Demonstrate the use of portable fire extinguishers.

- *Personal survival techniques* — Explain techniques for surviving at sea in case of ship abandonment.

Demonstrate how to locate and don life jackets and survival suits. References for survival training include *The Captain's Guide to Life Raft Survival* by Michael Cargal and *Survival Guide for the Mariner* by Robert J. Meuen (see Appendix B, References and Supplemental Readings).

- *Medical emergencies* — Explain how to take immediate action when encountering an accident or medical emergency. Demonstrate elementary first-aid techniques. A United States Coast Guard first-aid course is an available resource.

- *Doors* — Demonstrate the operation of fire doors and watertight doors, and explain their uses.

- *Hazardous materials* — Explain the hazards of any cargo or other hazardous materials on board, and show where they are located. Display and explain a sample material safety data sheet (MSDS), the manufacturer's document that gives information for a hazardous material. Show where the MSDSs are posted or stored for each material, and require that crew members read them.

- *Smoking* — Explain the proper disposal of smoking materials, and identify the areas where no-smoking regulations are observed.

- *Housekeeping* — Give procedures for good housekeeping routines and the reasons for observing them.

Recognition Programs

Vessel owners and operators must demonstrate their own continuing interest in a fire prevention program if they expect the people who operate their vessels to participate. Active participation, obvious concern for fire and life safety, and recognition of effort demonstrate such interest. Recognizing effort on an individual basis, by vessel or crew member, provides the incentive to maintain a good record. Generally awards include hats, jackets, individual and vessel certificates, coffee mugs, flashlights, safety patches, or monetary rewards.

 ## Safety Inspections

An important part of fire prevention is conducting safety inspections. The purpose is to find and eliminate fuels and ignition sources that could cause fires. A number of these possible fire causes were given earlier in this chapter. The elimination of most of these sources is not a technical matter but mainly common sense and good housekeeping practices.

Because many vessels are large and complex, the responsibility for safety inspections cannot rest with any individual or group of individuals. Instead, every crew member is an informal inspector. Crew members should check for fire hazards at all times (on and off duty), wherever they are on the vessel. This continual inspection is a matter of attitude and an extension of the *"one second for safety"* idea.

The master, chief officer, and chief engineer make a joint formal inspection of the entire vessel — from bow to stern and bilge to bridge — at least once each week. The formal inspection is systematic; a checklist is used to ensure that no area is overlooked. This checklist is used as a guide for informal inspections as well. The checklist can be customized to suit the needs of a particular vessel, depending on what facilities are provided. A sample checklist is given in Figure 2.23. Elements of the fire prevention program are addressed on this list, and it is also used as a reminder of necessary inspections. A careless check off of items on the list could mean overlooking vital elements of the fire prevention plan. More details on the inspection of fixed fire protection systems are provided in Chapter 8, Fixed Fire Suppression Systems.

Fire Prevention/Inspection Checklist

Part 1. Accommodation Areas		
Crew Quarters	**Pass**	**Fail**
1. Direct, uncluttered means of escape; at least two routes available		
2. General alarm system in good order		
3. Area free of combustible rubbish		
4. Area free of combustibles close to sources of heat		
5. Area free of overloaded electric circuits		
6. Area free of makeshift electrical wiring or repairs		
7. Electrical equipment properly grounded		
8. Fire extinguishers:		
a. In place and unobstructed		
b. Proper type and size		
c. Properly charged		
d. Date of last examination noted on inspection tag		
9. Noncombustible ashtrays:		
a. Adequate number		
b. Adequate size		
c. Properly placed		

NOTES:

Items not passing inspection:

Action required:

Other comments:

Galley	**Pass**	**Fail**
1. Area free of combustible rubbish		
2. Noncombustible receptacles with covers provided		
3. Galley hood and ducts:		
a. Free of grease accumulations		
b. Cleaning date recorded		
4. Fixed extinguishing system properly marked		

Galley	Pass	Fail
5. Fire extinguishers:		
a. In place and unobstructed		
b. Proper type and size		
c. Properly charged		
d. Date of last examination noted on inspection tag		
6. Area free of leaking pipes and fittings		
7. Area free of overloaded electrical outlets		
8. Electrical appliances in good repair		
9. Oven and ranges dry		
10. Oven free of cracks or crevices		
11. Oven burners secured		
12. Deep fryer secured and covered		
13. Ventilation ducts, dampers, and screens operational		

NOTES:

Items not passing inspection:

Action required:

Other comments:

Mess Rooms and Lounges	Pass	Fail
1. Noncombustible ashtrays:		
a. Adequate number		
b. Adequate size		
c. Properly placed		
2. Fire extinguishers:		
a. In place and unobstructed		
b. Proper type and size		
c. Properly charged		
d. Date of last examination noted on inspection tag		
3. Fireproof trash containers with covers provided		

NOTES:

Items not passing inspection:

Action required:	
Other comments:	
Submitted by:	
Signature:	Date:

Part 2. Deck Department		
Items	**Pass**	**Fail**
1. Decks:		
a. Free of combustible rubbish		
b. Free of oil and grease		
c. Free of leaking pipes and fittings		
d. Free of damaged or leaking containers		
2. Holds clean and dry before loading		
3. Cargo lights removed after loading		
4. Dangerous cargo stowed according to regulations		
5. Cargo stowed properly		
6. Dangerous cargo manifest and cargo stowage plan in order		
7. Fuel for lifeboats properly stored		
8. No-smoking signs posted		
9. Paints and flammables properly stowed		
10. Bos'n (boatswain) stores properly secured		
11. Ventilation ducts, dampers, and screens operational		

NOTES:

Items not passing inspection:

Action required:

Other comments:

Submitted by:

Signature: **Date:**

Part 3. Engineering Department

Items	Pass	Fail
1. Free of rubbish, waste, and oily rags		
2. Noncombustible receptacles with covers provided		
3. Ventilation ducts, dampers, and screens operational		
4. Decks and tank tops free of oil and grease		
5. Free of leaking pipes and fittings		
6. Out-of-service boilers free of oil accumulations		
7. No combustible liquids in open containers		
8. Paints and varnishes in proper storage room		
9. Dunnage in proper storage room		
10. Free of makeshift wiring		
11. Free of unsafe or makeshift stowage construction		
12. No unapproved electrical fixtures in paint lockers, battery rooms, etc.		
13. Warning signs posted:		
a. High Voltage – Keep Clear		
b. No Smoking		
14. Switchboard area free of obstructions		
15. Free of improper fusing or bridging		
16. Motors free of lint and dust		
17. Motors clear of combustible material		
18. Ladders unobstructed		
19. Oxygen, acetylene, and other gas cylinders stored upright and secured		
20. Fire extinguishers:		
a. In place and unobstructed		
b. Proper type and size		
c. Properly charged		
d. Date of last examination noted on inspection tag		

NOTES:

Items not passing inspection:

Action required:

Other comments:

Submitted by:

Signature: **Date:**

Part 4. Fire Protection Equipment		
Fire Main System	**Pass**	**Fail**
1. Hoses in place and free of cuts and abrasions		
2. Nozzles in place and applicator provided (if required)		
3. Valves unobstructed and easily operated		
4. Hose spanner in place		
5. Fire station properly marked		
6. Fire axe present		
7. Pumps tested for flow and pressure and ready for service		

NOTES:

Items not passing inspection:

Action required:

Other comments:

Fixed Fire Suppression Systems	**Pass**	**Fail**
1. System equipment or storage spaces free of debris or improper stowage		
2. Operating control valves unobstructed and in good operating condition		
3. Alarms and indicators in good order		
4. Operating controls set for proper operation		
5. Required amount of agent available and connected		
6. Pipes and fittings in good condition (no leaks)		
7. Discharge outlets in good condition		
8. Operating instructions posted		
9. Signs posted at all system alarms		
10. Hoses in place and free of cuts and abrasions		
11. Nozzles and equipment ready for use		
12. Foam containers free of leaks		
13. Valves and controls properly marked		
14. Monitor stations properly marked		

NOTES:

Items not passing inspection:

Action required:

Other comments:

Sprinkler Systems	Pass	Fail
1. Apparatus marked		
2. Control valves marked to indicate the protected compartments		
3. Spray heads in place, unobstructed, and clear		
4. Fire pumps operational		
5. Fresh-water pressure tank full		
6. Air compressor operational		
7. Check valves and control valve in proper positions and operate freely		
8. Sprinklers undamaged, properly rated, and free from obstructions		
9. Piping intact; free of visible leaks, rust, or corrosion		

NOTES:

Items not passing inspection:

Action required:

Other comments:

Emergency Equipment	Pass	Fail
1. Storage space properly marked		
2. Sets of personal protective equipment (PPE) in good condition and stored in widely separated, accessible locations		
3. The following items present for each set of PPE:		
a. Self-contained breathing apparatus (SCBA)		
b. Lifeline		
c. Explosionproof flashlight with spare batteries		
d. Protective coat and trousers		
e. Helmet		
f. Boots		
g. Gloves		
h. Protective hood		

Emergency Equipment		Pass	Fail
i.	Fire axe		
j.	Personal Alert Safety System (PASS) device		

NOTES:

Items not passing inspection:

Action required:

Other comments:

Submitted by:

Signature: **Date:**

Figure 2.23 Sample fire prevention/inspection checklist. Elements of the fire prevention program are addressed on this list, and it is also used as a reminder of necessary inspections.

Chapter 3

Vessel Types, Construction, and Arrangement

C rew members may think they know all about vessels because they are familiar with those on which they serve, but more complete vessel information is needed for fire and life safety implications. Tremendous advances have been made in worldwide standardization of maritime technology and terminology and vessel construction requirements. In many cases, however, great differences still exist in how vessels are classified and in the construction requirements that each classification must meet. Terminology may vary greatly depending on the nationality of the crew and the vessel itself. Finally, many construction features (that crew members may regard as quite normal) can either aid fire-fighting operations or hinder them. The layout of the vessel and locations of specific spaces (arrangement) also affect fire-fighting efforts. For example, crew members need to know which spaces surround a fire area or whether a void space is present in order to make an effective attack on a fire. The construction requirements discussed in this chapter and others found throughout this manual are based upon those given in the International Maritime Organization (IMO) *International Convention for the Safety of Life at Sea 1974 (SOLAS)* and amendments (1990, 1991), unless specified otherwise.

Many of the hazards associated with onboard fire fighting are related to the characteristics and construction of the vessel. Many vessels are basically constructed of steel, but many elements of a vessel's decks, bulkheads, compartments, and passageways are not. The lowest class of bulkhead construction (Class C) may use a noncombustible material such as

Marinite® structural insulation board. Composite elements such as aluminum and fiberboard, which may melt or burn readily, are also found. Regardless of the construction materials used, vessels with years of built-up layers of paint pose a fire hazard. Vessels are designed to carry different types of cargoes, and different cargoes may determine a specific shipboard layout. Mariners should understand the physical nature of their vessels and the type of fire protection afforded them. An understanding of the different types of vessels and how they are constructed enables a crew member to safely operate as part of an emergency response team aboard a vessel.

This chapter establishes a foundation for the remainder of the book concerning special features, arrangement configurations, and classification of vessel construction components and how they affect fire-fighting operations. It gives an overview of vessel types and classes, vessel construction elements, fire control plan symbols, and general arrangement features with the purpose of identifying where hazards may be encountered (holds, cargoes, engine rooms, etc.).

Vessel Classification

A number of classification societies exist that have written rules governing the construction of vessels. Some examples are the American Bureau of Shipping (ABS), Lloyd's Register of Shipping (LR), Bureau Veritas (BV), and Det Norske Veritas (DNV). The rules of the different societies vary somewhat in detail but are mainly very similar.

Vessels are classified to ensure that they are properly built, equipped, and maintained. Along with watertight integrity, other components are necessary: timely maintenance of the hull, machinery, fire prevention features, and fire suppression systems. Modifications made later must not compromise or reduce fire protection features. A vessel can be classified at any time during its service life if it meets the requirements of the society, but usually it is done at the drawing-board stage before construction. The classification is retained throughout the life of a vessel provided it is properly maintained and submitted for the required periodic surveys. Special, more intensive surveys are conducted at five-year intervals.

◆ Vessel Types

Vessels could be identified with almost an infinite number of terms. Just a brief list includes bulk cargo, container, roll-on/roll-off (ro/ro) vehicle carrier, tanker, passenger, ferry, tug, towboat, barge, mobile offshore drilling unit (MODU), and high-speed craft. However, for the purposes of fire protection, SOLAS designates vessels as being one of only three different types: passenger vessel, cargo vessel, and tanker.

Of course, the construction requirements for each different vessel type vary depending on other factors such as the number of passengers aboard a passenger vessel and the specific type of cargo ship such as a ro/ro. These specific requirements are dealt with as each related topic is discussed throughout this chapter. Specific fire fighting tactical procedures for each type of vessel are detailed in Chapter 12, Fire Fighting Strategies and Tactical Procedures.

The vessel type, function, construction, and arrangement present unique fire control and emergency situations to the mariner. The following sections review the six most common types of commercial vessels. There are many vessels that have combinations of these characteristics. Ferries may have both vehicles and an extensive number of passengers aboard. A LASH (lighter aboard ship) vessel may have dry cargo barges and tank barges on board that may present unusual fire-fighting situations.

Passenger Vessels

Passenger vessels obviously present an extreme life hazard potential. In an emergency, thousands of passengers may need to be evacuated to safe areas and protected. The availability of safe areas may be limited. The high number of vessel occupants increases the chances of fire from careless smoking and arson. A passenger vessel also presents the potential for large galley fires.

Passenger vessels are extremely compartmentalized, which reduces large open areas, but a trend toward large, open areas (atriums and casinos, for example) is evident in some vessels (Figures 3.1 a and b). Smoke, heat, and products of combustion from even a small fire can create a difficult fire-

Figures 3.1 a and b Passenger vessels (a) cruise ship. *Courtesy of Captain John F. Lewis.* (b) Hospital ship. *Courtesy of R. Wright/Maryland Fire and Rescue Institute/United States Coast Guard.*

fighting situation on the involved and adjacent decks. Accommodation spaces (cabins or berthing spaces) are small and numerous. Passengers may have to travel long distances from cabins via passageways and inclined ladders to reach safety. Search and rescue activities can be extensive, time-consuming, and labor-intensive. The extensive compartmentation also intensifies flooding problems, dewatering (removal of fire fighting water) procedures, vessel instability, and ventilation procedures.

Cargo Vessels

Cargo vessels have similar features and equipment, but many variations in vessel design, construction, appearance, and size exist (Figures 3.2 a and b). Hazardous material situations are commonly encountered on cargo vessels. Each of the cargo vessels (container, break bulk, bulk, and ro/ro) described in the following sections has unique characteristics based on its primary use.

Container Vessels

Container vessels have their own unique characteristics (Figure 3.3). They may have extensive electrical systems to power refrigerated container units. They often have belowdeck passageways that run fore and aft the entire length of the vessel. Although these passageways can allow protected movement under emergency conditions, the hatches to holds are often also in these passageways, making access even more complicated.

Access to a container in the middle of a stack or at the bottom of a hold can be very difficult. Containers in the holds may be under hatch covers that weigh over 20 tons (20.3 tonnes [t]) with high deck loads of containers on top. Often shore cranes must lift hatches. The cargo-handling equipment to move containers and hatch covers is usually not available when the vessel is underway or at anchorage. Most container vessels do not have cargo-handling equipment on board, so containers may be removed only at designated berths. Access to the interior of the hold may be through one or two narrow scuttles and down a vertical ladder.

Figures 3.2 a and b (a) *(top)* An example of a cargo vessel. Many variations exist. (b) *(right)* An example of a multipurpose cargo vessel (freight, bulk, ro/ro, etc.). *Both courtesy of Captain John F. Lewis.*

Figure 3.3 An example of a container vessel. *Courtesy of Captain John F. Lewis.*

Figure 3.4 This container vessel rolled more than 45 degrees in a typhoon and was struck by waves over 60 feet (18 m) high. Many containers and their lashings were broken, and several cargo holds were flooded. *Courtesy of Firefighter Aaron Hedrick, Seattle (WA) Fire Department.*

Figure 3.5 An example of a bulk carrier. *Courtesy of Captain John F. Lewis.*

Containers and stacks may collapse under fire and rough sea conditions (Figure 3.4). Refer to the manifest to determine the exact nature of the contents of containers involved in fire or collapse. A dangerous cargo manifest lists dangerous or hazardous cargo. Bills of lading for other containers might not specify their contents. A stowage plan lists container numbers, weights, and destinations. Exercise caution because contents may be incorrectly declared.

Break Bulk Cargo Vessels

Break bulk cargo vessels carry packages and pieces of cargo stacked in cargo holds. Break bulk cargo vessels have the same access and cargo-hold issues as container vessels. Cargo instability and shifting may be problems under emergency conditions. Class A fuels, Class B fuels, and toxic cargoes may be encountered. Wood dunnage used to brace cargoes may add to the fuel load. Fire may spread to materials in the center of the cargo, making access for extinguishment and overhaul difficult. Many smoldering, deep-seated fires have been encountered in certain types of cargo. Water-soaked cargo can swell and may burst the hull or cause vessel instability. Some cargo holds are refrigerated. A release of refrigeration gases can create oxygen-deficient and toxic atmospheres. Anhydrous ammonia, a common refrigerant, is toxic and explosive.

Break bulk vessels usually have cargo-handling gear on board that can move cargo and hatch covers. These types of vessels may also have mechanical hatch covers that make opening the space much easier, unless additional cargo is loaded on top of them.

Bulk Carriers

Bulk carriers carry loose, dry or granular materials that are poured into cargo holds (Figure 3.5). The unique problems bulk carriers present are usually associated with the cargo they carry. Cargoes of coal, grain, coke, and other loose, solid materials can create flammable dusts that present an explosion hazard similar to those at grain elevators. Loose, bulk materials can cause an engulfment hazard to people working in the holds. Certain cargoes have been known to self-heat or spontaneously ignite. Some may produce toxic or flammable gases.

Extensive deep-seated fires and fires difficult to reach have occurred in bulk carriers. Bulk cargoes soaked by fire streams or flooding can expand and cause structural damage to the vessel. Self-unloading equipment may be on board to move bulk cargo, and mechanical hatch covers that make opening the space easier are common.

Ro/Ro Vessels and Ferries

Ro/ro vessels and ferries have large, undivided areas that run the entire length of the vessel (Figures 3.6 a and b). Flooding can cause a loss of vessel stability due to shifting water caused by the free surface effect (tendency of liquid to remain level as a vessel leans). See Chapter 13, Shipboard Damage Control. Scuppers (drains) and freeing ports aid water falling on deck to flow overboard.

Vehicles carried on board have fuel in their tanks and batteries connected because they are driven on and off the vessel. A vehicle fire may be difficult to reach and extinguish because of congestion on the deck. Fire can spread quickly from vehicle to vehicle because of their close proximity. Fuel spills may cause fire extensions. Vehicles (such as trucks and trailers) may also carry hazardous cargoes.

Tankers

Tankers carry tremendous quantities of flammable and toxic liquids and gases (Figure 3.7). A small tanker (500 to 600 feet [152 m to 183 m] long) carries 16,000 tons (16 256 t) while a medium tanker carries 50,000 to 70,000 tons (50 800 t to 71 120 t). A supertanker carries 250,000 tons (254 000 t) or more. Fire, explosion, and environmental impact are much greater with tankers than with other types of vessels. Because tankers are designed to carry liquids, they have greater stability than other vessels. The center of gravity of the cargo is always as low as possible for a given load. Tankers are subdivided to minimize free surface effect. See Chapter 13, Shipboard Damage Control.

Due to the nature of cargo transfer and pumping systems, static electricity is more of an ignition problem and more of a consideration on tankers than on other vessels. To minimize ignition sources, the power supply for winches, windlasses, and pumps on tankers may be hydraulic or steam. Air-powered tools are commonly used, so air outlets may be readily available throughout the deck. Using inert gas systems on tankers also maintains nonexplosive atmospheres within the cargo tanks.

The movement of cargo for fire control and vessel stability may be easier because tankers are built to contain liquids and designed to move these liquids through pipes and pumps. Extensive pumping and piping systems, as well as heated liquids under pressure, may be encountered. Cryogenic, compressed, or refrigerated liquids may also be found and may present some unique and difficult emergency

Figures 3.6 a and b (a) *(top)* A roll-on/roll-off (ro/ro) vessel carries vehicles that are driven on and off the vessel. *Courtesy of R. Wright/Maryland Fire and Rescue Institute/ United States Coast Guard.* (b) *(bottom)* A ferry vessel carries vehicles and passengers. *Courtesy of Firefighter Luke Carpenter, Seattle (WA) Fire Department.*

Figure 3.7 An example of a tank vessel. *Courtesy of Captain John F. Lewis.*

situations. A tanker may also have a pump room — the location of many disastrous fires and explosions. Many tank vessel companies have special standard operating procedures (SOPs) or guidelines for dealing with pump-room emergencies.

Vessel Construction

Vessels are built to withstand the stresses expected in the conditions and trade for which they are intended. For example, a ship designed to transit northern waters in winter might have some degree of ice-breaking capability. The size and type of vessel, type of cargo, speed of loading and unloading (which put great stress on the structure), other features relevant to the vessel's intended use, and the weather in the areas in which the vessel will be working play a part in determining the construction design features. The scantlings (minimum thicknesses of frames and other structural members) are specified by the classification society and must be followed by the shipbuilder.

Discussion of various construction components and their fire protection implications are given in the sections that follow. Various classes of bulkheads and the several types of watertight and fire doors are described. Ventilation systems, electrical systems, and fuel systems are important factors in fire control. It is important that crew members know how to operate these systems along with knowing their construction elements.

Bulkheads

SOLAS and its amendments classify bulkheads and decks that form fire divisions into three different classes: A, B, and C. Information on fire-resistance ratings of bulkheads is found in SOLAS Chapter II-2, Construction – Fire Protection, Fire Detection and Fire Extinction. These classes are based essentially on construction methods and materials. An approved noncombustible material is one that neither burns nor gives off enough flammable vapors to self-ignite when heated to about 1,382°F (750°C). These classes are further divided into subclasses based on the temperature passed to the opposite side of the material over a period of time. The following sections summarize the SOLAS regulations on the construction of divisions formed by bulkheads and decks.

Class A Divisions

Class A divisions are those divisions formed by bulkheads and decks that comply with the following listed requirements. The port state authority may require a test of a prototype bulkhead or deck to ensure that it meets the requirements for integrity and temperature rise. Class A division requirements are as follows:

- Constructed of steel or other equivalent material (one that has structural and integrity properties equivalent to steel at the end of the standard fire test)

- Stiffened suitably

- Constructed so that they are capable of preventing the passage of smoke and flame to the end of the 1-hour standard fire test

- Insulated with approved noncombustible materials so that the average temperature of the unexposed side does not rise more than 282°F (139°C) above the original temperature, nor does the temperature at any one point (including any joint) rise more than 356°F (180°C) above the original temperature within the time listed:
  — Class A-60: 60 minutes
  — Class A-30: 30 minutes
  — Class A-15: 15 minutes
  — Class A-0: 0 minutes

Class B Divisions

Class B divisions are those divisions formed by bulkheads, decks, deckheads (overheads or ceilings), or linings that comply with the following listed requirements. The port state authority may require a test of a prototype division to ensure that it meets the requirements for integrity and temperature rise. Class B division requirements are as follows:

- Constructed of approved noncombustible materials (all materials entering into the construction and erection of Class B divisions are noncombustible, with the exception that combustible veneers are permitted provided they meet other requirements of SOLAS)

- Constructed so that they are capable of preventing the passage of flame to the end of the first half hour of the standard fire test

- Insulated so that the average temperature of the unexposed side does not rise more than 282°F (139°C) above the original temperature, nor does the temperature at any one point (including any joint) rise more than 437°F (225°C) above the original temperature within the time listed:
  — Class B-15: 15 minutes
  — Class B-0: 0 minutes

Class C Divisions

Class C divisions are constructed of approved noncombustible materials. They need not meet requirements relative to the passage of smoke and flame or limitations relative to the temperature rise. Combustible veneers are permitted provided they meet other requirements of SOLAS.

Doors

A person who is new to an emergency response team has probably never considered the implications that doors hold for fire fighting operations. Doors can be friends or foes — allies or enemies. *Watertight doors* are openings in bulkheads to allow movement between subdivisions below the main deck level. Above the main deck, *fire doors* allow movement between main vertical zones or rooms (subdivisions). The purpose of both watertight doors and fire doors is to contain flooding and prohibit the spread of smoke and fire. Include information on the kinds of doors, their locations, and how they operate during prefire planning.

Watertight Doors

Watertight doors are found throughout a vessel. Besides their intended use to section off portions of the vessel to contain flooding, watertight doors can also seal off a space that is involved in fire — helping to prevent the spread of fire or products of combustion. However, it is important to realize that these doors can also seal off fire fighting crews from a means of escape and at the same time completely sever a fire

hose. These doors are vital to the containment of fire, smoke, and water.

By their very nature, watertight doors can be under pressure from water or hot gases. Use extreme caution when opening watertight doors. Chock the doors open to keep them from closing when advancing fire hoses through them. Holding power-operated doors open may not always be possible because they are designed to close against weights of water and are very strong and heavy. Crew members need to distinguish the different types of watertight doors (*individually dogged, quick-acting,* and *power-driven*) and know how to operate them.

Individually dogged. Individually dogged (sometimes called *nonquick-acting*) watertight doors are manually opened and closed and are held in place by slip hinges and usually six (sometimes eight) locking levers or bolts called *dogs* (Figures 3.8 a and b). A dog

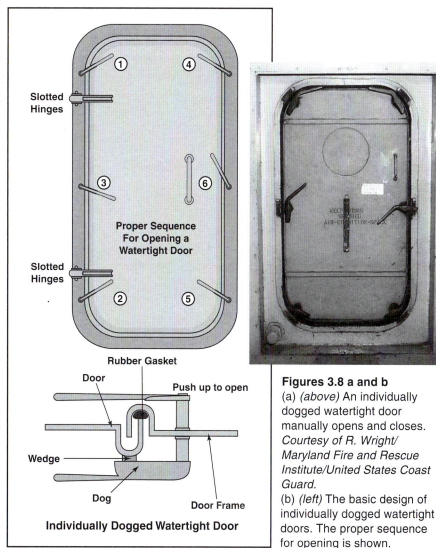

Figures 3.8 a and b
(a) *(above)* An individually dogged watertight door manually opens and closes. *Courtesy of R. Wright/ Maryland Fire and Rescue Institute/United States Coast Guard.*
(b) *(left)* The basic design of individually dogged watertight doors. The proper sequence for opening is shown.

wrench or extension (a short piece of pipe) is located both inside and outside the door (Figure 3.9). This dog wrench gives the user an appropriate extension to release or tighten the dogs. This tool is particularly useful when attempting to release a door that has pressure behind it or has been damaged by the heat of a fire. The recommended procedure for opening a door that may have pressure behind it is to first release the dogs on the hinged side of the door by pushing them up. The slip hinges allow the pressure to vent without the door swinging open uncontrollably. The dogs on the unhinged side can then be released when it is obviously safe to do so.

Quick-acting. All six to eight dogs used to open or close a quick-acting watertight door are activated by a central mechanism (single lever or handwheel) (Figures 3.10 a and b). A wheel activates all the dogs as it is turned. Rotate the mechanism only halfway at first to loosen the dogs and to vent any pressure behind the door, and then open the door completely after the pressure is relieved. Turn the wheel *counterclockwise* to open on one side and *clockwise* on the other. Hatches fitted with this mechanism usually have the inner side work counterclockwise because this direction is the more "natural" way to open.

Power-driven. Power-driven watertight doors are opened and closed by either electric or hydraulic motors that are remotely controlled. They can have either a horizontal or vertical (guillotine) opening (Figures 3.11 a–c). Each door can usually be opened or

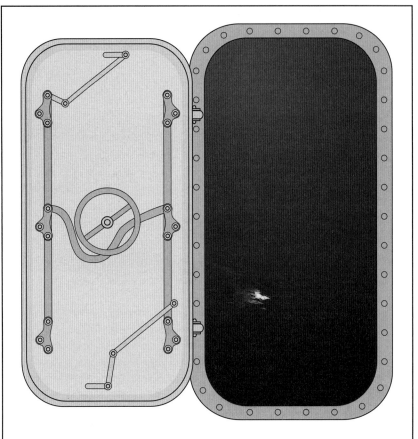

Quick-Acting Watertight Door

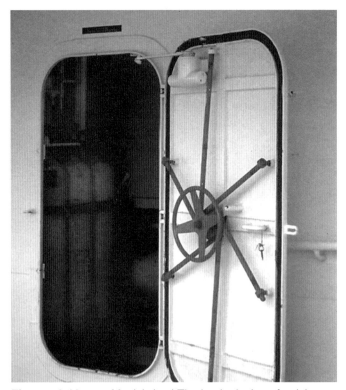

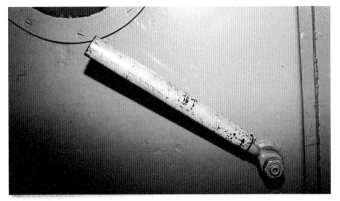

Figure 3.9 A dog wrench on a dog of an individually dogged watertight door. *Courtesy of Maritime Institute of Technology and Graduate Studies.*

Figures 3.10 a and b (a) *(top)* The basic design of quick-acting watertight doors. (b) *(bottom)* A quick-acting watertight door opens and closes with a single lever or handwheel. *Courtesy of David Ward.*

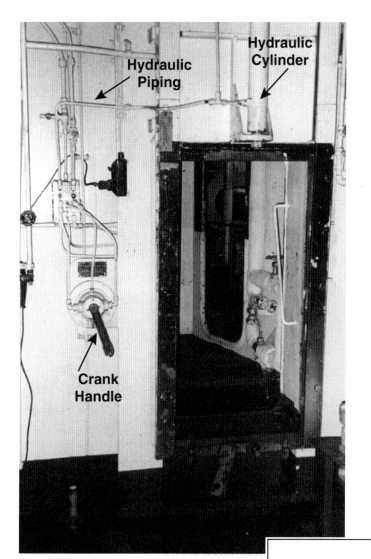

Hydraulic Piping

Hydraulic Cylinder

Crank Handle

Figure 3.11 b A sliding, power-driven watertight door. *Courtesy of R. Wright/Maryland Fire and Rescue Institute/ United States Coast Guard.*

Figure 3.11 a Power-driven watertight door (vertical opening). The entrance to the shaft alley tunnel is shown from the lower level of the engine room looking at the open door. The door is open (up). This hydraulically operated door is activated by the crank handle located to the left of the door. *Courtesy of Thomas F. Kiernan, Port Engineer, U.S. Coast Guard Chief Engineer/Steam, Norfolk, VA.*

Figure 3.11 c The basic design of a power-driven sliding watertight door.

Hydraulic Power Unit
to close/open door
(operated from Bridge)

Rack & Pinion Mechanism

Local Override
of Hydraulic System

Manual Operation

Sliding Watertight Door

closed by an electric switch or a hand crank or pump located at an operating station (Figure 3.12). Operating stations are usually located both nearby and in a remote location from the door (Figures 3.13 a and b). All doors may be opened and closed remotely from a control point such as the bridge. They must also be controlled locally from positions on either side of the bulkhead and from the deck above. Doors must be equipped with alarms that sound when doors are closing because anyone standing in their closing paths may be injured or killed. The mechanism is too powerful for a person to hold back the door. A horizontal door may operate slowly enough to allow a person to move out of the way, but the door's movement may be jerky or nonuniform in speed. Ensure that alarms are in working order.

> ## WARNING
>
> Do not attempt to pass through a power-driven door while the closing alarm is sounding. A person caught in the path of a closing power-driven door may be injured or killed.

Fire Doors

Openings in Class A bulkheads are required to be fitted with doors or shutters of an approved type. Port state regulations in accordance with IMO and SOLAS requirements for these closures are very stringent and intended to prevent compromising the integrity of the bulkhead. The seal of the door must also be approved, and gaskets are not permitted in achieving the seal. Double swinging and revolving doors are not permitted in a Class A bulkhead (Figures 3.14 a and b). See the source documents (SOLAS and port state regulations) for details of the requirements. Requirements include but are not limited to the following, depending on the size and intended use of the space:

- Approved closure construction materials (materials must match bulkhead class)

- Permissible size of the closure

- Arrangements for electrical closing of a door normally held open

- Indicators on the navigating bridge showing when a door is open

- Instructions for opening

- Amount and type of glass used for windows in doors (must be compatible with fire rating of bulkhead in which door is installed)

- Identification plate

Ventilation Systems

Ventilation is a factor in fire control. Ventilation systems can either remove the products of combustion by exhausting them to the atmosphere or prevent their spread by shutting down. Most vessels have procedures for ventilation shutdowns in various spaces so that in a fire response the airflow may be reduced or eliminated. Ventilation systems may

Figure 3.12 Close-up of the operating handle of the power-driven watertight door shown in Figure 3.11 a. Turning counterclockwise raises (opens) the door. Turning clockwise lowers the door to the closed position. *Courtesy of Thomas F. Kiernan, Port Engineer, U.S. Coast Guard Chief Engineer/ Steam, Norfolk, VA.*

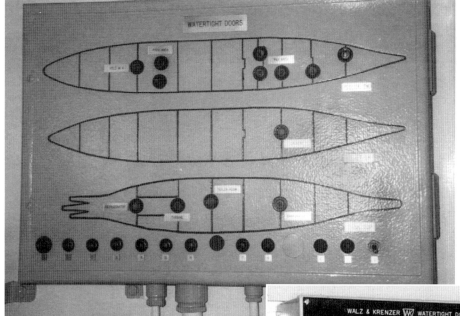

Figures 3.13 a and b Control board with lights and switches to operate watertight doors belowdeck. (a) *(left)* Control board on the bridge of a cruise vessel. *Courtesy of Robert E. (Smokey) Rumens.* (b) *(below)* Control panels for both watertight doors and fire doors on a hospital vessel. *Courtesy of R. Wright/ Maryland Fire and Rescue Institute/United States Coast Guard.*

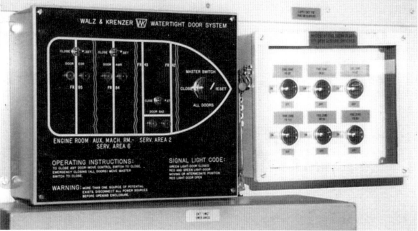

Figures 3.14 a and b (a) A fire door in the open position. *Courtesy of David Ward.* (b) A fire door in the closed position. *Courtesy of David Ward.*

provide exhaust (negative pressure) airflow or intake (positive pressure) airflow (see Chapter 9, General Fire Fighting Procedures). Intake and exhaust airflows can sometimes be reversed, but a delay may result before airflow begins again. Ventilation systems must be shut down in spaces where fixed fire protection systems are being discharged so that the extinguishing agents released are contained.

SOLAS regulations provide very specific requirements for construction of ventilation systems for both fresh air and conditioned air. Ventilation systems are either on fresh air intake or recirculated (conditioned) air. In an accommodation fire, a critical factor may be whether the air is being exhausted or recirculated because smoke may spread more quickly in the latter case. The requirements focus on the materials used in the construction of ducts, required locations for dampers and controls, and the provision of smoke detectors in ducts to provide for automatic shutdown of air-handlers in the event smoke enters the system (Figures 3.15 a–c).

When using heating, ventilating, and air-conditioning (HVAC) systems for fire control or smoke removal, first ensure that the systems (wiring, dampers, ducting, etc.) have not been damaged by the fire. A damaged blower motor could short-circuit and cause another fire in the system. Cool the ducts adjacent to the fire area by applying water streams to their surfaces. Especially, apply cooling streams wherever heat-activated smoke dampers are located. Crew members need a complete knowledge of the construction features and location of the ductwork route.

Electrical Systems/Equipment

Electricity is always a paramount concern to fire fighting crews. Electricity is vital to perform needed ventilation, communications, lighting, and other

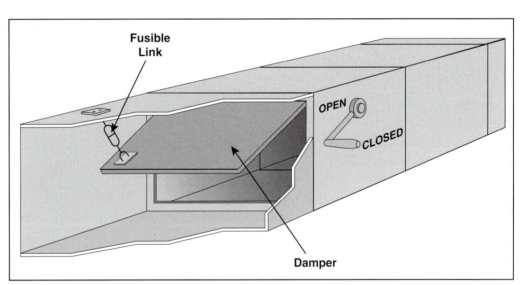

Figures 3.15 a–b (a) *(right)* Typical fire damper in a ventilation duct. (b) *(below)* Ventilation fire damper controls and general alarm bell. *Courtesy of R. Wright/Maryland Fire and Rescue Institute/ United States Coast Guard.*

Figure 3.15 c Box vents with dampers. *Courtesy of Maritime Institute of Technology and Graduate Studies.*

functions, but it can also be an extreme hazard to crew members and a potential source of ignition. During fire fighting, any energized system presents a serious shock hazard. It is crucial that crew members become familiar with the vessel's electrical systems and know how to isolate and control them within the fire area.

Electrical systems are routed throughout a vessel and penetrate bulkheads, so their routes must be monitored because they could allow fire to penetrate to the next compartment. According to regulations, the route of electrical cables must be fire-stopped but these are weak points when compared to the vessel's steel construction. Fire in cable trays are difficult to access and extinguish due to their locations and construction. The insulation gives off very toxic smoke even in light-smoke conditions.

Along with knowing about electrical systems and their hazards, use of the proper electrical equipment is also important. *Explosionproof* means that equipment is designed so as not to provide an ignition source in an explosive atmosphere. The equipment is enclosed in a case or container that can withstand an internal explosion of flammable mixtures and prevents internal sparks or flashes from igniting the surrounding flammable atmosphere. In addition, the external temperature of the enclosure container does not become high enough to ignite the surrounding atmosphere while the equipment is operating. Examples are electrical systems installed in paint lockers and flammable liquid rooms. The architect designs these systems into the vessel. Other examples are ventilation fans, lighting fixtures, and similar electrical equipment.

Very similarly, *intrinsically safe* means that by the very nature of its design, equipment or wiring is not capable of releasing sufficient electrical energy to cause the ignition of a flammable atmospheric mixture. Electronic equipment may be labeled intrinsically safe by design or by lab test. Either way, any damage to the case or mechanism renders the intrinsically safe designation void.

Explosionproof or intrinsically safe equipment is vitally important on tanker vessels and other hazardous materials carriers. These items are also important on other types of vessels, even passenger vessels, that also carry flammable, combustible, or otherwise hazardous cargo or materials. The propulsion fuel on virtually all vessels can create an atmosphere in which this type of equipment is required. Possible ignition sources such as light fixtures, electrical switches, electrical conduit, and ventilation fans must be explosionproof or intrinsically safe.

Vessel Arrangement Features

All vessels have common arrangement (layout) features that create certain fire control problems according to their type. For example, vessels can have large open spaces, high fuel loads, or areas where access is difficult. All vessels present a wide variety of confined-space situations. Belowdeck spaces are difficult to assess and may have hazardous atmospheres. Confined spaces such as paint lockers, ship stores and other storage areas, and battery compartments present fire hazards. For example, a cabin fire may heat a battery room and cause a release of hydrogen, which is explosive if ignited. Cargo holds, double bottoms, and the many types of tanks encountered can create oxygen-deficient, toxic, and flammable environments. But various vessels and spaces have unique characteristics that should be addressed as concerns for fire fighting. For example, general arrangement drawings showing side elevations would be useful for showing ventilation system operations during fire fighting. The following sections give brief overviews of many of the unique factors pertaining to cabin, engine room/machinery, cargo hold, and liquid cargo tank spaces.

Cabins

Cabins or accommodation spaces are usually small and loaded with a wide variety of Class A materials. Bedding, mattresses, clothing, paper, plastics, and foam rubber can produce large quantities of toxic products of combustion that can spread via ventilation systems throughout the involved deck or adjacent decks. Cabin bulkheads may be Class B construction (or sometimes Class C), so a fire may completely burn a whole section of cabins if it is not extinguished quickly. This burned area would then create a large space that could cause vessel instability if excessive fire fighting water is applied and not removed. Often only one passageway and two access doors to the cabin area of a vessel are available, making fire attack, search, and rescue difficult.

Engine Rooms/Machinery Spaces

The arrangement characteristics of engine rooms and machinery spaces present some of the most dangerous and difficult fire fighting environments on a vessel. Class B and Class C fires are the most likely types in these hazard areas because of the abundance of fuel oils, electricity, and high-pressure systems. A combination of hazards creates an even more hostile environment. Noise, steep ladders, slippery surfaces, open spaces in platforms and decks, and heights add to the already difficult situation. Some of the hazards found are as follows:

- *Class B fuels* — Many flammable liquids — diesel fuel, fuel oil, cleaning solvents, lubricating oils, and paints — are present.

- *Electricity (often high voltage)* — Tremendous amounts of electrical power, equipment, and circuits present ignition and shock hazards.

- *High-pressure fuel line systems* — A maze of steam, pneumatic, lubricating oil, and fuel piping systems and storage tanks are present.

- *High-temperature systems* — Engine casings/exhaust manifolds and steam (main propulsion or auxiliary) systems are some examples.

- *Moving machinery parts* — A few examples are pumps, piston rods, and flywheels.

- *Large, undivided machinery areas* — Smoke and heat from even a small fire can easily spread throughout this type of space. Access points are limited.

- *Difficult access* — Search and rescue operations are complicated by the congestion encountered in small engine rooms.

Cargo Holds

Cargo holds present large and deep undivided areas with limited access. Cargo holds can be large enough to make entry, fire control, and egress difficult in the time available on the standard 30-minute self-contained breathing apparatus (SCBA). Usually only one or two scuttles with vertical ladders are available to enter this type of space.

If cargo hatches are open or removed, a hold fire could have an unlimited supply of air and may spread rapidly. Open hatches may negate the use of fixed fire protection systems. However, the ability to open hatches may also allow fire control to be completed from outside the space.

A variety of fire fighting complications in cargo holds include the following:

- Cargo spaces may be loaded with a wide variety of Class A and Class B materials as well as other toxic materials.

- The seat of the fire may be difficult to reach.

- The fire may be deep-seated.

- Overhaul is usually difficult and extensive.

- Fire fighting water may damage adjacent cargo causing swelling of absorbent materials, vessel instability, and other unique problems.

- Cargo-handling gear may not be available to move materials, gain access to the fire, or assist in overhaul.

Liquid Cargo Tanks

Liquid cargo tanks present the same large-area and limited-access problems as cargo holds. Tremendous quantities of flammable and toxic cargoes can create explosive situations that can damage the structural integrity of the vessel, as well as compromise cargo transfer systems, gas inerting systems, and fixed fire fighting systems. Class B fires are common and require the use of fire fighting foam and unique fire control procedures. Reignition can be a constant concern. Large quantities of unburned products of combustion, which create smoldering coke deposits on the interior surfaces of tanks, are present in oxygen-deficient atmospheres. These coke deposits are a constant source of reignition and are very difficult to extinguish.

Fire Control Plan Symbols

To provide a universal method of diagramming ships, IMO has adopted a standardized set of symbols to use in fire control plans. These symbols concisely identify the arrangement elements of the ship, bulkheads and other construction features, fire fighting systems and control valves, ventilation controls, and so on. IMO encourages but does not require mariners to use these symbols (Figure 3.16).

Chapter 4

Safety Equipment

Crew members who fight fires require the best safety equipment available because of the hostile environment in which they perform these duties. Injuries are reduced or prevented and fire-fighting capability is enhanced if crew members have and use good equipment. However, providing and using quality protective equipment does not necessarily guarantee crew safety. All protective equipment has inherent limitations that must be recognized so that crew members do not overextend the item's range of protection. Extensive training in the use and maintenance of equipment is required to ensure that it provides optimum protection.

Any protective clothing, equipment, or gear worn by the mariner fighting fire is referred to as *personal protective equipment (PPE)*. Besides protective clothing (also known as *turnout gear, bunker gear,* or *structural fire fighting clothing*), other crucial pieces of protective equipment for engaging a fire are breathing apparatus and personal alert safety system (PASS) devices. In total, these pieces compose the full personal protective equipment (Figure 4.1). In addition, some situations require the use of environmental analysis instruments in order for crew members to determine whether an atmosphere is safe to enter.

Additional protective gear may be necessary to provide maximum protection in the many fire situations mariners may encounter. All crews should know the type of equipment needed for different exposures and have it readily available. The protective qualities of PPE vary; it is a vessel owner's responsibility to ensure that crew members are provided with acceptable protection. It is also the owner's responsibility to make sure that crew members use that protection.

Most standards governing maritime transportation require that vessels carry at least two complete sets of PPE. This minimal standard does not provide for any backup in fire fighting operations. It is accepted worldwide that at least two persons are required to operate each fire hose. The International Fire Service Training Association (IFSTA) recommends equipping each vessel with enough sets of PPE to outfit the personnel manning all attack and backup fire hoses. Optimally, the number of sets should correspond to the number of crew members on the vessel.

Figure 4.1 Mariner aboard ship in full personal protective equipment. *Courtesy of Maritime Institute of Technology and Graduate Studies.*

This chapter discusses most of the safety equipment that might be used by crew members who fight fires. Various types of protective attire and breathing apparatus, including limitations, storage requirements, training, donning/doffing techniques, changing and filling air cylinders, safety precautions, and emergency escape techniques are described. Skill sheets for various donning/doffing procedures and changing/filling air cylinders are given at the end of the chapter. Basic inspection, care, and maintenance procedures for breathing apparatus are discussed. PASS devices, personnel accountability systems, and environmental analysis instruments are also explained.

 ## Personal Protective Clothing for Fire Fighting

Personal protective clothing for fire fighting includes a helmet, turnout coat and pants, protective boots, interface components (protective hood and wristlets), gloves, and PASS device. Protective breathing equipment, which completes the full personal protective equipment ensemble, is discussed in a separate section later in this chapter. The water-resistant clothing commonly found on vessels (often referred to as a *marine firefighter suit*) consists of a water-resistant coat or jacket, rubber boots, a hard hat, and work gloves. This clothing is *only* acceptable for use in combating fires that are in the incipient phase or that can be extinguished with a handheld fire extinguisher. For larger fires, full PPE is required.

The quality of protective clothing must be thoroughly researched before selection and purchase. Purchase clothing that meets current applicable standards, for example, National Fire Protection Association (NFPA) 1971, *Standard on Protective Ensemble for Structural Fire Fighting.* A variety of protective clothing that has been tested to determine different characteristics is available. Crew members need to understand the design and purpose of the various types of protective clothing and be aware of each garment's limitations. This section discusses the types, limitations, and uses of the various garments.

Fire fighting clothing provides some of the best protection available for those engaging a fire on board vessels. This clothing was designed for protection from heat, moisture, and other common hazards associated with fire fighting activities. The purposes of the various components of personal protective clothing are as follows:

- *Helmet*—Protects the head from impact and puncture injuries as well as from scalding water; protects from both heat and cold (Figure 4.2). Ear covers and chin straps are required. The faceshield gives secondary protection of the face and eyes.

- *Protective hood* — Protects ears, neck, and portions of the crew member's face not covered by the helmet or coat; protects from exposure to extreme heat. Hoods are available in long (extending over shoulder and chest) or short styles.

- *Protective coat and trousers* — Protect trunk and limbs against cuts, abrasions, and burn injuries (resulting from heat) and provide limited protection from corrosive liquids; wristlets prevent materials from entering sleeves. Coats/trousers are composed of an outer shell, a moisture barrier, and a thermal barrier; all liners must be in place during fire fighting operations.

- *Gloves* — Protect the hands from cuts, puncture wounds, burn injuries, liquid absorption, and heat/cold. Gloves must fit properly and allow dexterity and tactile feel.

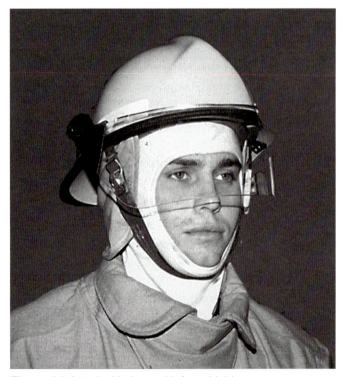

Figure 4.2 A typical helmet with faceshield.

- *Safety shoes or boots* — Protect the feet from burn injuries and puncture wounds; a stainless steel midsole plate provides puncture resistance. Protective boots are required for fire fighting. Sharing boots is unsanitary and not recommended. Safety shoes have safety toes and puncture-resistance soles and are used for other work activities.

- *Eye protection* — Protects the wearer's eyes from flying solid particles or liquids. Primary protection includes safety glasses and safety goggles; secondary protection includes helmet faceshields and breathing apparatus masks. Primary eye protection is required for any operation where flying particles or chemical splashes are likely. Eye goggles or faceshields may be part of the helmet or separate.

- *Hearing protection* — Limits noise-induced damage to the ears when loud noise situations cannot be avoided; includes earplugs, earmuffs, and intercom/ear protection systems. However, wearing hearing protection during actual fire fighting is generally impractical.

- *Personal alert safety system (PASS) device* — Provides life-safety protection by emitting a loud shriek if a crew member collapses or remains motionless for approximately 30 seconds; can also be activated manually. See Personal Alert Safety System Devices section.

All system components must be in place and working properly if a crew member is to receive full protection. Protective clothing protects the wearer from high temperatures, steam, hot materials, and most other common hazards of fighting fires on board. The garments discussed to this point *may not* qualify for use in hazardous materials (haz mat) incidents, because they are not rated to protect a crew member from chemical exposure. Refer to hazardous materials response references (material safety data sheets) to determine the suitability of the garments available to crew members when dealing with hazardous materials situations.

◆ Special Types of Personal Protective Equipment

Special types of high-temperature fire fighting clothing (approach, proximity, and fire entry suits) are available for situations such as fighting flammable liquid and gas fires (those involving high levels of radiated heat), to perform rescues, and to work in total flames for a short period. These garments are designed to protect crew members from conductive, convective, and radiant heat while working in close proximity to a fire. The three special types are discussed in the following sections along with their limitations and disadvantages.

Proximity and Approach Clothing

Both proximity and approach suits are used only in situations where the fire area is not physically entered (Figure 4.3). *Proximity clothing* is designed for short duration, close proximity exposures to flame and radiant heat temperatures as great as 1,500°F (816°C). *Approach clothing* is designed to protect crew members from radiant heat while approaching a fire. With approach clothing, radiant heat exposures up to 2,000°F (1 093°C) can be reached for brief periods of time (approximately 3 minutes or less). Some suits of these types may also afford a level of protection for exposure to steams, liquids, and weak chemicals. Each suit has a specific use and is not interchangeable. They are not designed to protect the wearer

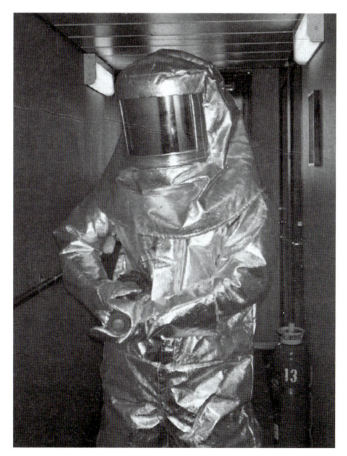

Figure 4.3 Special high-temperature fire fighting clothing. *Courtesy of R. Wright/Maryland Fire and Rescue Institute/ United States Coast Guard.*

against chemical hazards. Because these types of suits are typically used by airport crew members, personnel untrained in their use are not aware of their inherent limitations and may unknowingly subject them to environments for which they are not designed.

Various materials treated and coated with highly reflective aluminum are used in the construction of proximity and approach suits. Because the integrity of aluminized coatings is vulnerable to repeated flexing and scraping, aluminized overshells have been developed that can be stored separately. When needed, the shells easily slip over turnout coats and trousers, thus avoiding the abuse to the aluminized coating encountered in responses not requiring high-level protection. Combining the layers of additional clothing with the aluminized outer surface provides a high level of protection, but the ensemble restricts movement and the resulting heat buildup is fatiguing.

Entry Clothing

Entry clothing is designed for specialized work inside industrial furnaces and ovens. It is rarely used in fire fighting operations. Before using an entry suit, crew members must have special training in its donning, uses, and limitations. These special types of protective garments are normally all-enclosing (encapsulating), including hoods, gloves, and foot coverings. These suits are used particularly in chemical plants and other locations where expedient action is required to quickly close valves or save lives.

Limitations/Disadvantages

These types of special protective clothing pose several disadvantages: impaired mobility, limited vision, and loss of communication. Newly developed communication devices such as voice-activated radios help alleviate communication problems. Another limitation/disadvantage is that body heat cannot escape to the outside in these special suits. Thus, discomfort and fatigue quickly develop. Relief crews are necessary whenever this special clothing is used in an emergency setting. It is important that crew members who are expected to don and perform in this clothing practice using it regularly.

 ## Hazardous Materials Protective Clothing

Hazardous materials protective clothing is designed to protect a crew member from chemical exposure.

Regular and special high-temperature types of PPE may not give protection against the hazardous materials encountered because gases and some liquids can enter either through the clothing or through gaps. Four recognized levels of hazardous materials protection (A, B, C, and D) are discussed in the sections that follow. The appropriate level of protection is chosen based on the suspected threat posed by the incident and recommendations by various hazardous materials references and chemical manufacturers. Hazardous materials protective clothing *is not* suitable for fire fighting.

Level A

The highest level of protection against vapors, gases, mists, and particles is Level A, which consists of a fully encapsulating chemical entry suit with a full-facepiece self-contained breathing apparatus (SCBA) or a supplied air respirator (SAR) with an SCBA escape cylinder. A crew member must also wear boots with steel toes and shanks on the outside of the suit and specially selected chemical-resistant gloves for this level of protection. The breathing apparatus is worn inside (encapsulated within) the suit. To qualify as Level A protection, an intrinsically safe two-way radio is also worn inside the suit, often incorporating voice-operated microphones and an earpiece speaker for monitoring the operations channel.

Level B

Level B protection requires a garment (including SCBA) that provides protection against splashes from a hazardous chemical. Since the breathing apparatus is worn on the outside of the garment, Level B protection is not vapor-protective. It is worn when vapor-protective clothing (Level A) is not required. Wrists, ankles, facepiece and hood, and waist are secured to prevent any entry of splashed liquid. Depending on the chemical being handled, specific types of gloves and boots are donned. These may or may not be attached to the garment. The garment itself may be one piece or a two-piece hooded suit. Level B protection also requires the wearing of chemical-resistant boots with steel toes and shanks on the outside of the garment. As with Level A, chemical-resistant gloves and two-way radio communications are also required.

Level C

Level C protection differs from Level B in the area of equipment needed for respiratory protection. The same type of garment used for Level B protection is worn for Level C. Level C protection allows for the use of respiratory protection equipment other than SCBA. This protection includes any of the various types of air-purifying respirators. Crew members should not use this level of protection unless the specific hazardous material is known and its concentration can be measured. Level C equipment does *not* offer the protection needed in an oxygen-deficient atmosphere.

Level D

Level D protection does not protect the crew member from chemical exposure. Therefore, this so-called level of protection can only be used in situations where a crew member has no possibility of contact with chemicals. A pair of coveralls or other work-type garment along with chemical-resistant footwear with steel toes and shanks are all that is required to qualify as Level D protection.

 ## Protective Breathing Equipment

Respiratory protection is of primary concern to crew members because one of the major routes of exposure to the hazardous products of combustion is inhalation. Smoke, irritants, asphyxiants, poisons, and toxins are respiratory hazards crew members may encounter during an incident. Protective breathing equipment protects the face and lungs from these toxic products in addition to protecting from high temperatures. Inhaling heated gases can cause death from asphyxiation. Toxic and oxygen-deficient atmospheres (often found in confined spaces) can exist without fire. Failure to use breathing equipment in any of these situations can lead to injuries and deaths.

Breathing equipment is known by several names; the most common is *self-contained breathing apparatus (SCBA)*. Other names are *breathing apparatus (BA)* and *air packs*. Several categories of breathing equipment exist such as open-circuit SCBA, supplied-air breathing apparatus (SABA), air-purifying respirator (APR), and emer-

gency escape breathing device (EEBD). Each category of respiratory protection equipment is limited in its capabilities.

Besides describing categories and limitations, the following sections also discuss SCBA storage, donning/doffing steps, and training. Procedures are given for changing and filling air cylinders along with safety precautions and emergency escape techniques. Inspection, care, and maintenance procedures are also given.

Equipment Categories

The four basic categories of protective breathing equipment used by crew members are open-circuit SCBA, supplied-air breathing apparatus, air-purifying respirator, and emergency escape breathing device. Breathing apparatus is used on board for a variety of operational needs other than fire fighting, for example, confined-space entry. Knowledge of the various types and their limitations is important to crew members.

Open-Circuit Self-Contained Breathing Apparatus

Open-circuit SCBA is the most commonly used protective breathing apparatus (Figure 4.4). The air

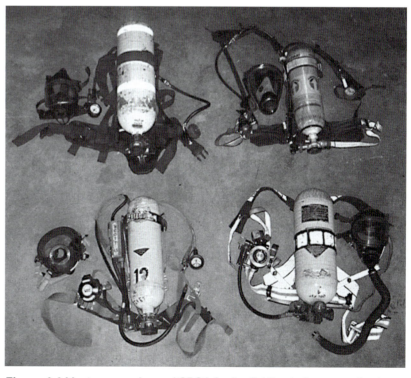

Figure 4.4 Most commonly used SBCA for fire fighting is the open-circuit type. Several examples are shown.

supply in an open-circuit SCBA is compressed breathing air. Exhaled air is vented to the outside atmosphere. The air supply duration of an open-circuit SCBA varies depending on the design of the unit, fitness of the wearer, amount of SCBA training by the wearer, amount of physical exertion required by the wearer, and self-confidence of the wearer. Most SCBA allow the performance of at least 15 to 20 minutes of heavy work. Some units may provide up to 45 minutes of actual duration under heavy-work conditions.

Before July 1, 1983, two types of open-circuit SCBA were available: demand and positive pressure. Since that date, demand units are no longer in compliance with regulatory requirements. *Any vessel crews still using demand apparatus must convert them to positive-pressure units as soon as possible.* The main reason for the change to positive pressure is the greater protection factor afforded by the positive-pressure units. Positive-pressure SCBA maintains a slightly increased pressure (above atmospheric) in the user's facepiece. This positive pressure significantly increases protection against contaminants entering the facepiece.

Numerous companies manufacture open-circuit SCBA, each with different design features or mechanical constructions. Certain parts such as cylinders and backpacks are often interchangeable; however, such substitution voids the certification of the unit by certifying agencies and is not recommended. Substituting different parts may also nullify warranties.

The four basic SCBA component assemblies are as follows:

- *Backpack and harness assembly* — Holds the air cylinder on a crew member's back as comfortably and securely as possible. Adjustable harness straps provide a secure fit. Waist straps distribute the weight of the cylinder to the hips. Removal of waist straps can void warranties.

- *Air cylinder assembly* — Includes air cylinder, valve, and pressure gauge (in some countries) (Figure 4.5). The cylinder constitutes the main weight of the breathing apparatus because it must be strong enough to safely contain the high pressure of the compressed air. Weights vary with each manufacturer.

- *Regulator assembly* — Includes high-pressure hose, low-pressure alarm, and pressure gauge (some

models). The regulator controls the flow of air to meet the respiratory requirements of the user. Some SCBA units have regulators that fit into the facepiece (Figure 4.6), others are on the person's chest or waist strap (Figure 4.7). All units have an audible alarm that sounds when cylinder pressure decreases to approximately one fourth of the maximum rated pressure of the cylinder.

- *Facepiece assembly* — Includes low-pressure hose (breathing tube) and exhalation valve on SCBA with harness-mounted regulators. The facepiece also includes other components that vary depending on the type, model, and manufacturer:

— Lens (viewing window)

— Connection to air supply

— Head harness

— Nosecup

— Speech diaphragm

Figure 4.5 The air cylinder constitutes the main weight of the SCBA.

Figure 4.6 This regulator connects directly to the facepiece.

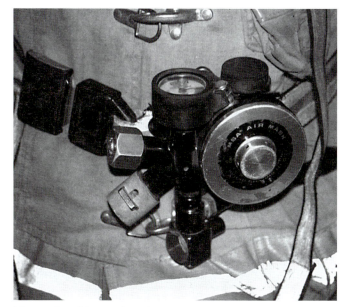

Figure 4.7 This regulator is attached to the waist strap.

Figure 4.8 Supplied air breathing apparatus is useful for extended duration operations.

Supplied Air Breathing Apparatus

Some special incidents (hazardous materials or rescues) often require a longer air supply than what can be obtained from standard open-circuit SCBA. In these situations, an airline attached to one or several large air cylinders is connected to an open-circuit facepiece, regulator, and escape (egress) cylinder (Figure 4.8). Airline equipment enables a crew member to travel limited distances, up to 300 feet (91 m), from the *regulated* air supply source. This type of respiratory protection enables a crew member to work for several hours without the encumbrance of a backpack. If greater mobility is needed, a crew member can also wear a standard SCBA with an airline option. The crew member can then temporarily disconnect from the airline supply, use the SCBA to provide breathing air, and perform necessary tasks beyond the range of the airline equipment.

More flexibility is allowed by a trolley-set configuration, which consists of two SCBA cylinders connected to a manifold to which 300 feet (91 m) of airline is attached. Considerable working time is allowed, specially in confined spaces, by alternating the two changing cylinders.

Any supplied air breathing apparatus must provide enough breathing air (usually 5 minutes) for the wearer to escape the atmosphere in the event the airline is severed. This requirement is usually accomplished by attaching a very small breathing cylinder, called an *escape cylinder,* to the airline unit. The five-minute escape cylinder is *not* intended for untethered work (detached from the airline). To perform untethered work, use a 30- or 60-minute SCBA that can be augmented by an airline.

The disadvantage to a supplied air breathing apparatus system is that the user's supply hose may become entangled while entering, exiting, and working in a hostile environment. The possibility of damage to the hose by cuts, abrasions, and heat damage in a fire area also exists.

Air-Purifying Respirator

An air-purifying respirator uses ambient air that is purified through a filter before inhalation. APRs enhance the mobility of a crew member because they do not require an air cylinder or backpack-type assembly. They are only worn in controlled atmospheres where the hazards present are known and sufficient oxygen is present. Three categories of air-purifying respirators are particulate-filtering, vapor- and gas-removing, and powered air-purifying.

> # WARNING
>
> Do not wear air-purifying respirators during emergency operations or in oxygen-deficient atmospheres. SCBA must be worn during these emergency situations.

Emergency Escape Breathing Device

Some vessels are equipped with emergency escape breathing devices for use in emergency escape from hazardous atmospheres (Figure 4.9). Several varieties are available: (1) stored-pressure, (2) oxygen-generating, or (3) filtering. EEBDs must *never* be used for fire fighting operations. They are designed to provide a limited amount of breathing time *for escape purposes only*. Some models are single-use only and cannot be refilled. On some vessels, one is stored above every seafarer's bunk and larger quantities are stored in various spaces (such as the engine room) throughout the vessel. Some older, obsolete units are unsafe and must be discarded or serviced by the manufacturer to make them safe for use.

Limitations

Awareness of the limitations associated with breathing apparatus allows a crew member to operate effectively. One way to become aware of equipment and personal limitations is to train in conditions as realistic as possible. For example, wear SCBA until the air cylinder is exhausted and record duration times, or

Figure 4.9 The emergency escape breathing device being demonstrated is the standard oxygen-generating type and is not yet inflated with oxygen. It can be used only once. *Courtesy of R. Wright/Maryland Fire and Rescue Institute/ United States Coast Guard.*

practice donning and doffing the SCBA in confined spaces. Limitation factors include the wearer, equipment, and air supply. The following sections highlight those limitation factors.

Wearer

Factors affecting the individual crew member's ability to use SCBA effectively include physical, medical, and mental limitations. These limitations are summarized as follows:

 ### Wearer Limitations
Physical

♦ **Physical condition** — *A wearer in sound physical condition can maximize the work performed and extend the air supply of a breathing apparatus as far as possible.*

♦ **Agility** — *Good agility can overcome the restrictions a breathing apparatus has on a wearer's movement and balance.*

♦ **Facial features** — *The shape and contour of a wearer's face and amount of facial hair affects the ability to get a good facepiece-to-face seal.*

Medical

- **Neurological functioning** — *Good motor coordination is necessary for performing tasks with breathing equipment, and a sound mind allows a wearer to handle emergency situations.*

- **Muscular/skeletal condition** — *Physical strength and size are necessary to effectively wear the breathing equipment and perform necessary tasks.*

- **Cardiovascular conditioning** — *Good heart and blood vessel conditioning prevents heart attacks, strokes, or other related problems during strenuous activity while wearing breathing apparatus.*

- **Respiratory functioning** — *Proper breathing maximizes a wearer's operating time while wearing breathing apparatus.*

Mental

- **Equipment training** — *Knowledge in every aspect of using breathing apparatus is necessary.*

- **Self-confidence** — *Belief in abilities has an extremely positive overall effect on the actions performed by a wearer.*

- **Emotional stability** — *Ability to maintain control in a high-stress environment reduces the chances of a wearer making serious mistakes.*

Equipment

Awareness of the limitations of the breathing apparatus equipment is necessary for crew members in order for them to operate properly. These limitations are as follows:

 Equipment Limitations

- **Limited visibility** — *Wearing the breathing apparatus facepiece reduces a wearer's peripheral vision, and facepiece fogging can reduce overall vision.*

- **Decreased communication capability** — *Wearing the breathing apparatus facepiece hinders voice communication.*

- **Increased weight** — *Depending on the model, breathing apparatus adds 20 to 35 pounds (9 kg to 16 kg) of weight, which decreases the wearer's mobility and causes fatigue.*

- **Splinting effect** — *The harness straps of the breathing apparatus give a splinting (immobilizing) effect, which reduces a wearer's mobility.*

Air Supply

Some air supply limitations are based on the breathing apparatus user while others are based on the actual supply of air in the cylinder. If training is conducted under realistic conditions and records are kept, the approximate time a crew member takes to exhaust a cylinder of air can be calculated. Factors that influence air supply include the following:

 Air Supply Limitations

- **Physical condition of user** — *A wearer in poor physical condition expends the air supply more quickly than one who is in good physical condition.*

- **Degree of physical exertion** — *The higher the physical exertion of a wearer, the faster the air supply depletes.*

- **Emotional stability of user** — *A wearer who becomes excited increases respirations and uses air faster than one who is calm.*

- **Condition of apparatus** — *Minor leaks and poor adjustment of regulators result in air loss.*

- **Cylinder pressure before use** — *If a cylinder is not filled to capacity, the amount of working time is reduced proportionately.*

- **Training and experience of user** — *Properly trained and highly experienced crew members can draw the maximum air supply from a cylinder.*

Storage

Store self-contained breathing apparatus units in a manner that makes them easily accessible when needed for emergency use. Although accessibility is important, do not make the units so accessible that they are subject to physical damage or contamination from the elements. Store them either in racks with a protective cover over them to prevent contamination and corrosion or in cases specially designed for storage (often provided with the unit). The key factor is to keep them in a protected environment so that they are readily available with other PPE items and fire fighting tools. Some good examples of proper storage are shown in Figures 4.10 a and b.

Answer the following questions when storing SCBA after use or during inspections to ensure serviceability when the units are needed:

- Is the air cylinder fully charged to its designed full pressure?

- Are all backpack and head-harness straps fully extended for ease of donning the equipment?

- Is the unit clean/free of foreign matter? Is the facepiece sanitary?

- Are all parts free of damage and excessive wear?

- Are all valves in the appropriate position? Most are usually placed in the OFF position, but some (depending on the model) need to be in the ON position.

- Are all hoses properly connected? This requirement may not apply to low-pressure hoses on many models.

- Is the air cylinder properly installed in the backpack harness?

Donning and Doffing

Several important factors are considered regarding the donning and doffing of SCBA. Several methods are possible, depending on how the apparatus is stored. Different types of facepieces also require different methods. The steps vary somewhat with different methods, and various manufacturers require different steps. It is impossible to list step-by-step procedures for each manufacturer's model in this book. Therefore, the information given in the following sections are general descriptions of the various techniques. Follow the manufacturer's instructions for donning and doffing particular breathing apparatus.

Donning and doffing breathing apparatus in the small spaces of a vessel's interior pose additional complications not faced by shoreside personnel. Crew members may be very close together, and movement of the vessel while donning/doffing can cause them to lose their balance and strike another person with the equipment. Two people may assist each other with donning and doffing procedures, but each person must be fully skilled in the techniques involved.

Donning Breathing Apparatus

Some donning methods described in the following paragraphs include the over-the-head method, the regular coat method, and donning from a compartment or backup mount. The steps needed to get the apparatus onto the body differ with each method, but once the apparatus is on the body, the method of

Figures 4.10 a and b Examples of good SCBA storage on board. (a) *Courtesy of David Ward.* (b) *Courtesy of R. Wright/Maryland Fire and Rescue Institute/United States Coast Guard.*

securing the unit is the same for any one model. Different steps are needed for securing different makes and models. Once the breathing apparatus has been donned, regardless of the method used, follow the manufacturer's instructions during operation.

Regardless of the SCBA model or method of donning, make the following precautionary safety checks when preparing to don an SCBA:

- *Air cylinder gauge* — Check that the cylinder is full (Figure 4.11).

- *Regulator gauge and cylinder gauge* — Ensure that they read within approximately 10 percent of the same pressure. A crew member needs to be familiar with the equipment, and know the amount of pressure that constitutes a 10 percent difference.

- *Harness assembly and facepiece* — Ensure that all straps are fully extended (Figure 4.12).

- *Regulator valves* — Check that they are in the proper positions.

Once these checks are complete, the protective breathing apparatus may be donned using the most appropriate method. Position the breathing apparatus for donning. Adjust protective clothing appropriately: Put on the protective hood and pull it back, button the turnout coat, and turn the collar up so that the shoulder straps of the apparatus do not hold the collar down. Skill Sheet 4-1 describes the general procedures for donning full protective clothing and SCBA (see end of chapter).

Over-the-head method. Arrange the backpack straps so that they do not interfere with grasping the cylinder or backplate. Raise the harness assembly overhead. Slide the arms into their respective harness shoulder strap loops as the SCBA slides down the back (Figure 4.13). The step-by-step procedures for donning a SCBA using the over-the-head method are given in Skill Sheet 4-2 at the end of the chapter.

Regular coat method. Don the SCBA like a coat by putting one arm at a time through the shoulder strap loops. Arrange the unit so that *either* shoulder strap can be grasped for lifting (Figure 4.14). The step-by-step procedures for donning a SCBA using the regular coat method are given in Skill Sheet 4-3.

An alternate method, the crossed-arms coat method, can also be used. Arrange the equipment so that both shoulder straps can be grasped for lifting. See IFSTA's **Self-Contained Breathing Apparatus** manual for the step-by-step procedures.

Figure 4.11 Check the cylinder gauge to make sure that the air cylinder is full when preparing to don SCBA.

Figure 4.12 Fully extend the harness assembly and facepiece straps when preparing to don SCBA.

Figure 4.13 Over-the-head donning method. *Courtesy of Maritime Institute of Technology and Graduate Studies.*

Figure 4.14 Regular coat donning method. *Courtesy of Maritime Institute of Technology and Graduate Studies.*

Compartment or backup mount. Breathing apparatus stored in a closed compartment can be ready for rapid donning by using any number of methods. A mount on the inside of a compartment allows a crew member to don the breathing apparatus by backing up to it. The backup mount provides quick access. However, the height of the compartment must be appropriate for crew members to use this method. If a breathing apparatus unit is mounted too high, remove it from the mount and don using the over-the-head or coat method. See IFSTA's **Self-Contained Breathing Apparatus** manual for the procedures for donning SCBA using the backup method.

Donning the Facepiece

A facepiece cannot be worn loosely or it will not seal against the face properly. An improper seal may permit toxic gases to enter the facepiece or untimely loss of the air supply. Do not rely solely on tightening facepiece straps to ensure proper facepiece fit. A facepiece tightened too much is uncomfortable or may cut off circulation to the face. Fit each crew member with an SCBA facepiece that conforms properly with the face shape and size. Many SCBA are available with different-sized facepieces. Nosecups, if used, must also properly fit the person.

Several other conditions affect facepiece fit, for example, missing dentures or long hair. Long hair, sideburns, or beards can interfere with the outer edges of the facepiece, thus preventing contact and a proper seal with the skin. Many vessel companies and masters require that crew members be clean shaven. The temple pieces of eyeglasses also prevent the proper sealing of the facepiece; therefore, eyeglasses cannot be worn by a crew member outfitted in a SCBA. Some manufacturers offer special prescription lens holders that are installed inside the facepiece. Another remedy is to obtain a facepiece with a lens that is made to the wearer's corrective prescription.

Even though various styles are available from manufacturers, the uses and donning procedures for facepieces are essentially the same. When storing facepieces, fully extend the straps for donning ease and to keep the facepiece from becoming distorted. A facepiece may be packed in a case or stored in a bag or coat pouch. General steps for donning all SCBA facepieces are as follows:

General Facepiece Donning Steps

Step 1: Do not allow hair to come between the skin and the sealing surface of the facepiece.

Step 2: Center the chin in the chin cup, and center the harness at the rear of the head.

Step 3: Tighten facepiece straps by pulling them evenly and simultaneously to the rear. Pulling them outward to the sides may damage them and prevent proper engagement with the adjusting buckles. Tighten the lower straps first, then the temple straps, and finally the top strap if there is one.

Step 4: Check the facepiece for proper seal and operation by sealing the cylinder connection in the palm of the hand. Then inhale slowly to ensure that the mask is sealed. Check that the exhalation valve is functioning properly, all connections are secure, and the donning mode switch is in the proper position if present.

Step 5: Check for positive pressure by gently breaking the facepiece seal by inserting two fingers under the edge of the facepiece. Air moves past the fingers if the seal is correct. If no air movement is felt, remove the unit and have it inspected.

Step 6: Wear the protective hood over the facepiece harness or straps. Cover all exposed skin, but do not obscure vision.

Step 7: Wear the helmet with all straps secured.

Before donning, it is important to recognize some of the important variations in facepieces. For example, different models from the same manufacturer may have a different number of straps to tighten the head harness. The shape and size of lenses may vary. Most importantly, however, different regulator locations require different donning procedures. The regulator is either attached to the facepiece or mounted on the harness (waist belt or shoulder strap). Descriptions of these two donning procedures are given in the following paragraphs.

Harness-mounted regulator. The facepiece for an SCBA with a harness-mounted regulator has a low-pressure hose, or breathing tube, attached to the facepiece with a clamp or threaded coupling nut. Before donning the facepiece, pull the protective hood back and down so that the face opening is around the neck. Turn up the collar of the turnout coat. The procedures for donning a facepiece having a low-pressure hose or harness-mounted regulator are given in Skill Sheet 4-4.

Facepiece-mounted regulator. Before donning the facepiece, pull the protective hood back and down so that the face opening is around the neck. Turn up the collar of the turnout coat (Figure 4.15). Depending upon the style of helmet, it may be necessary to don the helmet first and allow it to rest on the shoulder. An

Figure 4.15 Before donning the facepiece, pull the protective hood back and down so that the face opening is around the neck. Turn up the collar of the turnout coat.

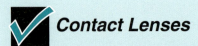

Contact Lenses

The advisability of wearing contact lenses with SCBA facepieces is questionable. Some authorities do not permit this practice, but others allow it if the user demonstrates successful use. Some users report that they have no problems with wearing contact lenses with SCBA, but others report that the lenses can be lost due to the pressure that the seal places on the facial area. Other problems encountered are debris from inside the facepiece becoming lodged behind the contact lens and irritation of the eyes due to the confined airflow causing a drying effect.

alternate method is to leave the helmet on while donning the backpack, then loosen the chin strap and allow the helmet to rest on the air cylinder while donning the facepiece. The procedures for donning a facepiece with a facepiece-mounted regulator are given in Skill Sheet 4-5.

Doffing Breathing Apparatus

Doffing techniques differ for different types of SCBA, and the location of the regulator (harness-mounted or facepiece-mounted) also determines the doffing procedures used. Skill Sheet 4-6 describes the steps for doffing SCBA with a harness-mounted regulator, and Skill Sheet 4-7 describes the steps for doffing SCBA with a facepiece-mounted regulator. When a crew member is in a safe atmosphere and SCBA is no longer required, the following steps generally apply to all SCBA when doffing:

General SCBA Doffing Steps

Step 1: *Disconnect the flow of air from the regulator to the facepiece.*

Step 2: *Disconnect the low-pressure hose from the regulator or remove the regulator from the facepiece, depending on the type of SCBA.*

Step 3: *Remove the facepiece.*

Step 4: *Remove the backpack assembly while protecting the regulator.*

Step 5: *Close the cylinder valve.*

Step 6: *Relieve pressure from the regulator in accordance with manufacturer's instructions.*

Step 7: *Extend all straps.*

Step 8: *Refill and replace the cylinder.*

Step 9: *Clean and disinfect the facepiece.*

Step 10: *Inspect the unit, and store properly.*

Changing Air Cylinders

With care and caution, a crew member can change an air cylinder at the location of an emergency so that the equipment can be used again as soon as possible. Doff the unit using one of the procedures described earlier. Obtain a full air cylinder and have it ready. Changing cylinders can be either a one- or two-person job. Skill Sheet 4-8 describes the one-person method for changing an air cylinder. When there are two people, the person with an empty cylinder bends or kneels to position the cylinder so that the other person can easily change it (Figures 4.16 and 4.17). Mark and remove cylinders that are out of service.

Training

Breathing apparatus is a device that protects one of the body's most vital organs: the lungs. To simply have a working knowledge of wearing and using breathing apparatus is not adequate. A crew member must be skillful and comfortable when using the apparatus. Crew members must understand the operation and uses of breathing apparatus intimately. They must learn to work in total darkness with confidence while wearing breathing apparatus and have the know-how to escape should the unit fail.

Simply reading and understanding the information given in this chapter is not enough to qualify an individual to use breathing apparatus. Training has no substitute. Crew members need training by a knowledgeable authority to acquire the competence and ability to safely perform emergency duties while wear-

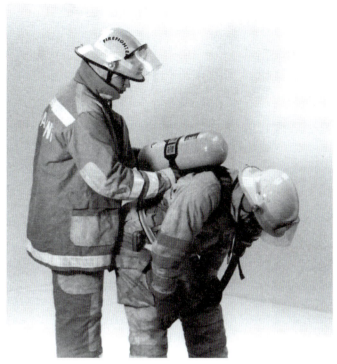

Figure 4.16 One person slides a full cylinder into the backpack assembly while the other person braces to remain steady.

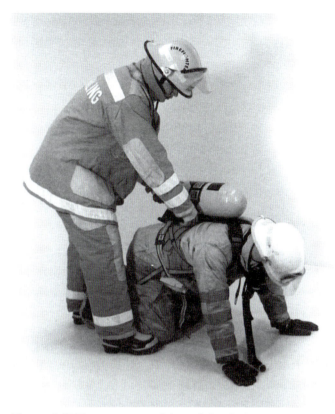

Figure 4.17 The person receiving a full cylinder may choose to kneel while the cylinder is being replaced.

ing breathing apparatus. After training is complete, thoroughly drill crew members in the proper use and operation of breathing apparatus on a regular basis. The skill of using breathing apparatus is learned through extensive, repetitive practice. Follow the instructions provided here, the recommendations of the equipment manufacturer, and the policies established for the emergency response team.

Filling Air Cylinders

Air cylinders are made from a variety of materials, including steel, aluminum, and a composite of aluminum wrapped in a fiberglass outer layer. Air cylinders are filled from either a cascade system (a series of at least three, 300 cubic-foot [8 490 L] cylinders) (Figure 4.18) or directly from a compressor purification system if available. Purchase air for refilling cylinders from a reliable shoreside source. Vessel air or air from compressors on board most vessels do not have the necessary filtration systems to ensure breathable air. Whichever means of filling air cylinders is used, always refer to the cylinder manufacturer's refilling procedures for safety measures and recommended practices. Educate all crew members who fill air cylinders in these procedures,

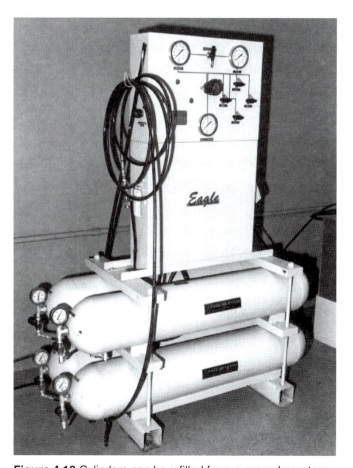

Figure 4.18 Cylinders can be refilled from a cascade system.

and train them on the refilling process. The following safety precautions always apply when refilling air cylinders:

- Put cylinders in a shielded fill station to contain fragments should the cylinder rupture.

- Fill slowly to prevent overheating the cylinder shell. A water bath over the cylinders is often used as an additional safety feature to reduce heating.

- Fill cylinder completely to the correct pressure, but do not overpressurize. Air cylinders are typically rated at either 2,216 psi (15 279 kPa) {153 bar} or 4,500 psi (31 026 kPa) {310 bar}.

Skill Sheet 4-9 gives a sample recommended procedure for filling air cylinders from a cascade system. The procedures for filling air cylinders from a compressor/purifier are much the same as those for a cascade system. The one exception is that there is usually a pressure relief adjustment that prevents the overpressurization of the cylinder (Figure 4.19). Set the pressure control to not exceed the rated pressure of the cylinder. For compressor/purifiers not equipped with a pressure relief valve, maintain constant watch on the fill pressure to determine when the cylinder is full.

Safety Precautions

Fire fighting is a strenuous, demanding activity. Although protective gear is designed to protect crew members, it can work against them at the same time. The basic required protective coat accumulates body heat and hinders movement, increasing fatigue. These conditions intensify when the additional weight of breathing apparatus is added. Be alert to the signs and symptoms of heat-related conditions that can occur under these conditions. Observe the following precautions for maximum safety when using breathing apparatus:

- Maintain physical fitness.

- Monitor personal fatigue, and rest when necessary.

- Be aware of varying air-supply durations.

- Ensure that the area is free of contamination before removing breathing apparatus.

- Work in groups of two or more (buddy system).

- Crawl rather than walk upright in areas where visibility is obscured.

Emergency Escape Techniques

The emergencies created by the malfunction of protective breathing apparatus can be overcome in several ways. Crew members should practice, *practice*, and **practice** controlled breathing when using breathing apparatus. However, when air supply is low or problems arise, emergency breathing techniques may be required. One technique is skip breathing: inhale (as during regular breathing), hold the breath as long as it takes to exhale, and then inhale once again before exhaling. Take normal breaths, and exhale slowly.

Although a regulator usually works as designed, it can malfunction. One method of using SCBA when the regulator becomes damaged or malfunctions is to open the bypass valve. During normal SCBA operation, the mainline valve is fully open while the bypass valve is fully closed. If needed in an emergency, a crew member can close the mainline valve and open the bypass valve to provide a flow of air into the facepiece. The crew member closes the bypass valve after taking a breath and then opens it again for each breath.

If the facepiece fails, it may be possible to breathe directly from the low-pressure hose (if it can be disconnected from the facepiece) or the regulator. In either event, the crew member must make sure that the hose or regulator opening is held close to the mouth to avoid breathing fire gases when inhaling. Know the equipment thoroughly, and follow the manufacturer's recommendations.

In any emergency, the conservation of air and immediate withdrawal from the hazardous atmosphere

Figure 4.19 Cylinders can also be filled from a compressor purification system.

are of the utmost importance. The following is a list of suggestions that can effectively resolve an emergency escape situation:

- Do not panic! Panicking causes rapid breathing that uses more valuable air. Suggestions:
  — Control breathing while crawling.
  — Communicate with others if possible.
  — Follow established emergency procedures if equipment fails.
- Stop and think. How did you get to where you are?
  — Downstairs? Upstairs?
  — Left turns? Right turns?
- Stop and listen. What do you hear?
  — Noises from other crew members?
  — Hose and equipment operation?
  — Sounds that indicate the location of fire?
- Activate the PASS device.
- Use different methods to find a way out. Suggestions:
  — Follow the fire hose out if possible.
  — Move in a straight line to nearest bulkhead.
  — Move in one direction once in contact with the bulkhead (make all left-hand turns or all right-hand turns — do not mix them).
  — Call for directions; call out or make noise so other crew members can give assistance.
  — Break through a bulkhead, if possible, to escape.

Inspection, Care, and Maintenance

Breathing apparatus units require proper care and inspection before and after each use to provide complete protection. Clean and sanitize units immediately after using. Moving parts that are not clean may malfunction. An unclean facepiece can spread germs and may contain an unpleasant odor. An air cylinder with less air than prescribed is inefficient if not useless.

Thoroughly wash facepieces with warm water containing any mild commercial disinfectant. Rinse with clear, warm water (Figure 4.20). Give the exhalation valve special care to ensure proper operation. Inspect

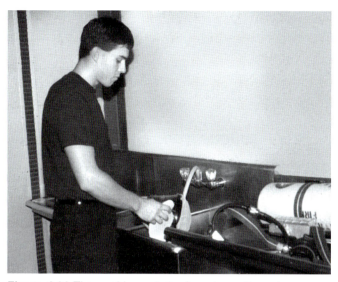

Figure 4.20 Thoroughly wash the facepiece after using.

the air hose for cracks or tears. Dry the facepiece with a lint-free cloth or air dry (do not use paper towels because they can scratch the lens).

Proper care also includes making a daily inspection if possible or at least weekly. Use the suggested checklist shown in Figure 4.21. Monthly inspections include removing the equipment from service and checking all components for deterioration. Check the operation of all gauges, valves, regulator, exhalation valve, and low-air alarm. See if leaks occur around valves and air cylinder connections.

Annual maintenance, testing, and repairs require the expertise of factory-certified technicians and is done in accordance with manufacturer's recommendations. Air cylinders are stamped or labeled with the date of manufacture and the date of the last hydrostatic test (either 3 or 5 years according to composition of cylinder) to determine the strength of the shell.

◆ Personal Alert Safety System Devices

Personal alert safety system devices are worn anytime entry is made into a hazardous area. They assist in locating downed or disoriented rescuers. PASS devices must conform to the standards set forth in NFPA 1982, *Standard on Personal Alert Safety Systems (PASS) for Fire Fighters.* These mechanisms, about the size of a small box of cereal, are worn on a crew member's SCBA or coat and operate by producing a loud, audible signal whenever body movement stops for more than 30 seconds or when the rescuer manually

Daily/Weekly SCBA Inspection Checklist

	Yes	No
◆ *Air cylinder is full.*	❏	❏
◆ *All gauges work. Cylinder gauge and regulator gauge read within approximately 10 percent of the same pressure.*	❏	❏
◆ *Low-pressure alarm is in working condition.*	❏	❏
◆ *All hose connections are tight and free of leaks.*	❏	❏
◆ *Facepiece is clean and in good condition.*	❏	❏
◆ *Harness system is in good condition, and straps are in the fully extended position.*	❏	❏
◆ *All valves are operational.*	❏	❏

NOTES:

Items Not Passing Inspection:

Action Required:

Other Comments:

Submitted by:

Signature: **Date:**

Figure 4.21 Daily/weekly SCBA inspection checklist.

triggers the alarm (Figure 4.22). Even in zero visibility conditions, fellow rescuers can follow the loud tone to locate a downed crew member.

The automatic function of a PASS device may not work well in rough seas because the movement of the vessel may be mistaken by the device as the movement of the crew member. Another possibility is that in many loud machinery spaces, the PASS device may not be heard by other crew members. These considerations are currently being studied by the maritime industry. Certainly it is better to rely on PASS devices at sea than not use any safety system at all.

PASS devices can save lives, but they must be used and maintained properly. Turn the device on, and test it before entering a hazardous space. Conduct training classes on techniques to use when attempting to rescue a lost crew member. Tests performed by the Mesa (Arizona) Fire Department show that locating a person in poor visibility conditions can be more difficult than expected even with the loud shriek of a PASS device. One reason (in a shoreside setting) was that the sound reflected off walls, ceilings, and floors making the device difficult to locate. Noise from SCBA operation and muffled hearing because of protective hoods added to the difficulty. Rescuers had a tendency to sidestep established search procedures when they thought they could tell the location of the alarm sound. The Mesa tests resulted in the following recommendations for using PASS devices:

• Make sure that the PASS devices selected are approved by a competent authority and meet the requirements of NFPA 1982.

Figure 4.22 PASS devices can save lives.

- Test PASS devices at least weekly, and maintain them in accordance with manufacturer's instructions.

- Conduct practical and regular training with PASS devices under realistic conditions.

- Train crew members to always turn on and test the devices before entering a hazardous atmosphere.

- Train rescuers to listen for distress sounds by stopping in unison, controlling breathing, and lifting hood or earflaps away from ears.

- Turn off the PASS device when a downed crew member is located or communications will be impossible.

◆ Personnel Accountability Systems

Too many emergency personnel have died because they were not discovered missing until it was too late. Flashover and backdraft may trap or injure fire fighting crew members. SCBAs can malfunction or run out of air. Crew members can get disoriented or lost. Use an on-duty list or other system, such as a tag system, and a head count at every incident to account for all personnel involved in the incident. Ensure that the master or emergency team leader has a list or some means of identifying the location of all crew members.

On-duty lists allow officers to know exactly who is operating on the scene, but officers must be in constant contact with crew members by head count, communication, or sight. Keep one copy of the on-duty list at the forward command location, and maintain one copy on the bridge. By keeping track of head counts, it is possible to determine whether someone is missing. For shipboard applications, a completed muster list is the best starting place to identify what personnel are aboard and, of those personnel, which are on duty.

A simple tag system aids in accounting for crew members during emergencies. Equip crew members with a personal identification (ID) tag. Crew members leave their tags at a given location or with a designated person upon entering the danger area. Attach tags to a control board or personnel ID chart for quick reference. Crew members collect their tags upon leaving the emergency zone. By using this system, it is possible for officers to know exactly who is operating in the emergency area.

An SCBA tag system provides even closer accountability for crew members inside a danger area. Each SCBA is provided with a tag containing the name of the user and the unit's air pressure. Crew members give their tags to a designated officer upon entering the danger area. The officer records time of entry and expected time of exit. This officer also does a brief check to ensure that all personal protective equipment is in place and being used properly. This system provides complete accountability for those inside a hazardous area. Crew members leaving the danger area retrieve their tags so that the accountability person knows who is safely outside and who is still inside. Send in relief crews before the estimated time of the sounding of the low-pressure alarms. Figure 4.23 illustrates one form of a control and accountability system used during an emergency response.

Each vessel must develop its own system of personnel accountability. Standardize the system so that it is used at every emergency incident and during training. The system can be as simple or complex as the vessel needs. Ensure that all crew members are

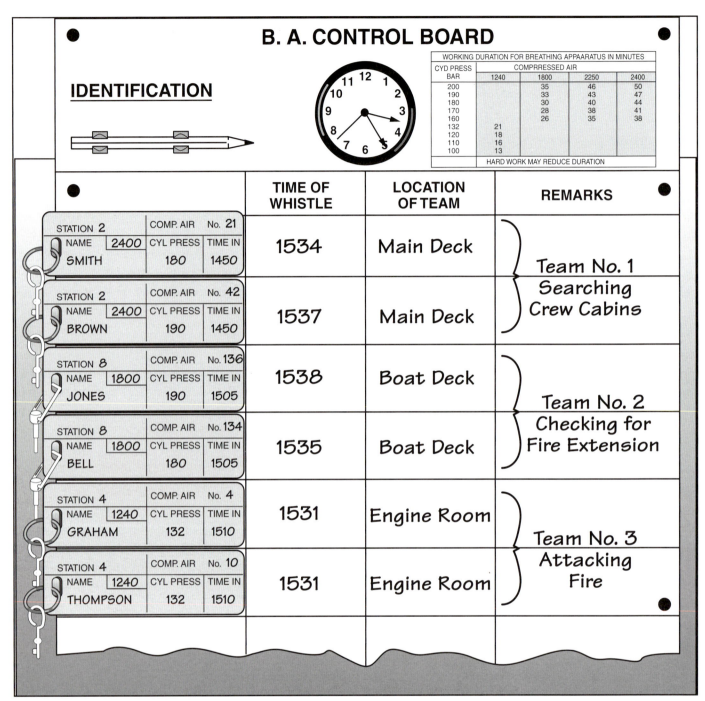

IDENTIFICATION	TIME OF WHISTLE	LOCATION OF TEAM	REMARKS
STATION 2 COMP. AIR No. 21 NAME 2400 CYL PRESS TIME IN SMITH 180 1450	1534	Main Deck	Team No. 1 Searching Crew Cabins
STATION 2 COMP. AIR No. 42 NAME 2400 CYL PRESS TIME IN BROWN 190 1450	1537	Main Deck	
STATION 8 COMP. AIR No. 136 NAME 1800 CYL PRESS TIME IN JONES 190 1505	1538	Boat Deck	Team No. 2 Checking for Fire Extension
STATION 8 COMP. AIR No. 134 NAME 1800 CYL PRESS TIME IN BELL 180 1505	1535	Boat Deck	
STATION 4 COMP. AIR No. 4 NAME 1240 CYL PRESS TIME IN GRAHAM 132 1510	1531	Engine Room	Team No. 3 Attacking Fire
STATION 4 COMP. AIR No. 10 NAME 1240 CYL PRESS TIME IN THOMPSON 132 1510	1531	Engine Room	

B. A. CONTROL BOARD

WORKING DURATION FOR BREATHING APPAARATUS IN MINUTES				
CYD PRESS BAR	COMPRRESSED AIR			
	1240	1800	2250	2400
200		35	46	50
190		33	43	47
180		30	40	44
170		28	38	41
160		26	35	38
132	21			
120	18			
110	16			
100	13			
HARD WORK MAY REDUCE DURATION				

Figure 4.23 SCBA wearers are timed from the time they turn on their air supplies and enter the scene. Locations are noted along with the anticipated time of the sounding of low-pressure alarms. *Adapted from the B.A. Control Board of The Fire Service College, Moreton-in-Marsh, England.*

familiar with the system. Accountability is vital. If the emergency team leader does not know who is in the emergency area and where, it is impossible to determine who and how many may be trapped or injured. If the master knows exactly how many people are on the scene at all times and where they are operating, it is much easier to manage a incident. Refer to Chapter 11, Emergency Response Process, for more information.

◆ Environmental Analysis Instruments

Monitoring instruments are designed to determine specific concentrations of hazardous materials in the environment. These instruments are routinely used to determine oxygen deficiency, the presence of flammable or explosive atmosphere, or toxicity. They may also be required before, during, and after a fire on board a vessel to assess the presence and degree of hazard within a compartment. Entry into a fire area also presents challenging visual conditions. Dense smoke and a lack of lighting give crew members visual conditions that the eyes, flashlights, or lanterns cannot overcome.

Low-light enhancement devices are not designed for interior fire operation, but they may be useful for night operations on deck. Typically, a binocular or monocular is equipped with low-light gathering optics with electronic circuitry to enhance the available light.

Most instruments operate on the same basic principle, with an ON/OFF mechanism and an audible or visible alarm. Follow the manufacturer's instructions for each different analyzing instrument. Multiple readings from different areas within a space may be necessary to make an accurate assessment of the environment. Make every attempt to take readings from outside the space through monitoring holes or access ports. Above all, properly train the operators of these instruments so that the results are reliable.

In general, the instruments are sensitive to the environment and easily damaged by dropping, bumping, or extreme temperatures. They require regular maintenance and frequent *calibration:* the process of adjusting a monitoring instrument so that its readings correspond to actual, known concentrations of a given material. The manufacturer initially calibrates instru-

ments. Consider having the ability to calibrate instruments on board without having to return the instrument to the manufacturer. Conditions for calibration include atmosphere, temperature, atmospheric pressure, etc. The maintenance requirements of various instruments vary widely from one instrument to another, from one manufacturer to another, and from one model to another. Follow the manufacturer's maintenance instructions and calibration specifications on each individual instrument.

The instruments most likely to be used when responding to a fire are oxygen analyzers, pyrometers, and thermometers (Figure 4.24). More than one instrument may be required to develop a complete picture of the hazards involved. Instruments may vary greatly in design and operation from one make or model to another. The monitoring instruments most used in the maritime industry are briefly described as follows:

- *Oxygen analyzers* — Measure the concentrations of oxygen in the air and typically reveal both oxygen-enriched and oxygen-deficient atmospheres (Figure 4.25).

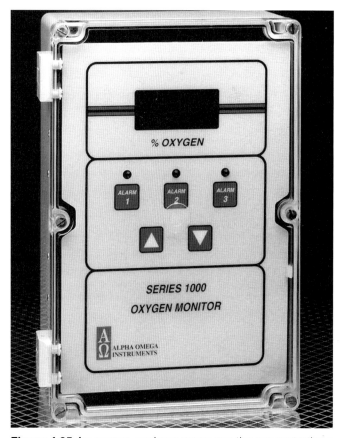

Figure 4.25 An oxygen analyzer measures the concentrations of oxygen in the air. *Courtesy of Alpha Omega Instruments Corporation, Cumberland, RI.*

Figure 4.24 Various temperature-measuring devices are available. A pyrometer indicates the presence of heat, and a thermometer indicates the intensity of heat present. *Courtesy of R. Wright/Maryland Fire and Rescue Institute/United States Coast Guard.*

- *Pyrometers* — Indicate the presence of heat, and often the relative intensity, although not specific temperatures.

- *Thermometers* — Indicate the intensity of heat present expressed in degrees Fahrenheit (°F) or degrees Celsius (°C).

- *Infrared detectors or scanners* — Detect fires in concealed spaces by converting radiation energy (heat) into a displayed electrical signal. These devices can detect heat from hidden fires. They work by measuring and comparing surface temperatures. Ranges of the devices vary, but most detect heat within 10 feet (3 m) of the source. Infrared devices are often used for victim search or locating the seat of a fire in smoke.

- *Thermal imaging devices/cameras* — These devices use very high-tech electronic and optical systems and allow an emergency responder to see in zero visibility by converting heat patterns into clear, visible images that are viewed on a screen. They can convert temperatures as low as 0.1°F (-17.7°C) radiated from objects and humans at a range of 600 feet (183 m). The units are useful not only for interior victim search and fire detection but also for scanning the water's surface. The equipment can be handheld or mounted directly to a crew member's helmet. Many manufacturers offer optional video capabilities that allow transmission of the thermal image to a location outside the search area (such as a command post) where they are recorded and viewed remotely (Figure 4.26).

- *Explosimeters or combustible gas detectors* — Measure the percentage of the lower flammable (explosive) limit (LFL) or the presence of a flammable vapor expressed in parts per million (ppm).

- *Multi-gas detectors* — Detect several (two to twenty or more) gases in one instrument. They are most commonly used in addition to instruments that check for oxygen deficiency and the presence of an explosive atmosphere. These devices are very accurate but must be diligently maintained. *CAUTION: Failure to detect one of the gases for which this instrument is designed, does not indicate the absence of a hazard.*

- *Toxicity monitoring devices* — Detect the presence and measure the concentration of specific toxic chemicals in the air. The type of instrument may range from simple colorimetric tubes to sophisticated instruments.

Figure 4.26 Thermal imaging devices/cameras allow emergency responders to see in zero visibility. *Courtesy of R. Wright/Maryland Fire and Rescue Institute/United States Coast Guard.*

- *Colorimetric indicator tubes* — Measure the concentration of a specific gas or vapor in the air. They consist of small glass tubes filled with different reagents that react with the gas being tested. A sample of air is drawn up through the tube, usually by a vacuum device. A change in the color or appearance inside the tube indicates specific chemicals are present in the air.

- *Specific chemical monitors* — Detect either a large group of chemicals or a specific chemical. Manufacturers can configure these monitors to detect almost any chemical. These monitors are more accurate than detector tubes.

- *Radiation monitors* — Detect alpha, beta, or gamma radiation sources; some measure the accumulated radiation exposure. Examples of radiation monitors include Geiger counters, survey meters, and dosimeters.

- *Personal monitoring devices* — Detect and record the exposure of a worker to potential hazards. Some models can be set to sound an alarm if a worker is exposed to a predetermined level of the hazard. The level indicated may be either an acute exposure or a chronic exposure. The radiation dosimeter is the most common personal monitoring device. Two common types of radiation dosimeters are the film badge and the pocket dosimeter.

Donning Personal Protective Equipment

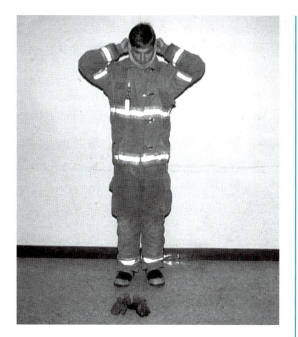

Step 1: Don protective coat, trousers, and boots.

Step 2: Pull protective hood down around the neck.

Step 3: Place gloves in a readily accessible location.

Step 4: Position SCBA ready for donning.

Step 5: Check the cylinder gauge to make sure that the air cylinder is full (at least 90 percent).

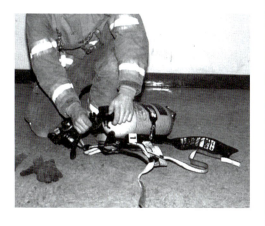

Step 6: Open the cylinder valve slowly, and listen for the audible alarm as the system pressurizes.

Step 7: Verify the operation of the low air supply warning alarm.

NOTES:
- If the audible alarm does not sound, or if it sounds but does not stop, place the unit out of service.

- On some styles of SCBA, the audible alarm does not sound when the cylinder valve is opened. Know the operation of each particular unit.

Step 8: Check the regulator gauge and the cylinder gauge to ensure that they register within 10 percent of the same pressure.

Step 9: Don the SCBA in accordance with manufacturer's recommendations:

- Secure all straps.

- Properly position the facepiece.

- Check the exhalation valve.

- Check facepiece seal.

- Attach low-pressure tube to the regulator, or attach the regulator and air line to the facepiece depending on the style of SCBA.

- Check donning mode switch if present.

- Activate the airflow.

- Activate the PASS device.

Step 10: Place the hood and helmet in proper position for operations.

Step 11: Don gloves.

Donning Self-Contained Breathing Apparatus

Over-the-Head Method

Step 1: Crouch or kneel at the cylinder valve end of the unit.

Step 2: Check the unit.

— Check the cylinder gauge to make sure that the air cylinder is full (at least 90 percent).

— Open the cylinder valve slowly and listen for the audible alarm as the system pressurizes. Then, open the cylinder valve fully.

NOTES:

• If the audible alarm does not sound, or if it sounds but does not stop, place the unit out of service.

• On some styles of SCBA, the audible alarm does not sound when the cylinder valve is opened. Know the operation of each particular unit.

— Check the regulator gauge and the cylinder gauge to ensure that they register within 10 percent of the same pressure.

— Check donning mode switch if present.

NOTES:

• For units with facepiece-mounted regulators, leave the cylinder valve open and the unit in the donning mode.

• If the unit is positive pressure only, refer to the manufacturer's instructions concerning the cylinder valve.

Step 3: Spread the harness straps out to their respective sides.

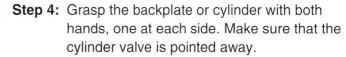

Step 4: Grasp the backplate or cylinder with both hands, one at each side. Make sure that the cylinder valve is pointed away.

NOTE: There should be no straps between the hands.

Step 5: Lift the cylinder, and let the regulator and harness hang freely.

Step 6: Raise the cylinder overhead, and let the elbows find their respective loosened harness shoulder strap loops. Keeping elbows close to the body, tuck chin and grasp the shoulder straps as the SCBA begins to slide down the back. Let the straps slide through the hands as the backpack lowers into place.

Step 7: Lean forward to balance the cylinder on the back and partially tighten the shoulder straps by pulling them outward and downward.

NOTE: It is sometimes necessary to lean forward with a quick jumping motion to properly position the SCBA on the back while tightening the straps.

Step 8: Continue leaning forward, and fasten the chest buckle if the unit has a chest strap.

NOTE: Depending upon a person's build, it may be more comfortable to fasten the chest buckle before tightening the shoulder straps.

Step 9: Fasten and adjust the waist strap until the unit fits snugly.

Step 10: Don the facepiece (see Skill Sheets 4-4 and 4-5).

Regular Coat Method

Step 1: Perform Steps 1 through 3 of the over-the-head method.

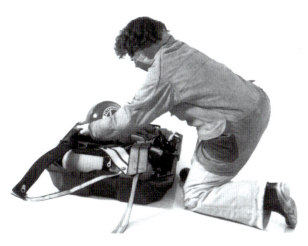

Step 2: Position the upper portion of the straps over the top of the backplate.

NOTES:

• This positioning makes the straps less likely to fall, and the arms can go through the straps with less difficulty.

• This procedure is for those harnesses having the regulator attached to the left side of the harness. For these that have the regulator mounted on the right,

— Grasp the right strap with the right hand.

— Don the backpack following the instructions in the next steps, but using alternate method directions.

Step 3: At the top of the harness, grasp the left strap with the left hand; grasp the lower portion of the same strap with the right hand.

NOTE: When kneeling at the cylinder valve end, the left harness strap will be to the right.

Alternate Method:

Step 3a: Grasp the top of the left shoulder strap with the left hand; grasp the regulator with the right hand.

Step 4: Lift the unit; swing it around the left shoulder and onto the back. Both hands should still be grasping the shoulder strap.

Alternate Method:

Step 4a: Lift the unit; swing it around the left shoulder and onto the back, maintaining control of the regulator with the right hand.

Step 5: Continue to hold the strap with the left hand, release the right hand, and insert the right arm between the right shoulder strap and the backpack frame.

Alternate Method:

Step 5a: Transfer the regulator to the left hand.

— Insert the right arm through the right shoulder strap.

— Grasp the end of the waist strap with the right hand.

— Loosely connect the waist strap.

Step 6: Lean slightly forward to balance the cylinder on the back.

Step 7: Tighten the shoulder straps by pulling them outward and downward.

Step 8: Continue leaning forward, and fasten the chest buckle if the unit has a chest strap. Tighten the shoulder straps further if necessary.

Step 9: Fasten and adjust the waist strap until the unit fits snugly.

Step 10: Don the facepiece (see Skill Sheets 4-4 and 4.5).

Donning the Facepiece

Harness-Mounted Regulator

Step 1: Grasp the head harness with the thumbs inserted through the straps from the inside, and spread the webbing.

Step 2: Push the top of the harness up the forehead to remove hair that may be present between the forehead and the sealing surface of the facepiece.

Step 3: Center the chin in the chin cup and position the harness so that it is centered at the rear of the head.

Step 4: Tighten the harness straps by pulling them evenly and simultaneously to the rear (not outward). Tighten the lower straps first, then the temple straps, and finally the top strap if there is one.

NOTE: Pulling the straps outward, to the sides, may damage them and prevents proper engagement with the adjusting buckles.

Step 5: Check the facepiece seal.

— Exhale deeply.

— Seal the end of the low-pressure hose with a bare hand.

— Inhale slowly (not deeply).

— Hold the breath for 10 seconds.

Step 5a: Adjust or redon the facepiece if the facepiece does not collapse tightly against the face or if there is evidence of leaking.

NOTE: Inhaling very quickly temporarily seals any leak and gives a false sense of a proper seal.

Step 6: Check the exhalation valve.

— Inhale.

— Seal the end of the low-pressure hose with the palm of a hand.

— Exhale.

Step 6a: Keep the low-pressure hose sealed, press the facepiece against the sides of the face, and exhale to free the valve if the exhalation does not go through the exhalation valve. If the exhalation valve does not become free, remove the facepiece from service.

NOTE: Use caution when exhaling against a sealed facepiece in order to prevent discomfort and possible damage to the inner ear from exhaling forcefully.

Step 7: Put on the helmet, first inserting the low-pressure hose through the helmet's chin strap. Rest the helmet on the shoulder until the SCBA is completely donned.

NOTE: Helmets with straps that completely disconnect may be donned as a last step.

Step 8: Connect the low-pressure hose to the regulator.

— Turn the donning switch to the PRESSURE, USE, or ON position if the unit has one.

— Open the mainline valve if the unit does not have a donning switch.

Step 9: Check for positive pressure. Gently break the facepiece seal by inserting two fingers under the edge of the facepiece. Air should move past the fingers.

Step 9a: Remove the unit and have it checked if air movement is not felt.

Step 10: Pull the protective hood into place, making sure that all exposed skin is covered and that vision is unobscured. Check to see that no portion of the hood is located between the facepiece and the face.

Step 11: Place the helmet on the head and tighten the chin strap.

Alternate Method: Wear the helmet while donning the SCBA. After donning the backpack,

- Loosen the chin strap.

- Allow the helmet to rest on the air cylinder or on the shoulder.

- Don the facepiece.

- Lift the helmet back onto the head and tighten the chin strap when the facepiece straps have been tightened and the hood is on.

Donning the Facepiece

Facepiece-Mounted Regulator

Step 1: Grasp the head harness with the thumbs inserted through the straps from the inside, and spread the webbing.

Step 2: Stabilize the facepiece with one hand, and use the other hand to remove hair that may be present between the forehead and the sealing surface of the facepiece.

Step 3: Center the chin in the chin cup and position the harness so that it is centered at the rear of the head.

Step 4: Tighten the harness straps by pulling them evenly and simultaneously to the rear (not outward). Tighten the lower straps first, then the temple straps, and finally the top strap if there is one.

NOTE: For two-strap harnesses, tighten the neck straps, then stroke the harness firmly down the back of the head. Retighten the straps as necessary.

Step 5: Attach the regulator to the facepiece if the regulator is separate from the facepiece by positioning it firmly into the facepiece fitting. Lock it into place.

NOTE: This procedure varies, depending upon the make of SCBA. Always follow the manufacturer's instructions.

Step 6: Check the facepiece seal.

— Make sure that the donning switch is in the DON position (positive pressure off).

— Inhale slowly (not deeply).

— Hold the breath for 10 seconds.

NOTE: The mask should collapse against the face. There should be no sound of airflow and no inward leakage through the exhalation valve or around the facepiece.

Alternate Method: Check the facepiece seal:

• Close the cylinder valve.

• Continue to breathe slowly until the mask collapses against the face.

• Hold the breath for 10 seconds.

• Reopen the cylinder valve if the mask draws up to the face and no leaks are detected.

• Adjust or redon the facepiece if there is evidence of leaking.

NOTE: Use care with this method because it uses some of the air supply.

Step 7: Check the exhalation valve. When exhaling during Step 6, make sure that the exhalation goes through the exhalation valve and not to the edges of the facepiece.

Step 7a: Press the facepiece against the sides of the face, and exhale to free the valve if the exhalation does not go through the exhalation valve. If the exhalation valve does not become free, remove the facepiece from service.

NOTE: Use caution when exhaling against a sealed facepiece in order to prevent discomfort and possible damage to the inner ear from exhaling forcefully.

Step 8: Check for positive pressure. Gently break the facepiece seal by inserting two fingers under the edge of the facepiece. Air should move past the fingers.

Step 8a: Remove the unit and have it checked if air movement is not felt.

Step 9: Pull the protective hood into place, making sure that all exposed skin is covered and that vision is unobscured. Check to see that no portion of the hood is located between the facepiece and the face.

Step 10: Put the helmet back on the head and tighten the chin strap. Be sure to get the helmet strap under the chin.

NOTE: Helmets with a breakaway strap can be donned at this point.

Alternate Method: Leave the helmet on while donning the backpack, then loosen the chin strap and allow the helmet to rest on the air cylinder while donning the facepiece.

Doffing Self-Contained Breathing Apparatus

Harness-Mounted Regulator

Step 1: Close the mainline valve and disconnect the low-pressure hose from the regulator.

Step 1a: Make sure that the donning switch is in the donning mode If the unit has one.

Step 2: Take off the helmet or loosen it and push it and the hood back off the head.

Step 3: Loosen the facepiece harness strap buckles. Either rub them toward the face or lift the buckles slightly to loosen them.

Step 4: Take off the facepiece, and extend the harness straps fully.

Step 5: Unbuckle the waist belt and fully extend the adjustment.

Step 6: Disconnect the chest buckle if the unit is so equipped.

Step 7: Lean forward; release shoulder strap buckles and hold them open while fully extending the straps.

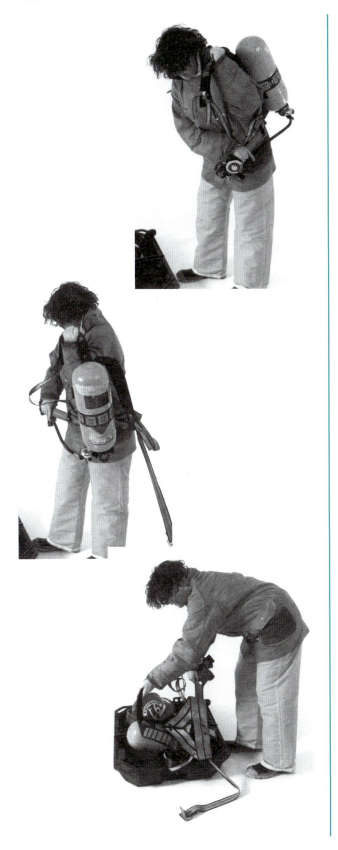

Step 8: Grasp the shoulder straps firmly with the respective hands. Slip off the shoulder strap from the shoulder opposite the regulator, and remove the arm from the shoulder strap.

Step 9: Grasp the regulator with the free hand, allow the other strap to slide off the shoulder, and lower the SCBA to the ground. Do not drop the regulator or allow it to strike anything.

Step 10: Close the cylinder valve, and relieve the excess pressure from the regulator.

Step 10a: Reconnect the regulator if it has been removed from the facepiece. Hold the facepiece against the face, and breathe until the pressure is depleted.

Alternate Method: Open the mainline valve and allow the excess pressure to vent. Do not use the bypass valve to relieve excess pressure.

Step 11: Remove the facepiece from the regulator and extend the straps fully.

Step 12: Prepare the facepiece for inspection, cleaning, sanitizing, and storage.

Step 13: Inspect and store unit properly.

Doffing Self-Contained Breathing Apparatus

Facepiece-Mounted Regulator

Step 1: Take off the helmet or loosen it and push it and the hood back off the head.

Step 2: Turn off the positive pressure or place in donning mode if the unit has a donning switch.

Step 3: Disconnect the regulator from the facepiece, depending upon the make of SCBA and manufacturer's instructions.

Step 4: Loosen the facepiece harness strap buckles. Either rub them toward the face or lift the buckles slightly to loosen them.

Step 5: Take off the facepiece, and extend the harness straps fully.

Step 6: Unbuckle the waist belt and fully extend the adjustment.

Step 7: Disconnect the chest buckle if the unit is so equipped.

Step 8: Attach the regulator to the harness clip if the unit is so equipped or control the regulator by holding it while performing the next steps.

Step 9: Lean forward; release shoulder strap buckles and hold them open while fully extending the straps.

Step 10: Grasp the shoulder straps firmly with the respective hands. Slip off the shoulder strap from the shoulder opposite the regulator, and remove the arm from the shoulder strap.

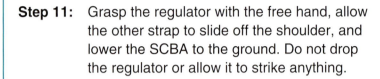

Step 11: Grasp the regulator with the free hand, allow the other strap to slide off the shoulder, and lower the SCBA to the ground. Do not drop the regulator or allow it to strike anything.

Step 12: Close the cylinder valve and relieve the excess pressure from the regulator.

Step 12a: Reconnect the regulator to the facepiece. Hold the facepiece against the face, and breathe until the pressure is depleted.

NOTE: Do not use the bypass valve to relieve excess pressure.

Step 13: Prepare the facepiece for inspection, cleaning, sanitizing, and storage.

Step 14: Inspect and store unit properly.

Changing an SCBA Air Cylinder

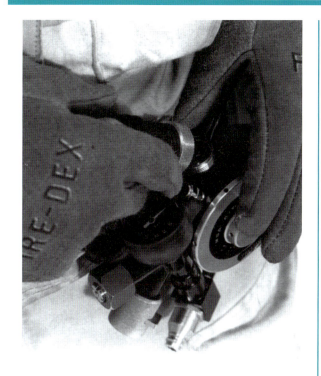

One Person

Step 1: Disconnect the regulator from the facepiece or disconnect the low-pressure hose from the regulator.

Step 2: Close the cylinder valve on the used cylinder and release the pressure from the high-pressure hose.

NOTES:

• On some units, the pressure must be released by breathing down the regulator or opening the mainline valve. Refer to the manufacturer's instructions for the correct method for the particular unit.

• The high-pressure coupling will be difficult to disconnect if the pressure is not released.

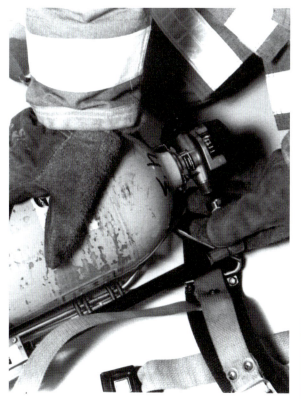

Step 3: Disconnect the high-pressure coupling from the cylinder.

Step 3a: Repeat Step 2 if more than hand force is required to disconnect the coupling.

Step 4: Lay the hose coupling on the deck, directly in line with the cylinder outlet, as a reminder so that the replacement cylinder can be aligned correctly and easily.

NOTE: Be sure that grit or liquids do not enter the end of the unprotected high-pressure hose prior to attaching it to the cylinder outlet valve.

Step 5: Release the cylinder clamp and remove the empty cylinder.

Step 6: Place the new cylinder into the backpack, position the cylinder outlet, and lock the cylinder into place.

NOTE: For some cylinders, it may be necessary to rotate the cylinder one-eighth turn to the left; this protects the high-pressure hose by lessening the angle of the hose and preventing twisting.

Step 7: Check the cylinder valve opening and the high-pressure hose fitting for debris and the condition of the O-ring.

— Clear any debris from the cylinder valve opening by quickly opening and closing the cylinder valve or by wiping the debris away.

— Replace the O-ring if it is distorted or damaged.

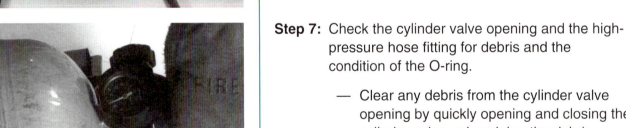

Step 8: Connect the high-pressure hose to the cylinder valve opening.

NOTE: Do not overtighten; hand-tightening is sufficient.

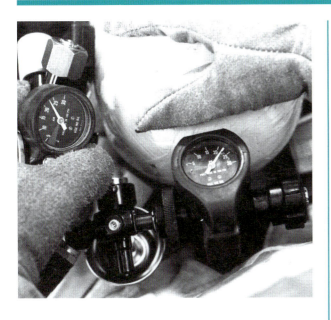

Step 9: Open the cylinder valve and check the gauges on the cylinder and the regulator.

NOTES:

- Both gauges should register within 10 percent of the same pressure.

- Some units require that the mainline valve on the regulator be opened in order to obtain a gauge reading. Seal the regulator outlet port by placing one hand over it. On a positive-pressure regulator, the port must be sealed for an accurate regulator gauge reading.

Filling an SCBA Air Cylinder

From a Cascade System

Step 1: Check the hydrostatic test date of the SCBA air cylinder.

Step 2: Inspect the SCBA air cylinder for damage such as deep nicks, cuts, gouges, or discoloration from heat.

Step 2a: Remove the cylinder from service If it is damaged or is out of hydrostatic test date, and tag it for further inspection and hydrostatic testing.

CAUTION: Never attempt to fill a cylinder that is damaged or out of hydrostatic test date.

Step 3: Place the SCBA air cylinder in a fragment-proof fill station.

Step 4: Connect the fill hose to the cylinder.

NOTE: If the fill hose has a bleed valve, make sure that the bleed valve is closed.

Step 5: Open the SCBA air cylinder valve.

Step 6: Open the valve at the fill hose, the valve at the cascade system manifold, or the valves at both locations if the system is so equipped.

NOTE: Some cascade systems may have a valve at the fill hose, at the manifold, or at both places.

Step 7: Open the valve of the cascade cylinder that has the least pressure but has more pressure than the SCBA cylinder.

NOTE: The airflow from the cascade cylinder must be slow enough to avoid "chatter" of the connecting lines or excessive heating of the cylinder being filled.

Step 8: Watch to see that the cylinder gauge needle rises slowly by about 300 to 600 psi (2 068 kPa {21 bar} to 4 137 kPa {41 bar}) per minute.

NOTE: A hand should be able to rest on the SCBA cylinder without undue discomfort from the heating of the cylinder.

Step 9: Close the cascade cylinder valve when the pressures of the SCBA cylinder and the cascade cylinder equalize.

Step 9a: Open the valve on the cascade cylinder with the next highest pressure if the SCBA cylinder is not yet full.

Step 9b: Repeat Steps 8 and 9 until the SCBA cylinder is completely full.

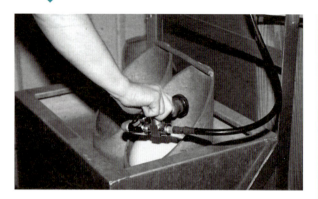

Step 10: Close the valve or valves at the cascade system manifold and/or fill line if the system is so equipped.

Step 11: Close the SCBA cylinder valve.

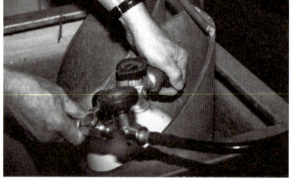

Step 12: Open the hose bleed valve to release excess pressure between the cylinder valve and the valve on the fill hose.

CAUTION: Failure to release excess pressure could result in O-ring damage.

Step 13: Disconnect the fill hose from the SCBA cylinder.

Step 14: Remove the SCBA cylinder from the fill station.

Step 15: Return the SCBA cylinder to proper storage.

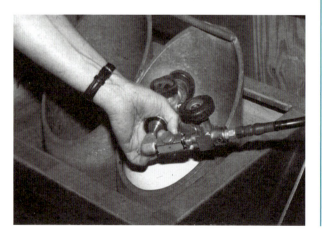

Chapter 5

Portable and Semiportable Fire Extinguishers

The fire extinguisher is one of the most common fire protection appliances in use today. Extinguishers of various types are required by all agencies that regulate marine operations, and they are found on all vessels. Seafarers are always educated to sound the alarm before taking any other action when a fire is discovered, but a portable fire extinguisher is a good first line of defense for small fires. Approximately 90 percent of marine fires are small enough to be extinguished with a portable extinguisher when discovered early. In many cases, a portable extinguisher can extinguish a fire in much less time than it would take to deploy a fire hose. In any case, sound the alarm before attempting extinguishment, no matter how small the fire.

It is important that crew members are knowledgeable about the different types of fire extinguishers and their correct operation. This chapter discusses the various types of fire extinguishers (both portable and semiportable) that crew members are likely to encounter. Almost all types are found in either the portable or semiportable variety. For marine purposes, an extinguisher is classified as portable if its total gross weight is 55 pounds (25 kg) or less. Semiportable types are commonly wheeled units. Another type of semiportable extinguisher is the hose-reel system, which may use carbon dioxide (CO_2), dry chemical, or halogenated extinguishing agents. Requirements for portable fire extinguishers are set forth in Chapter II-2 of the *International Convention for the Safety of Life at Sea* (SOLAS) regulations from the International Maritime Organization (IMO).

This chapter also discusses North American and European fire extinguisher rating systems. Information on selecting the proper extinguisher, operating procedures, maintenance requirements and procedures, inspection guidelines, and general extinguisher refilling procedures are also addressed. Information about damaged and obsolete extinguishers is given. Skill sheets describing operating fire extinguishers and refilling cylinder procedures are given at the end of the chapter.

 Sandbox

A common, though often forgotten, fire-fighting item on board vessels is the sandbox. It is usually found in the engine room, and its operation is very simple: Spread the sand with the scoop provided. The sand is effective on small Class A, Class B, Class C, and on some Class D fires. The sandbox is low maintenance and reliable in operation.

Courtesy of Captain John F. Lewis.

Extinguisher	Type	Agent	Fire Class	Size	Stream Reach	Discharge Time
Pump-Tank Water	Hand-carried; backpack	Water	A only	1½–5 gal (6 L to 19 L)	30–40 ft (9 m to 12.2 m)	45 sec to 3 min
Stored-Pressure Water	Hand-carried	Water	A only	1¼–2½ gal (5 L to 10 L)	30–40 ft (9 m to 12.2 m)	30–60 sec
Aqueous Film Forming Foam (AFFF)	Hand-carried	Foam	A & B	2½ gal (10 L)	20–25 ft (6 m to 7.6 m)	Approximately 50 sec
Halon 1211*	Hand-carried; wheeled	Halon	B & C	*Hand-carried:* 2½–20 lb (1.1 kg to 9 kg) — *Wheeled:* to 150 lb (68 kg)	8–18 ft (2.4 m to 5.5 m) — 20–30 ft (6 m to 10.7 m)	8–18 sec — 30–44 sec
Halon 1301	Hand-carried	Halon	B & C	2½ lb (1.1 kg)	4–6 ft (1.2 m to 1.8 m)	8–10 sec
Carbon Dioxide	Hand-carried	Carbon dioxide	B & C	2½–20 lb (1.1 kg to 9 kg)	3–8 ft (1 m to 2.4 m)	8–30 sec
Carbon Dioxide	Wheeled	Carbon dioxide	B & C	50–100 lb (23 kg to 45 kg)	8–10 ft (2.4 m to 3 m)	26–65 sec
Dry Chemical	Hand-carried stored-pressure; cartridge-operated	Sodium bicarbonate, potassium bicarbonate, ammonium phosphate, potassium chloride	B & C	2½–30 lb (1.1 kg to 14 kg)	5–20 ft (1.5 m to 6 m)	8–25 sec

Continued on next page

Table 5.1 (Continued)
Operational Characteristics of Portable/Wheeled Fire Extinguishers

Extinguisher	Type	Agent	Fire Class	Size	Stream Reach	Discharge Time
Multipurpose Dry Chemical	Hand-carried stored-pressure; cartridge-operated	Monoammonium phosphate	A, B, & C	2½–30 lb (1.1 kg to 14 kg)	5–20 ft (1.5 m to 6 m)	8–25 sec
Dry Chemical	Wheeled; ordinary or multipurpose	Sodium bicarbonate, etc.	A, B, & C	75–350 lb (34 kg to 159 kg)	Up to 45 ft (13.7 m)	20 sec to 2 min
Dry Powder	Hand-carried; wheeled	Various, depending on metal fuel (this description is for sodium chloride plus flow enhancers)	D only	*Hand-carried:* to 30 lb (14 kg) *Wheeled:* 150 lb & 350 lb (68 kg & 159 kg)	4–6 ft (1.2 m to 1.8 m)	28–30 sec

* Rating: Those larger than 9 lb (4 kg) capacity have small Class A ratings (1-A to 4-A).

Types of Portable Fire Extinguishers

Many different types of portable fire extinguishers exist, and the following sections highlight some of the common extinguishers encountered in any setting, marine or otherwise. Brief application techniques are given for each type. See Selecting and Using Portable Fire Extinguishers section for general application steps. Various extinguishing agents are used with the different types: water, foam, halogenated hydrocarbons, carbon dioxide, dry chemicals, and dry powders. Some extinguishers are stored-pressure or cartridge-operated types. Although some of the extinguisher types described are rarely seen on marine vessels, they are included in the interest of completeness. See Table 5.1 for a list of operational characteristics of a number of portable and wheeled fire extinguisher types.

Stored-Pressure Water Extinguishers

Stored-pressure water extinguishers, also known as *air-pressurized water* (APW) extinguishers, are used only on Class A fires (Figures 5.1 a and b). These extinguishers are useful for all types of small Class A fires and are often applied to confined hot spots during overhaul operations. They are available in sizes of 1¼ to 2½ gallons (5 L to 10 L). Under normal conditions, their stream reach is 30 to 40 feet (9 m to 12.2 m). The discharge time is 30 to 60 seconds, depending on the size of the unit. These extinguishers need protection against freezing if they are exposed to temperatures less than 40°F (4°C). Add an antifreeze solution to the water. Stored-pressure extinguishers are very useful in incipient fire situations involving the Class A combustibles found in most accommodation spaces.

The water in a stored-pressure extinguisher is discharged by a compressed gas (either air or nitrogen) that is stored in the tank with the water. A gauge located on top of the tank shows when the extinguisher is properly pressurized (Figure 5.2). With this type of extinguisher, the pressure is ready to release the extinguishing agent at any time. When the shutoff device is opened, a stream of water expels through the hose.

Carry a stored-pressure water extinguisher to a fire in an upright position. Hold the hose in one hand and the shutoff device in the other. When water is needed,

Figures 5.1 a and b
(a) *(right)* A stored-pressure water fire extinguisher. *Courtesy of Captain Mark Turner, Syndicated Management Services Ltd.* (b) *(below)* Cutaway view of a stored-pressure water fire extinguisher.

Figure 5.2 Pressure gauge for stored-pressure water fire extinguisher clearly showing operable range.

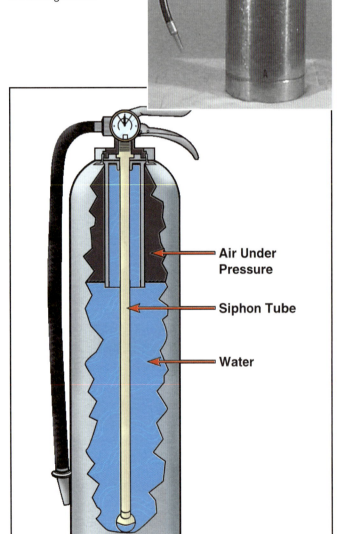

- Air Under Pressure
- Siphon Tube
- Water

Figure 5.3 Aqueous film forming foam (AFFF) fire extinguisher.

squeeze the shutoff handle and direct the water at the target.

Aqueous Film Forming Foam (AFFF) Extinguishers

Another type of extinguisher that is found on marine vessels is the aqueous film forming foam (AFFF) extinguisher (Figure 5.3). AFFF extinguishers are suitable for use on Class A and Class B fires. They are particularly useful in combating fires or suppressing vapors on small liquid fuel spills. The most common AFFF extinguishers are 2½ gallons (10 L) in size. Their stream reach is 20 to 25 feet (6 m to 7.6 m). The discharge time is about 50 seconds. These extinguishers need protection against freezing if they are exposed to temperatures less than 40°F (4°C). Add an antifreeze solution that is compatible with the AFFF concentrate to the AFFF/water.

AFFF extinguishers are either stored-pressure or cartridge-operated types. Stored-pressure AFFF extinguishers are very similar to the stored-pressure water extinguishers. The main difference is that in addition to water, the tank contains a proportionate amount of AFFF concentrate. The other primary difference is that an AFFF extinguisher has a special aspirating nozzle that provides better quality foam than a regular nozzle.

The water/AFFF solution is expelled by compressed air stored in the tank with the solution. When the shutoff device is opened, the solution expels through the hose and aspirating nozzle where air mixes with the solution and forms the finished foam. When applying the foam, prevent disturbing the foam layer by either letting the foam gently "rain" down on the fuel surface or deflecting it off an object. Avoid splashing liquid fuels. This special foam has the ability to make water float on fuels that are lighter than water. The vapor seal that is created extinguishes the flame and prevents reignition. The foam also has good wetting and penetrating properties on Class A fires.

The cartridge-operated AFFF extinguisher is equally as effective as the stored-pressure type but is slightly different. Only water is stored in the cylinder, but a cartridge located in the hose holds a solid form of AFFF. When activated, water flows and mixes with the solid AFFF to make the foam solution. Application of the water/AFFF solution is similar to that described for the stored-pressure AFFF extinguisher.

Halogenated Agent Extinguishers

Halon is a generic term for halogenated hydrocarbons: a chemical compound that contains carbon plus one or more elements from the halogen series (fluorine, chlorine, bromine, or iodine). An extinguisher using a halogenated agent is a very effective extinguisher because the vapor is nonconductive and inherently clean. In spite of their effectiveness in extinguishing fires in flammable and combustible liquids and electrical equipment, halogenated agents are very expensive and toxic under fire conditions. They have been found to significantly contribute to the depletion of the earth's ozone layer. Halogenated extinguishing agents are included in the *Montreal Protocol on Substances that Deplete the Ozone Layer,* which requires a complete phaseout of the production of halogens by 2000 (later changed to 1994). The United

States stopped producing halogens at the end of 1993, and research was begun on possible alternative extinguishing agents. Many users are replacing halon with more traditional agents that are environmentally sound and less expensive.

Nonetheless, halon is a very effective extinguishing agent, and these extinguishers are still found. The user often misunderstands how halogenated agents extinguish fires, but research indicates that they interrupt the chain reaction of the combustion process. The most common halon agents for fire extinguishing purposes are Halon 1211, Halon 1301, and Halon 1202.

Halon 1211 Extinguishers

Halon 1211 (bromoclorodifluoromethane) portable extinguishers are intended primarily for Class B and Class C fires (Figure 5.4). However, Halon 1211 extinguishers greater than 9 pounds (4 kg) in capacity also have a low Class A rating (See Fire Extinguisher Rating Systems section). Hand-carried extinguishers are found in sizes from 2½ to 20 pounds (1.1 kg to 9 kg). The hand-carried extinguishers have a stream reach of about 8 to 18 feet (2.4 m to 5.5 m) with a total discharge time of 8 to 18 seconds. Halon 1211 extinguishers do not require freeze protection.

Halon 1211 is stored in the extinguisher as a liquefied compressed gas. Nitrogen is added to give Halon 1211 added pressure when discharged. When the shutoff handle is squeezed, Halon 1211 is released in a clear liquid stream, giving it greater reach ability than a gaseous agent, but the stream may be affected by wind when operated outside.

Carry Halon 1211 extinguishers by the top handle. Make the initial application close to the fire, and direct the discharge at the base of

Figure 5.4 A Halon 1211 hand-carried fire extinguisher.

the flames. Continue application even after the flames are extinguished. Best results are obtained on flammable liquid fires if the discharge is directed to sweep the flames from the burning surface. Apply the agent first at the near edge of the fire and gradually progress forward while moving from side to side.

Halon 1301 Extinguishers

Halon 1301 (bromotrifluoromethane) is most commonly found in fixed fire protection systems (see Chapter 8, Fixed Fire Suppression Systems), but it is occasionally found in small portable fire extinguishers. These portable extinguishers are comparable to the Halon 1211 extinguishers with a few exceptions. They are limited to 2½ pounds (1.1 kg), and the stream is discharged in a nearly invisible gaseous form. Because a gas stream has less reach than a liquid stream, the effective reach of a Halon 1301 extinguisher is only about 4 to 6 feet (1.2 m to 1.8 m).

Halon 1202 Extinguishers

Halon 1202 (dibromodifluoromethane) is another halon product occasionally found in extinguishers, but it is not widely used. It is considered the most effective of the halons but is also the most toxic.

Halon Replacements

Although the production of halon fire fighting agents has been banned worldwide for several years, regulatory restrictions on the use and maintenance of present units do not exist. The International Maritime Organization's Subcommittee on Fire Protection has done considerable work in this area but has not yet developed international standards for halon replacement. The United States allows these units to remain in service and be maintained. The restrictive factor is cost. The cost of halons per pound (½ kg) is rising to the point that refill and replacement costs justify replacement of the entire system.

Halon replacement gases, technology, and equipment are being developed worldwide. In the United States, any replacement agent must be submitted to the United States Environmental Protection Agency (EPA) for evaluation through its Significant New Agent Program (SNAP). Water mist and steam extinguishment systems are currently under testing for vessel application by the United States Coast Guard (USCG) and United States Navy.

The EPA has approved many new agents, however, it appears that almost all these agents require twice the volume of halon. Some of these agents are halocarbons (perfluorobutane, trifluoromethane, and halotron for example), inert gases (argon and nitrogen for example), or various blends (inergen and argonite for example). New agents are continually being evaluated, and it is expected that the list of agents will continue to grow.

Carbon Dioxide Extinguishers

Carbon dioxide extinguishers are effective in extinguishing Class B and Class C fires. They are not effective on reactive metals because these substances decompose the CO_2. The primary extinguishing characteristic of CO_2 is smothering. Carbon dioxide displaces 20 to 30 percent of the oxygen in the air, making an environment unsuitable for combustion.

Hand-carried units are available in sizes from 2½ to 20 pounds (1.1 kg to 9 kg). Because their discharge is in the form of a gas, they have a limited reach of only 3 to 8 feet (1 m to 2.4 m). Total discharge time ranges from 8 to 30 seconds, depending on the size of the extinguisher (Figure 5.5). CO_2 extinguishers do not require freeze protection, but they should be stowed

Figure 5.5
A portable carbon dioxide fire extinguisher. *Courtesy of Captain Mark Turner, Syndicated Management Services Ltd.*

at temperatures below 130°F (54°C). At higher temperatures, the internal pressures can build to unsafe levels.

Carbon dioxide is stored as a liquefied compressed gas (850 psi at 70°F or 5 861 kPa {59 bar} at 21°C). The carbon dioxide is stored under its own pressure and is ready for release at any time. Little dry ice crystals or "snowflakes" usually accompany the gaseous discharge. These flakes sublime into a gaseous form shortly after discharge. Even though CO_2 is expelled at -110°F (-79°C), the cooling effect is minimal when compared to an equal amount of water because CO_2, a gas, does not have the heat-absorbing capabilities of water.

Carry carbon dioxide extinguishers by the top handle, and use in an upright position. Units 15 pounds (6.8 kg) or larger are grounded to the deck to avoid shock from static electricity. When the shutoff handle is squeezed, the carbon dioxide discharges in a gaseous form through a horn on the end of a hose or short metal fitting. Approach from the upwind side. Point the discharge horn at the base of the fire. Continue application even after the flames are extinguished to prevent a possible reflash of the fire. On flammable liquid fires, best results are obtained when the discharge from the extinguisher is employed to sweep the flames from the burning surface. Apply the discharge first at the near edge of the fire and gradually progress forward, moving the discharge cone very slowly from side to side.

As a CO_2 extinguisher nears depletion, an audible change in the sound of discharge is quite noticeable. Practice often with carbon dioxide extinguishers in order to become familiar with the operation and behavior of the extinguishing capabilities and the change in sound pitch.

CAUTION: Use caution in small, confined areas. Carbon dioxide displaces the oxygen in a space.

A "frost" residue may form on the extinguisher nozzle horn. Contact with the skin could cause frostbite.

Dry Chemical Extinguishers

Dry chemical extinguishers are among the most common portable fire extinguishers in use today. The terms dry chemical and dry powder are often interchanged. In North America, *dry chemical* agents are for use on Class A-B-C fires and/or B-C fires, whereas *dry powder* agents are for Class D fires only. The British refer to *dry chemical* agents as *powder*. Two basic types of dry chemical extinguishers are available: ordinary/regular and multipurpose. Ordinary/regular dry chemical extinguishers are rated for Class B and Class C fires. Some commonly used ordinary dry chemicals are sodium bicarbonate, potassium bicarbonate (Purple K®), potassium chloride (Super K®), and potassium bicarbonate/urea base (Monnex®). Multipurpose dry chemical extinguishers are rated for Class A, Class B, and Class C fires. Commonly used multipurpose dry chemicals are ammonium phosphate (commonly called monoammonium phosphate) and barium sulfate.

Dry chemical agents are corrosive, so other agents may be preferable in some cases. It may be better to use halon-type agents or carbon dioxide on electrical fires to prevent costly cleaning around open electrical contacts. Dry chemical will not extinguish fires in materials that produce their own oxygen.

Hand-carried dry chemical fire extinguishers are available in sizes from 2½ to 30 pounds (1.1 kg to 14 kg) in both the stored-pressure and cartridge-operated forms. The stream reach under normal conditions is 5 to 20 feet (1.5 m to 6 m); however, it is easily affected by wind. The total discharge time is 8 to 25 seconds. Dry chemicals require no freeze protection, but in the cartridge-operated extinguisher a dry nitrogen cartridge is used when the extinguisher is subject to freezing temperatures (Figure 5.6).

The stored-pressure type is similar in design to the pressurized-water extinguisher. The agent storage tank is maintained under a constant pressure. This pressure is commonly in the area of 200 psi (1 379 kPa) {14 bar}. Cartridge-operated extinguishers maintain a separate agent tank and pressure cylinder (Figure 5.7). The agent tank is not pressurized until a plunger is pushed to release the gas from the cartridge. Both types of extinguishers primarily use nitrogen as the pressurizing gas, although CO_2 is used in some models.

During manufacture, the chemicals are mixed with small amounts of additives that prevent them from caking, and this mixing allows the agent to discharge easily. Avoid mixing or contaminating ordinary/regular agents with multipurpose agents and vice versa. The dry chemicals presently used as agents are

Figure 5.6 Portable dry chemical fire extinguishers (cartridge-operated [left] and stored-pressure [right]).

Figure 5.7 Cutaway of a cartridge-operated dry chemical fire extinguisher.

nontoxic when swallowed but may cause breathing problems if inhaled. When outside, approach the fire from the upwind side to avoid inhaling the discharge cloud.

The methods of operating a stored-pressure or cartridge-operated extinguisher are slightly different. To operate the stored-pressure type, simply pull the pin and squeeze the handle to discharge the agent. To operate the cartridge type, first remove the hose from its storage position. In the stored position, the hose prevents the activation plunger from being accidentally pushed. Once the hose is removed, tilt the top of the extinguisher away from the body and other people, and depress the activation plunger.

Regardless of which type of extinguisher is used, operation expels a cloud of dry chemical. Control the discharge by using the shutoff valve. Best results are obtained by attacking the near edge of the fire and progressing forward while moving the nozzle in a rapid, side-to-side sweeping motion. With a multipurpose agent, apply the discharge to the burning surface to coat the hot surface even after the flames are extinguished to prevent possible reflash.

A dry chemical extinguisher is most effective when the stream is discharged at the proper distance from the base of the flames. Although fire fighting crew members have a tendency to approach a fire as close as possible, closer is not always better. Read the manufacturer's label for the recommended distance, and then apply the agent so that the stream is beginning to break into a "cloud" at the near edge of the flames. It takes practice to master this technique. Expel extinguishers that are scheduled for refilling on practice fires. Dry chemical extinguishers are refilled after any use, no matter how much agent is used. Incorporate onboard maintenance cycles for these extinguishers into training drills so that a portion of them is used for each drill. Use and refill dry chemical extinguishers at least once a year.

Dry Powder Extinguishing Agents

Normal extinguishing agents generally are not used on metal (Class D) fires. Specialized techniques and dry powder extinguishing agents have been developed to control and extinguish metal fires. A given dry powder agent does not, however, necessarily control or extinguish *all* metal fires. Some agents are valuable in suppressing fires in several metals; others are useful for combating only one type of metal fire. Some of the agents are applied by a hand shovel or scoop, others by portable fire extinguishers designed for use with dry powders.

Hand-carried dry powder extinguishers have a capacity of 30 pounds (14 kg). Stream reach is from 4 to 6 feet (1.2 m to 1.8 m). The discharge time ranges from 28 to 30 seconds. Most of these extinguishers use a sodium chloride agent. Flow enhancers and a thermoplastic material are added to the sodium chloride to enhance crusting after the material is discharged onto a fire.

Apply dry powder to a depth that adequately covers the fire area and provides a smothering blanket. Apply the agent gently on metal fires to avoid breaking any crust that forms over the burning metal. If the crust is broken, the fire may flare and expose more raw materials to combustion. Apply additional agent as necessary to cover any hot spots that develop. Leave the material undisturbed, and do not attempt disposal until the mass has cooled. Avoid scattering the burning metal. If the burning metal is on a combustible surface, first cover the fire with powder, then spread a 1- or 2-inch (25 mm to 51 mm) layer of powder nearby, and shovel the burning metal onto this layer (add more powder as needed). Refer to the manufacturer's recommendations for uses and special techniques for extinguishing fires in various combustible metals.

 ## Types of Semiportable Fire Extinguishers

Because of their size, semiportable fire extinguishers utilize various methods to achieve mobility. Some are wheeled or placed on carts. Others have hose reels that unwind to attack a fire. Most of these units use carbon dioxide, dry chemical, or foam agents. Some hose-reel systems also use halogenated agents. Semiportable fire extinguishers are used in areas of high hazard where portable fire extinguishers are insufficient and fixed extinguishing systems are not possible because of the lack of an enclosure (such as near purifiers or adjacent to boilers).

Wheeled Units

Semiportable units are mounted on wheels for relative ease of movement, although their locations on board a vessel rarely permit easy movement. They must be secured to prevent movement during passage. Generally the smaller semiportable extinguishers are wheeled because of the difficulty in moving a large wheeled object in rough seas.

The advantages of wheeled units over portable ones include their greater capacity of agent and available

reach because of the hoses and/or lance-type nozzles. Wheeled units usually contain foam, carbon dioxide, or dry chemical. Some dry powder fire extinguishers for Class D fires also come in wheeled models in 150- and 350-pound (68 kg and 159 kg) sizes. Large Halon 1211 extinguishers are found as wheeled units weighing up to 150 pounds (68 kg). Sometimes 10-gallon (38 L) water extinguishers are wheeled units.

Carbon Dioxide

Carbon dioxide wheeled units are similar to the hand-carried units except that they are considerably larger. These units are used only on Class B and Class C fires. Wheeled units range in size from 50 to 100 pounds (23 kg to 45 kg). Because of their size, they have a longer stream reach: 8 to 10 feet (2.4 m to 3 m) under normal conditions. The discharge time is usually 26 to 65 seconds.

Wheeled units are commonly found on board vessels and at shoreside facilities. The principle of operation for these larger units is the same as it is for the smaller CO_2 extinguishers. They are wheeled to a fire and operated according to the manufacturer's instructions. They have a short hose, typically less than 15 feet (4.6 m), which is deployed before use.

Dry Chemical

Dry chemical wheeled units are similar to the hand-carried units but on a larger scale (Figure 5.8). They

Figure 5.8 A wheeled dry chemical fire extinguisher. *Courtesy of Captain Mark Turner, Syndicated Management Services Ltd.*

are used for Class A, Class B, and Class C fires. Wheeled units range in size from 75 to 350 pounds (34 kg to 159 kg). They are capable of shooting a stream up to 45 feet (13.7 m). The total discharge time ranges from 20 seconds to 2 minutes. They are usually equipped with hoses that are 50 to 100 feet (15 m to 30.5 m) long.

Operating the wheeled dry chemical extinguisher is similar to operating the hand-carried, cartridge-type dry chemical extinguisher. The extinguishing agent is kept in one tank, and the pressurizing gas is stored in a separate cylinder. Turning a handwheel on top of the gas tank or activating a quick-pressurization device releases the gas into the agent tank. The agent tank is totally pressurized within seconds. Hold the nozzle securely, and be ready for a substantial nozzle reaction. Apply the agent in the same way as described for the hand-carried extinguishers.

On Class A fires, direct the discharge at the burning surfaces to cover them with chemical. Intermittently direct the chemical discharge on any glowing areas when the flames are extinguished. Maintain a careful watch for hot spots that may develop. Apply additional agent to those surfaces as necessary to adequately coat them with the extinguishing agent.

Aqueous Film Forming Foam

Aqueous film forming foam wheeled units are commonly found in machinery spaces (Figure 5.9). These units are large versions of the hand-carried AFFF extinguishers. Because they have a greater capacity, the discharge time and stream reach are greater. Read the manufacturer's instructions for specific discharge time, stream reach, and operating procedures of the AFFF wheeled units. Like the smaller units, they need protection against freezing if they are exposed to temperatures less than 40°F (4°C). Add an antifreeze solution that is compatible with the AFFF concentrate to the AFFF/water.

Hose-Reel Systems

Hose-reel systems are much the same as wheeled units, except for size. Hose-reel type systems usually have a much larger agent capacity and thus are too heavy to "wheel" to a fire. They consist of a length of hose that can be spooled from the reel or storage tank to reach a fire, while all other components remain fixed in place. Most hose-reel systems use carbon dioxide, dry chemical, or halogenated agents.

Carbon Dioxide Hose-Reel Systems

Carbon dioxide hose-reel systems are often employed in engine rooms and in machinery spaces containing electrical equipment. The most common system configuration consists of two CO_2 cylinders, a reel for the hose, a ½-inch (13 mm) diameter hose that is 50 to 75 feet (15 m to 23 m) long, and a discharge horn with a shutoff valve. Manually activate the system by operating the control lever on top of one of the cylinders. It is not necessary to open both cylinders because the pressure increase from opening one cylinder automatically causes the other to open. Once the system activates, unwind the necessary length of hose to reach the fire. Discharge the agent by opening the control valve on the horn. Apply the agent in the same manner as described for other CO_2 extinguishers.

Figure 5.9 A wheeled foam fire extinguisher. *Courtesy of R. Wright/Maryland Fire and Rescue Institute/United States Coast Guard.*

Dry Chemical Hose Systems

Dry chemical hose systems typically consist of a storage tank or vessel that contains the agent, high-pressure nitrogen cylinders to provide propellant, a length of rubber hose, and a nozzle with a shutoff valve. This type of system is usually installed in areas where Class B and Class C fires are likely. Activate the system by pulling the release mechanism on the head of the nitrogen cylinder. A remote-cable activating device is used in some systems where the cylinders are housed in a separate space. In some installations, multiple hose stations with corresponding activation points on the system are available. Once activated, the nitrogen gas agitates the agent and propels it through the hose. The entire length of the hose, usually looped on a rack, is deployed to assure a quality and constant flow. Apply the agent by opening the shutoff valve on the nozzle.

Halogenated Agent Hose-Reel Systems

Halogenated agents are also used in hose-reel systems. These systems are very similar to the carbon dioxide variety and are intended for combating Class B and Class C fires. As with the CO_2 system, the halogenated agent is stored under pressure in one or two cylinders that are equipped with release devices. Activation of either of the cylinders automatically opens the other. As soon as the system activates, spool the hose from the reel to the fire, and apply the gaseous agent as needed.

◆ Fire Extinguisher Rating Systems

In an effort to make this manual as international as possible, the extinguisher rating systems used in both North America and Europe are discussed. In general, fire extinguishers are classified according to how they are used on the classes of fire (see Chapter 1, Fire Science and Chemistry). It is important for crew members to learn about the system that is used for the extinguishers on board their vessel.

North American Extinguisher Ratings

In North America, portable fire extinguishers are classified according to their intended use on the four classes of fire (A, B, C, and D) (Figure 5.10). In addition

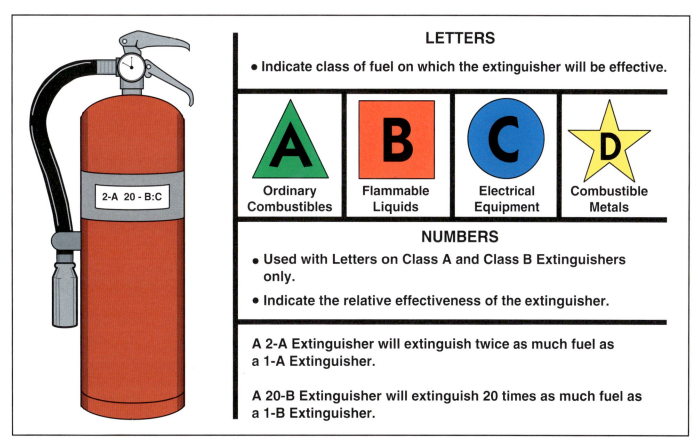

LETTERS

- Indicate class of fuel on which the extinguisher will be effective.

A Ordinary Combustibles	B Flammable Liquids	C Electrical Equipment	D Combustible Metals

NUMBERS

- Used with Letters on Class A and Class B Extinguishers only.
- Indicate the relative effectiveness of the extinguisher.

A 2-A Extinguisher will extinguish twice as much fuel as a 1-A Extinguisher.

A 20-B Extinguisher will extinguish 20 times as much fuel as a 1-B Extinguisher.

(extinguisher label: 2-A 20 - B:C)

Figure 5.10 North American fire extinguisher rating symbols and pictographs. Fire extinguishers are classified according to their intended use.

to the letter classification, Class A and Class B extinguishers also receive a numerical rating. Multiple letters or numeral-letter ratings are used on extinguishers that are effective on more than one class of fire. Extinguishers are not classified according to their effectiveness, but by size only. The numerical rating system is based on tests conducted by Underwriters Laboratories Inc. (UL) and Underwriters Laboratories of Canada (ULC). These tests are designed to determine the extinguishing potential for each size and type of extinguisher.

Extinguishers for use on Class C fires receive only the letter rating because Class C fires are essentially Class A or Class B fires involving energized electrical equipment. The Class C designation just confirms that the extinguishing agent is nonconductive. Class D extinguishers, likewise, do not contain a numerical rating. The effectiveness of the extinguisher on Class D metals is detailed on the faceplate.

Class A Ratings

Class A portable fire extinguishers are rated from 1-A through 40-A, depending on their size. A 1-A rating requires 1¼ gallons (5 L) of water. A 2-A rating requires 2½ gallons (10 L) or twice the 1-A capacity. Therefore, a dry chemical extinguisher rated 10-A is equivalent to five 2½-gallon (10 L) water extinguishers.

To receive a 1-A through 6-A rating, an extinguisher must be capable of extinguishing the three following types of Class A fires: wood crib, wood panel, and excelsior. Although all three are Class A combustibles, each presents a substantially different type of burning. Portable fire extinguishers that are tested for larger than a 6-A rating are only subjected to the wood crib test.

When tests are conducted on Class A fires, there is little difference in the size of fire extinguished by expert operators and nonexpert operators. According to the rating board, application techniques are not as important on a Class A fire as they are on a Class B fire.

Class B Ratings

Extinguishers suitable for use on Class B fires are classified with numerical ratings ranging from 1-B through 640-B. The test used by UL to determine the rating of Class B extinguishers consists of burning a flammable liquid in square steel pans. The rating is based on the approximate square-foot (square meter) area of a flammable liquid fire that a nonexpert operator can extinguish. An expert operator conducts the tests; however, the numerical rating of the extinguisher is applied under the assumption that a nonexpert or untrained operator will use the extinguisher. To determine the amount of fire a nonexpert operator can extinguish, a working rating was calculated to be 40 percent of the fire area an expert operator could consistently extinguish in the tests.

Review a common-sized extinguisher such as the multipurpose extinguisher rated 4-A 20-B:C to better understand the rating system. This extinguisher will extinguish a Class A fire that is four times larger than a 1-A fire, approximately 20 times as much Class B fire as a 1-B extinguisher, or a deep-layer flammable liquid fire of a 20-square-foot (2 m²) area. It is also safe to use on fires involving energized electrical equipment.

Class C Ratings

No fire-extinguishing capability tests are specifically conducted for Class C ratings. In assigning a Class C designation, the extinguishing agent is only tested for electrical nonconductivity. If the agent meets the test requirements, the Class C rating is given in conjunction with a rating previously established for Class A and/or Class B fires.

Class D Ratings

Test fires for establishing Class D ratings vary with the type of combustible metal being tested. The following factors are considered during each test:

- Reactions between the metal and the agent

- Toxicity of the agent

- Toxicity of the fumes (products of combustion) produced

- Time of burnout of the metal versus time of extinguishment

When an extinguishing agent is determined safe and effective for use on a combustible metal, operation instructions are included on the faceplate of the extinguisher, although no numerical rating is given. Class D agents cannot be given a multipurpose rating for use on other classes of fire.

Multiple Markings

Combinations of the letters A, B, and/or C or the symbols for each class identify extinguishers suitable for more than one class of fire. The three most common combinations are Class A-B-C, Class A-B, and Class B-C. There is no extinguisher with a Class A-C rating. A new fire extinguisher is labeled with its appropriate marking.

Two systems of labeling portable fire extinguishers are available. The first system uses specifically colored geometric shapes with the class letter shown within the shape: green triangle (Class A fires), red square (Class B fires), blue circle (Class C fires), and yellow star (Class D fires) (Figure 5.11). The second system is a pictograph labeling system that is designed to make the selection of the most appropriate fire extinguishers more effective and easier (Figure 5.12). The pictograph system also emphasizes when not to use an extinguisher on certain types of fires (Figures 5.13). Regardless of which system is used, it is important that the markings are clearly visible.

European Extinguisher Ratings

The Europeans classify and rate fire extinguishers in much the same way as North Americans with some exceptions. The primary difference is that types of fires are classified slightly different in Europe. No great difference exists between what North Americans refer to as Class A, Class B, and Class D fires and the European classifications, but other classes are different (Figure 5.14). Some European countries have their own differences. Because of the desire to associate all of Europe with one set of fire codes, these classifications will likely change in the future. What crew members must understand is the types of fires and what extinguishers they have available to combat the different

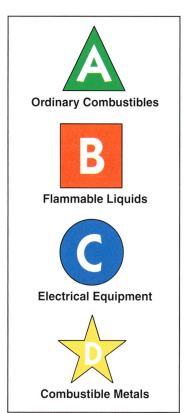

Figure 5.11 Alphabetical/geometric symbols for the North American classes of fire.

Figure 5.12 Pictographs for the North American classes of fire.

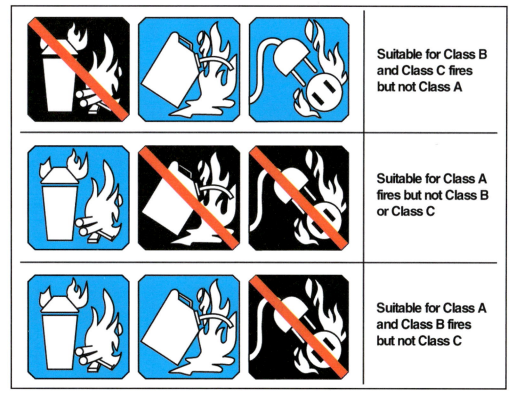

Figure 5.13 Pictographs showing the North American classes of fire for which an extinguisher is not suitable.

Fire Class	Extinguishing Principles
CLASS **A** **Fires involving solid materials usually of organic nature in which combustion normally takes place with the formation of glowing embers. Wood, paper, textiles, etc.**	Water cooling or combustion inhibition. ***Extinguisher:*** water and all purpose powder (excellent); foam.
CLASS **B** **Fires involving liquids or liquefiable solids. Burning liquids, oil, fat, paint, etc.**	Flame inhibiting or surface blanketing and cooling. ***Extinguisher:*** all purpose powder, dry powder, and foam (excellent); carbon dioxide.
CLASS **C** **Fires involving gases.**	Flame inhibiting. ***Extinguisher:*** all purpose powder, dry powder, and carbon dioxide.
CLASS **D** **Fires involving metals, magnesium, sodium, titanium, zirconium, etc.**	Exclusion of oxygen and cooling. ***Extinguisher:*** metal powder (excellent).
CLASS **E** **Fires involving electrical hazards.**	Flame inhibiting. ***Extinguisher:*** carbon dioxide (excellent); all purpose powder and dry powder.

Figure 5.14 European classes of fire and extinguishing principles.

types. The sections that follow give a general summary of the European method of classifying extinguishers.

The new European standard specifies that all extinguishers be colored red with a panel of up to 10 percent of the body area in another color denoting the contents: water (red), foam (cream), dry powder (blue), carbon dioxide (black), and halon type (green) (Figure 5.15). The traditional British method colored the entire extinguisher according to its contents. Mariners will likely find many extinguishers of the traditional coloring method in use for some time.

Class A Ratings

The European Class A rating is for fires involving solid materials, normally of an organic nature. Extinguishers certified for these types of fires are evaluated with a series of wood crib test fires. The cribs are of standard height and width but have varying lengths. The numerical value is based upon the largest standard wooden crib test fire that it can extinguish. A rating of 54A means that it is capable of extinguishing a standard Class A test fire 5.4 meters (17.7 ft) in length.

Figure 5.15 Fire extinguishers showing new European standard: all red with bands of agent colors.

Class B Ratings

The European Class B rating is for fires involving flammable and combustible liquids, but not flammable gases. The numerical value indicates the largest flammable liquid tray fire that can be extinguished by the tested size of extinguisher, using specified procedures. The flammable liquid used for the test fires is one of a range of aliphatic hydrocarbons, floating on a volume of water. The rating reflects the total volume of liquid in the tray. One example is an extinguisher rated as 34B. This extinguisher is certified as being capable of extinguishing a fire in a tray containing 34 liters (9 gal) of liquid.

Classes C, D, and E Ratings

The European Class C rating indicates the extinguisher's effectiveness for fires involving flammable gases. Presently no performance rating for Class C fires exists. Dry chemical, halon-type, and carbon dioxide extinguishers are typically suitable for gas fires, but great caution is necessary. Always turn off the gas supply before attempting extinguishment. Otherwise, the risk of reignition in an explosive manner is a likely threat.

Currently no performance standard for Class D extinguishers for fires involving metals exists. Some countries use a Class E category for energized electrical equipment or wiring fuels.

British Markings

The British standard of extinguisher markings is based on the extinguishing agent rather than the fire classification. Portable extinguishers in Britain have traditionally been colored according to their contents as shown in the following chart:

Extinguishing Medium	Extinguisher Color
Water	Signal Red
Foam	Pale Cream
Dry Powder	French Blue
Carbon Dioxide	Black
Halon Type	Emerald Green

Many of these extinguishers are still found, but new extinguishers are colored red with a panel of the body area in a color denoting the contents according to the

new European standard. Figure 5.16 shows which extinguisher to use for each class of fire.

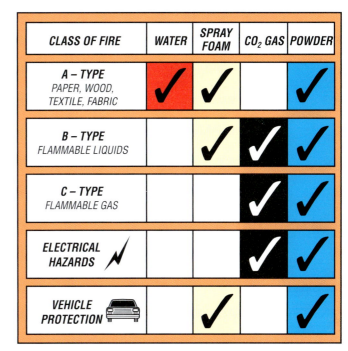

Figure 5.16 Chart showing which extinguisher to use in Britain for the classes of fire.

German Markings

The German standard provides for the use of pictographs to denote the type of fire for which a given extinguisher is approved, which is similar to the current method of marking extinguishers in North America. The pictographs are described in the following chart:

Fire Class	Symbol Description
CLASS A Ordinary Combustibles	Wood logs with a flame and ash
CLASS B Flammable Liquids	Square gas can
CLASS C Flammable Gases	Gas stove burner with flame
CLASS D Metals	A grinding wheel with shavings dropping off
CLASS E Electrical	Electrical wall outlet

German Fire Extinguisher Symbols

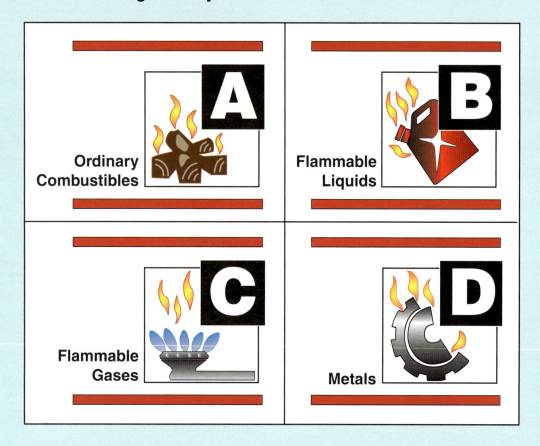

Ordinary Combustibles — **A**

Flammable Liquids — **B**

Flammable Gases — **C**

Metals — **D**

 ## Selecting and Using Portable Fire Extinguishers

Selection of the proper portable fire extinguisher depends on the following factors:

- Classification of the burning fuel
- Ratings of the extinguisher
- Hazards to be protected
- Severity of the fire
- Atmospheric conditions
- Number of crew members available
- Ease of handling the extinguisher
- Any life hazard or operational concerns

Choose extinguishers that minimize the risk to life and property, yet are effective in extinguishing a fire. Address these considerations when choosing a fire extinguisher to mount in a particular area. For example, it is unwise to place dry chemical extinguishers with a corrosive agent in areas where highly sensitive computer equipment is located. In these particular areas, halon-type or carbon dioxide extinguishers are better choices.

Portable extinguishers come in many shapes, sizes, and types. Become familiar with the detailed instructions found on the labels of extinguishers even though the operating procedures of each type are similar. In an emergency, every second is important; therefore, know the general instructions applicable to most portable fire extinguishers. General operating procedures for virtually all portable extinguishers, except the cartridge-operated dry chemical or dry powder extinguishers, are easily remembered by using the acronym *P-A-S-S*. See Skill Sheet 5-1 at the end of the chapter for steps. Skill Sheet 5-2 shows how to operate cartridge-operated dry chemical or dry powder extinguishers.

Australian Portable Fire Extinguisher Guide

Fire Protection Association Australia		Portable Fire Extinguisher Guide					Fire Protection Association Australia Website www.fpaa.com.au
		CLASS A	CLASS B	CLASS C	CLASS E	CLASS F	CLASS D For fire involving combustible metals use special purpose extinguisher
Two colour schemes for fire extinguishers exist	EXTINGUISHANT	Wood Paper Plastics	Flammable & Combustible Liquids	Flammable Gases	Electrically Energised Equipment	Cooking Oils and Fats	
PRE 1999 / FROM 1999							
	WATER	YES	NO	NO	NO	NO	Dangerous if used on flammable liquid, energised electrical equipment and cooking oils/fat fires
	WET CHEMICAL	YES	NO	NO	NO	YES	Dangerous if used on energised electrical equipment
	FOAM	YES	YES	NO	NO	LIMITED	Dangerous if used on energised electrical equipment
	POWDER	YES (ABE) / NO (BE)	YES (ABE) / YES (BE)	YES (ABE) / YES (BE)	YES (ABE) / YES (BE)	NO (ABE) / LIMITED (BE)	Look carefully at the extinguisher to determine if it is an BE or ABE unit as the capability is different
	CARBON DIOXIDE	LIMITED	LIMITED	LIMITED	YES	LIMITED	Not suitable for outdoor use
	VAPORISING LIQUID	YES	LIMITED	LIMITED	YES	NO	Check the characteristics of the specific extinguisher agent

LIMITED indicates that the extinguisher is not the agent of choice for the class of fire, but that it may have a limited extinguishing capability.
Solvents such as alcohol or acetone mix with water and therfore require special foam
Green text indicates the class or classes in which agent is most effective

©FPA Australia ACN 005 366 576

Courtesy of Fire Protection Association Australia.

Quickly check an extinguisher before using it in an emergency to ensure that it is operable. Inspect the following components:

- **External condition** — No apparent damage
- **Hose/nozzle** — In place
- **Weight** — Feels as though it contains agent
- **Pressure gauge (if available)** — In operable range

Carry modern extinguishers to a fire in an upright position. Make sure that the fire is within reach before discharging the extinguishing agent. Otherwise, the agent is wasted. Small extinguishers require closer approach to a fire because they have less stream reach than larger extinguishers.

 ## Inspecting, Maintaining, and Refilling Portable Fire Extinguishers

Servicing of portable fire extinguishers is the responsibility of the vessel owner or operator, but some responsibilities (simple inspection and refilling) may be delegated to crew members. If proper procedures are followed and the appropriate agents are used, three types of extinguishers are easily refilled by crew members (stored-pressure water, stored-pressure AFFF, and cartridge-operated dry chemical). Maintenance is usually performed by trained service technicians employed by extinguisher distribution and service companies. Maintenance provides maximum assurance that a portable extinguisher will operate effectively and safely. Maintenance procedures include a thorough examination of the three basic components of a portable extinguisher: mechanical parts, extinguishing agents, and expelling means. Extinguisher maintenance is performed either whenever an inspection uncovers a need for maintenance/repairs/replacement or at specified intervals. The National Fire Protection Association (NFPA) requires a thorough inspection of portable fire extinguishers by qualified personnel at least once a year.

It is best to inspect fire extinguishers monthly to ensure that they are accessible and operable. Verify that extinguishers are in their designated locations,

Fire Extinguisher Inspection Checklist

	Yes	No
◆ Extinguisher is in the proper location and accessible.	☐	☐
◆ Extinguisher is properly mounted.	☐	☐
◆ Water and AFFF extinguishers have proper freeze protection.	☐	☐
◆ Discharge nozzle or horn is free of obstructions, cracks, or dirt or grease accumulations.	☐	☐
◆ Operating instructions on the extinguisher nameplate are legible.	☐	☐
◆ Lock pins and tamper seals are intact.	☐	☐
◆ Extinguisher is full of agent and/or fully pressurized (check pressure gauge, weight the extinguisher and cartridge [cartridge-operated extinguishers], or inspect the agent level).	☐	☐
◆ Hose and associated fittings are in good condition.	☐	☐
◆ Extinguisher is free of corrosion (check bottom of extinguishers that are exposed to outside air).	☐	☐
◆ Dry chemical agents are loose and soft (turn extinguisher upside down once a month and tap with rubber mallet to loosen agent).	☐	☐
◆ Inspection tag contains date of previous inspection, maintenance, or refilling.	☐	☐

NOTES:

    Items Not Passing Inspection:

    Action Required:

    Other Comments:

Submitted by:

Signature: **Date:**

Figure 5.17 Fire extinguisher inspection checklist.

have not activated, and show no obvious physical damage or condition that prevents their operation. When inspecting fire extinguishers, consider three important factors that determine their value: (1) serviceability, (2) accessibility, and (3) user's ability to operate them. The procedures on the checklist given in Figure 5.17 are part of every fire extinguisher inspection. If any of the items listed are deficient, remove the extinguisher from service and repair as required. Replace the extinguisher with one that has an equal or greater rating. Remove a CO_2 extinguisher that is found deficient in agent weight by 10 percent or more, and replace with a full unit of equal size. The full and empty weights are stamped on the heads of carbon dioxide extinguishers for this purpose. Be sure to consider a loss of 10 percent of the weight of the agent and *not* 10 percent of the gross weight of the unit, which is substantially greater. Ensure that dry chemical extinguishers are not subjected to temperatures above 140°F (60°C) because additives may melt at higher temperatures.

Vessel owners/operators have a responsibility to keep accurate and complete records of all maintenance/repairs and inspections, including the month, year, type of maintenance/repair, and date of the last recharge/refill. In addition to these central records, tags are affixed to each portable extinguisher indicating the date of its last inspection. Identify types and locations of extinguishers by numbering or some other means.

Some testing procedures are also necessary. Portable extinguishers are pressure vessels and require hydrostatic testing to determine the strength of the shell at specified intervals. The period between hydrostatic tests is determined by the authority having jurisdiction and depends upon the type of extinguisher. Carbon dioxide hose assemblies require a conductivity test. Replace hoses that are nonconductive. Hoses must be conductive to prevent the generation of static electricity.

When refilling an extinguisher, refer to the manufacturer's specifications for detailed instructions. Determine that no physical damage exists by carefully examining the extinguisher. When refilling multiple extinguishers, keep the parts for each extinguisher together, separate from the others. Do not use parts from one extinguisher on another. If a part is needed, use an exact manufacturer's replacement. The following steps for refilling the stored-pressure water, stored-pressure AFFF, and cartridge-operated dry chemical extinguishers are general procedures only.

Refilling a Stored-Pressure Water Extinguisher

Step 1: Squeeze the discharge handle until all water or air expels from the hose to exhaust the extinguisher's pressure. *CAUTION: Relieve all pressure from the extinguisher before proceeding.*

Step 2: Loosen the retaining nut that holds the valve and hose assembly in place on the extinguisher cylinder.

Step 3: Remove the valve assembly and pickup tube from the cylinder.

Step 4: Pour any remaining water from the cylinder, and ensure no foreign objects are inside.

Step 5: Refill the cylinder to the designated fill line with fresh, clean water. Add antifreeze solution if needed.

Step 6: Reinstall/replace any gaskets or other parts that require replacement. Lubricate the gasket with petroleum jelly or a similar lubricant, and reinstall the valve and pickup tube assembly.

Step 7: Tighten the valve assembly nut securely.

Step 8: Recharge the unit with clean pressurized air introduced through the air chuck assembly. Point the top of the unit away from people. Stop introducing air when the pressure gauge indicates fully charged or is in the green zone (approximately 100 psi [689 kPa] {7 bar}).

Step 9: Check for leaks around the valve assembly by using soapy water. If a leak is discovered, do not attempt to eliminate it under pressure. Identify the cause of the leak, discharge the unit as in Step 1, and repeat the refilling procedure.

Step 10: Replace the retaining pin in the valve handle to prevent accidental discharge, and secure with an appropriate seal device.

Step 11: Record the date of refilling on the tag or label; record all details in the central records.

Refilling a Stored-Pressure AFFF Extinguisher

Step 1: Squeeze the discharge handle until all water or air expels from the hose to exhaust the extinguisher's pressure. CAUTION: Relieve all pressure from the extinguisher before proceeding.

Step 2: Loosen the retaining nut that holds the valve and hose assembly in place on the extinguisher cylinder.

Step 3: Remove the valve assembly and pickup tube from the cylinder.

Step 4: Pour any remaining AFFF/water from the cylinder, and ensure no foreign objects are inside.

Step 5: Refill the cylinder to the designated fill line with fresh, clean water. Add antifreeze solution if needed. Ensure the antifreeze solution is compatible with the foam agent.

Step 6: Add the correct amount of foam concentrate.

Step 7: Reinstall/replace any gaskets or other parts that require replacement. Lubricate the gasket with petroleum jelly or a similar lubricant, and reinstall the valve and pickup tube assembly.

Step 8: Tighten the valve assembly nut securely.

Step 9: Recharge the unit with clean pressurized air introduced through the air chuck assembly. Point the top of the unit away from people. Stop introducing air when the pressure gauge indicates fully charged or is in the green zone (approximately 100 psi [689 kPa] {7 bar}). CAUTION: Use extreme care because of high pressures.

Step 10: Check for leaks around the valve assembly by using soapy water. If a leak is discovered, do not attempt to eliminate it under pressure. Identify the cause of the leak, discharge the unit as in Step 1, and repeat the refilling procedure.

Step 11: Replace the retaining pin in the valve handle to prevent accidental discharge, and secure with an appropriate seal device.

Step 12: Record the date of refilling on the tag or label; record all details in the central records.

Refilling a Cartridge-Operated Dry Chemical Extinguisher

Step 1: Invert the unit, direct the nozzle downwind and away from people, and squeeze the valve handle on the nozzle to release all remaining pressure from the unit.

Step 2: Upright the unit and squeeze the valve handle on the nozzle again to see if all pressure is relieved. CAUTION: Relieve all pressure from the extinguisher before proceeding.

Step 3: Remove the protective cover over the pressure cartridge.

Step 4: Unscrew the pressure cartridge from the unit. The cartridge threads can be either left-handed or right-handed, depending on the manufacturer.

Step 5: Remove the cap on top of the extinguisher cylinder with the hands or specially designed tools. These caps are installed only hand-tight and do not require excessive force to remove.

Step 6: Blow residual agent from the hose.

Step 7: Add the correct dry chemical agent for the extinguisher rating to the fill mark (or determine proper weight). *CAUTION: Do not use either a different agent or a similar agent from a different manufacturer. Some dry chemicals are slightly acidic and some are slightly alkaline. Intermixing of agents or residue can cause chemical reaction and pressure buildup inside the extinguisher.*

Step 8: Remove any agent in the threads of both the cylinder and cap with a small, clean, soft-bristle toothbrush. Ensure that the safety vent hole is not obstructed.

Step 9: Reset the indicator pin if so equipped.

Step 10: Replace the gasket after applying a light coat of petroleum jelly or a similar lubricant. Replace the cap on the top of the cylinder, and tighten hand-tight or to manufacturer's specifications.

Step 11: Secure the actuator or safety before installing cartridges.

Step 12: Replace the spent cartridge with a new one, tightening by hand only. *CAUTION: Use only the appropriately sized cartridge for the unit.*

Step 13: Replace the protective cover and return the hose to its stored position to prevent the plunger from accidentally being depressed. *NOTE: Some models may also use a pin to prevent the plunger from being depressed. In such cases, replace the pin and secure it with an appropriate seal device.*

Step 14: Record the date of refilling on the tag or label; record all details in the central records.

Damaged/Obsolete Portable Extinguishers

Knowing that a safe and reliable portable extinguisher is available gives crew members a first line of defense against beginning fires and a complement to other fire suppression systems. Information about damaged and obsolete portable extinguishers is important to crew members because using either one is dangerous and could cause serious injury to the operator and others. Portable fire extinguishers can become damaged by fire, corrosion, or other abuse. Out-of-date extinguishers that are no longer suitable for use are still found on some vessels.

Damaged Extinguishers

Remove any leaking, corroded, or otherwise damaged extinguisher shells or cylinders from service. Discard or return them to the manufacturer for repair (Figure 5.18). Damaged portable extinguishers can fail at any time. If an extinguisher shows only slight damage or corrosion and it is uncertain whether it is safe to use,

send it to the manufacturer or a qualified testing agency for hydrostatic testing. Crew members who are trained to perform extinguisher maintenance can replace leaking hoses, gaskets, nozzles, and inner chambers.

CAUTION: Never attempt to repair the shell or cylinder of a portable fire extinguisher. Contact the manufacturer for instructions.

Obsolete Extinguishers

Occasionally, crew members may find old portable fire extinguishers that are not familiar. In most cases, these extinguishers are no longer recommended for use; therefore, it is important to recognize obsolete extinguishers. *Remove all obsolete extinguishers from service, and replace with new ones.*

The most common obsolete extinguishers found are the types designed to operate after being inverted. U.S. manufacturers stopped producing inverting-type extinguishers in 1969. Extinguishers made of copper or brass with soft soldered or riveted cylinders were discontinued at the same time (Figures 5.19 a–c). The major dangers and disadvantages of these extinguishers include the following:

- Once activated, the extinguisher cannot be turned off.

- The extinguishing agent is corrosive.

- If the discharge hose is blocked, the extinguisher can build to pressures in excess of 300 psi (2 068 kPa) {21 bar} and explode.

- The extinguisher can explode if the cylinder is damaged or corroded.

Some obsolete extinguishers that crew members may find include the following:

- *Soda-acid* — When this extinguisher is inverted, acid from a bottle within the cylinder mixes with a soda-and-water solution and produces a gas that expels the liquid. The pressure on an acid-corroded shell has exploded many soda-acid extinguishers.

- *Inverting foam* — This extinguisher looks similar to the soda-acid type. It contains two solutions that produce foam as well as an expellent gas when mixed.

- *Cartridge-operated water* — Besides inverting, this extinguisher requires bumping on the deck to puncture a CO_2 cylinder. The pressure of the gas released from the cartridge expels the water. This extinguisher has a tendency to explode when pressurized.

- *Liquid carbon tetrachloride* — Some of these extinguishers resemble pump-type insecticide sprayers. When carbon tetrachloride comes in contact with heat, it releases a highly toxic phosgene gas.

Figure 5.18 A damaged extinguisher — discard or return to the manufacturer for repair.

Figures 5.19 a–c Obsolete extinguishers — remove from service and replace with new types. (a) *(above left)* Soda acid agent. (b) *(above middle)* Riveted cylinder. (c) *(right)* Carbon tetrachloride agent.

Operating a Portable Fire Extinguisher

P-A-S-S Procedure

Step 1: *P* **Pull** the pin that keeps the handle from being pressed at the top of the extinguisher. The plastic or thin wire inspection band that retains the pin in position breaks easily as the pin is pulled.

Step 2: *A* **Aim** the nozzle or outlet toward the base of the fire. Some hose assemblies are clipped to the extinguisher body. Release the hose, and point.

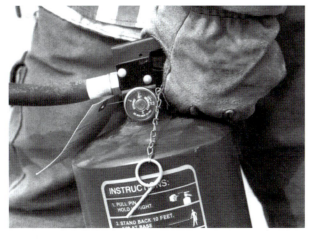

Step 3: *S* **Squeeze** the handle above the carrying handle to discharge the agent.

— Try a very short test burst to ensure proper operation before approaching the fire.

— Release the handle to stop the discharge at any time.

Step 4: *S* **Sweep** the nozzle back and forth at the base of the flames to distribute the extinguishing agent.

— Watch for smoldering hot spots or possible reflash of flammable liquids.

— Make sure that the fire is completely extinguished.

— Back away from the fire area.

Operating a Portable Fire Extinguisher

Cartridge-Operated Extinguishers Dry Chemical/Dry Powder

Step 1: Select the appropriate extinguisher based on the size and type of fire.

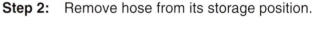

Step 2: Remove hose from its storage position.

Step 3: Position to one side of the extinguisher and depress the activation plunger.

Step 4: Point the nozzle or horn in a safe direction, and discharge a very short test burst to ensure proper operation.

Step 5: Carry the extinguisher to within stream reach of the fire.

Step 6: Aim the nozzle or horn toward the material that is burning.

Step 7: Squeeze the carrying handle and the discharge handle together to start the flow of agent. Release the handle to stop the flow.

Step 8: Sweep the nozzle back and forth. Start at the near edge of the fire and move forward while sweeping the nozzle from side to side until the fire is extinguished.

NOTE: Do not plunge **dry chemical** into flammable liquid fires.

Step 9: Watch for smoldering hot spots or possible reignition of flammable liquids. Make sure that the fire is completely extinguished. Be prepared to reapply agent if reignition occurs.

Step 10: Back away from the fire area.

Chapter 6

Water and Foam Fire Fighting Equipment

In the previous chapter we discussed portable and semiportable fire extinguishers. Familiarity with a great deal of other manual fire fighting equipment is needed in order to ensure successful performance in responding to fire emergencies. This chapter covers the basic information on other types of equipment beginning with fire hose and couplings, fire hose appliances and tools, and various nozzles. Different nozzles produce different types of fire streams. A *fire stream* is a stream of water or other extinguishing agent after it leaves a nozzle until it reaches the desired target. Fire streams can reduce high temperatures and provide protection to crew members and exposures. Choosing the proper fire stream for the particular emergency fire situation is an important factor in effectively extinguishing fires. Along with information about water fire fighting equipment and fire streams, a discussion of the properties and uses of fire fighting foams, foam equipment, foam generation, and foam application techniques is given. Much of the fire fighting equipment (hose, nozzles, etc.) is available at fire stations located throughout a vessel. Safety equipment such as protective clothing and self-contained breathing apparatus (SCBA) are kept elsewhere, sometimes in a easily accessible room or emergency equipment locker.

◆ Fire Stations

A *fire station* is a place on board a vessel where fire hoses and other fire fighting equipment are stored ready for immediate use (Figures 6.1 a–c). It is important that all required fire fighting equipment is kept in designated, accessible places. The number of fire stations on board a vessel depends on its size and type. Small vessels may have two or three, while large vessels may have two on each interior deck and six on the main deck. Fire stations are numbered sequentially in a uniform pattern so that crew members can recall the location of a given station by its number. Fire stations generally contain the following equipment:

- *Fire hose* — Carries extinguishing agent from water source; stowed on a rack or hose reel in the open where it is readily visible

- *Wrench(es), hose tools, and/or appliances (adapters)* — Complete fire hose connections and layouts

- *Nozzles* — Discharge extinguishing agent

- *Hydrant valve/outlet and cap* — Accesses fire main water supply

- *Marine strainer (self-cleaning)* — Filters debris from water that might clog nozzles

Some optional equipment that may be found at a fire station include portable fire extinguishers, axes, and damper keys (to open or close ventilation dampers in an emergency). Fire stations may be equipped with gated wyes so that several fire hoses can deploy from one location.

Fire stations are painted red for high visibility. Fire hose are stored so that they can be put into service easily and quickly. However, this visibility makes them vulnerable to misuse and damage. Misuse is any use other than the intended purpose. The hydrant valve or piping can be damaged if it is used as a cleat for tying a line. Hydrant valve stems can also be damaged during the handling of cargo or the moving of heavy

Figures 6.1 a–c Several examples of fire stations on vessels: (a) Ferry engine room. *Courtesy of Captain John F. Lewis.* (b) Galley. (c) Outside on deck. *Photographs b and c courtesy of R. Wright/Maryland Fire and Rescue Institute/United States Coast Guard.*

materials through passageways. If cargo is carried on deck, stow it so that it does not block access to the fire station hydrant outlet.

Fire stations are located so that the water streams from at least two separate hydrant outlets overlap. Fire hoses access the fire main water supply by connecting to hydrant outlets. See Chapter 8, Fixed Fire Suppression Systems, for information on fire main systems. Most regulations require that hydrants supply water to every part of the vessel normally accessible to persons aboard while the vessel is underway and to any cargo space when the vessel is empty. At least two streams of water from separate outlets must reach all spaces. At least one of these streams must be from a single length of hose. Hydrants are positioned near the entrances to protected spaces. See *International Convention for the Safety of Life at Sea (SOLAS)*, Chapter II-2, Regulation 4, for details.

A fire station hydrant outlet (Figure 6.2) has two major components: a control valve and the hose connection (either 1½ or 2½ inches [38 mm or 64 mm]) with appropriate threads or other connecting features. The connection may be equipped with a marine

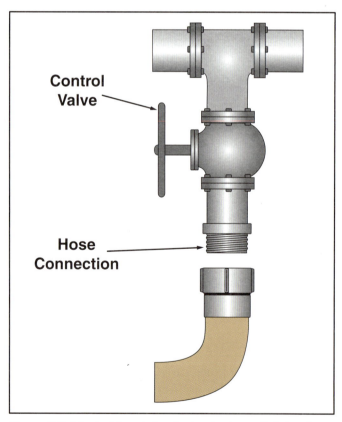

Figure 6.2 A typical fire station hydrant valve outlet showing the control valve and hose connection.

strainer. According to SOLAS regulations, fire hydrant outlets must have the following features:

- Each fire hydrant outlet has a valve that allows the removal of fire hose while pressure is in the fire main system.

- The fire hydrant outlet may be in any position, from horizontal to pointing vertically downward, to minimize twisting or kinking of the fire hose.

- The threads or other connecting feature on the fire hydrant outlet conform to the requirements of the regulating authority. Use a standardized connecting system to allow all approved fire hose to attach to the hydrant outlet.

- On interior hydrant outlets in certain passenger vessels, a gated wye may connect two 1½-inch (38 mm) fire hoses to a 2½-inch (64 mm) outlet.

- Fire hoses are connected to hydrant outlets at all times in interior locations on passenger vessels carrying more than 36 passengers.

◆ Fire Hose

Fire hose is a flexible tube that carries water or other extinguishing agent under pressure from a source of supply to a point of discharge. Fire hose must be flexible and watertight and have a smooth lining and durable covering. *Fire hose couplings* connect the various lengths of fire hose. Fire hose is the most used item in fire fighting activities. Do not use it for other purposes. The following sections discuss fire hose construction materials, sizes, uses, care and maintenance, testing, and safety. In addition, the proper care of fire hose couplings is important to efficient fire fighting operations. The various hose appliances and tools discussed allow flexibility in hose layout arrangements.

Fire Hose Materials and Sizes

Fire hose is classified by the material from which it is constructed and its size. The measurement of the internal diameter of the hose (for example, 1½ inches [38 mm]) is the size of the hose. Construction materials come in several grades and degrees of quality and may be susceptible to deterioration and wear. Fire hose is manufactured in different configurations such as single-jacket, double-jacket, woven-jacket, and rubber-covered. Some of the major fibers or materials used in the construction of the outer jacket of fire hose are cotton, nylon, rayon, vinyl, rubber blends, and polyester fibers. Fire hose must withstand relatively high pressures and transport water with a minimum loss in pressure. It must be flexible enough to store in rolls or load onto a hose rack without occupying excessive space.

Each size of fire hose is designed for a specific purpose. Fire hose is most commonly cut and coupled into *lengths* (also known as *sections*) of 50 or 100 feet (15 m or 30.5 m) for convenience of handling and replacement, but other lengths are available. Sections are coupled together to produce a continuous fire hose. The common sizes of fire hose used in vessel fire fighting operations and their applications are as follows:

- *Utility hose: ¾ inch (19 mm)* — Hose reels; semiportable fire extinguisher wheeled units; allowed for interior use on some small vessels

- *Attack fire hose: 1½ to 1¾ inches (38 mm to 44 mm)* — Combines water flow with suitability

- *Supply hose: 2½ inches (64 mm)* — Moves volumes of water to the fire scene; used as attack fire hose for large fires

Couplings

Fire hose couplings are designed to connect and unconnect hose in a short time with little effort. They are made of durable materials, generally alloys in varied percentages of brass, aluminum, or magnesium. These alloys make the coupling more durable and easier to attach to the hose. Marine couplings are usually made of brass because it does not rust. Numerous types of hose couplings are used around the world. Examples are the threaded and Storz types, expansion, quarter turn, pin lug, rocker lug, spring-loaded lug, nakajima, snap (sometimes called the Jones snap), and instantaneous (quick connect) (Figure 6.3).

Figure 6.3 Example of Storz couplings. *Courtesy of Captain Mark Turner, Syndicated Management Services Ltd.*

The *swivel* and the *expansion-ring* are two types of gaskets used with fire hose couplings. The swivel gasket makes the connection watertight when the couplings are connected. The expansion-ring gasket is used at the end of a fire hose where it expands into the shank of the coupling. These two gaskets are not interchangeable; their thickness and width are different.

Protecting Fire Hoses and Couplings

Fire hoses and couplings are subjected to many potential sources of damage during fire fighting. All parts of the fire hose coupling are susceptible to being damaged, bent, or crushed. Usually, little can be done at fires to protect fire hose from injury. The most important factor relating to the life of fire hose is the care it gets after fires, in storage, and on hose racks. Drain excess water from fire hose after each use. Although moisture is good for the interlinings, accumulations of water in the hose can cause damage. Properly test fire hose if any damage is suspected. Pressure test fire hose at least annually (quarterly is recommended). See Fire Hose Testing section. Be aware of the sources of damage to fire hose and couplings and procedures for proper care and protection. Sources of damage and proper care procedures are given in the sections that follow.

Sources of Fire Hose Damage

The life of fire hose is considerably dependent upon how well it is protected against mechanical injury, excess heat, organic damage, and chemical contacts. Ways to prevent these damages are described in the following paragraphs.

Mechanical. Some common mechanical injuries are worn places, rips, abrasions, and cracked interlinings. To prevent these damages, the following practices are recommended:

- Avoid laying or pulling fire hose over rough, sharp edges or objects.

- Do not lay fire hose over equipment tracks, steam pipes, etc. Prevent mobile equipment from running over fire hose or pinching it.

- Close nozzles and valves slowly to prevent *water hammer* (force created by the rapid deceleration of water; often heard as a distinct sharp clank).

- Change position of bends in hose when restowing.

- Avoid excessive pressures on fire hose.

- Avoid subjecting wet fire hose to freezing temperatures.

- Avoid stowing fire hose where the vessel's movement underway or seawater on deck could cause chafing.

Thermal. Exposing fire hose to excessive heat or fire can char, melt, or weaken the fabric and dry the rubber lining. A similar drying effect may occur to interlinings when fire hose is hung vertically to dry for a longer period of time than is necessary or dried in intense sunlight. To prevent thermal damage, the following practices are recommended:

- Protect fire hose from exposure to excessive heat or fire when possible.

- Keep the outside of woven-jacket fire hose dry.

- Avoid drying fire hose on a hot deck.

- Prevent fire hose from contacting engine exhaust systems or steam line piping.

- Use covers on fire hose boxes at fire stations that are exposed to sunlight.

Organic. Organic damage such as mildew and mold may occur on woven-jacket fire hose when moisture remains on the outer surfaces. Mildew and mold cause decay and the consequent deterioration of the hose. Rubber-jacket fire hose is not susceptible to mold and mildew damage. Some fire hose is chemically treated to resist mildew and mold, but such treatment is not always completely effective. Some methods of preventing mildew and mold on woven-jacket hose are as follows:

- Remove wet woven-jacket fire hose from fire stations after a fire, and replace it with dry hose.

- Remove woven-jacket fire hose from racks at least every 30 days. Inspect, clean (if necessary), and reload.

- Run water through woven-jacket fire hose every 90 days to prevent drying and cracking of the rubber lining.

Chemical. Chemicals and chemical vapors can damage the rubber lining and often cause the lining and jacket to separate. When hose is exposed to petroleum products, paints, acids, or alkalis, it may weaken to the point of bursting. Runoff water from a fire may carry foreign materials that can damage fire hose.

Clean fire hose as soon as practical after exposure to chemicals or chemical vapors. Some recommended good practices are as follows:

- Scrub fire hose thoroughly, and brush traces of acid contacts with a solution of baking soda and water. Baking soda neutralizes acids.

- Remove fire hose periodically from racks, wash with freshwater, and dry thoroughly.

- Avoid laying fire hose on decks where oil or other chemicals have spilled.

- Dispose of fire hose properly if it has been exposed to hazardous materials and cannot be decontaminated.

- Flush fire hose with freshwater (if available) after using seawater, foam, or other agents.

Fire Hose Care

The life span of fire hose is appreciably extended when it is cared for properly. Washing, drying, and storing are very important functions. The following paragraphs highlight proper washing, drying, and storing procedures for fire hose.

Washing. The method used to wash fire hose depends on its type. Hard-rubber hose and rubber-jacket collapsible hose require little more than rinsing with freshwater if available (avoid seawater if possible). Use a mild soap if necessary. Most woven-jacket fire hose requires more care than the hard-rubber or rubber-jacket collapsible hose. Thoroughly brush accumulations of dust and debris from woven-jacket fire hose after use. If dirt cannot be removed by brushing, wash and scrub the fire hose with freshwater. When fire hose has been exposed to oil, wash with a mild soap or detergent, making sure that the oil is removed completely. Rinse thoroughly, using freshwater if available. Use common scrub brushes or brooms with streams of water.

Drying. The method used to dry fire hose depends on its type. Hard-rubber and rubber-jacket collapsible hose may be placed back on racks while still wet with no ill effects. However, woven-jacket fire hose requires thorough drying before being rolled for storage or stowed onto hose racks. Lay woven-jacket fire hose flat on a clean area of the main deck. Avoid excessive temperatures and prolonged exposure to direct sunlight.

Storing. Roll extra lengths of fire hose not required at fire stations, and store in a suitable location. Place storage shelves or cabinets in a clean, well-ventilated space that is easily accessible when replacement fire hoses are needed.

Coupling Care

Couplings are less likely to receive damage during common usage when they are connected; however, they can be bent or crushed if run over by mobile equipment. When a coupling is severely bent, the only safe and recommended solution is to replace it. Some simple rules for the care of fire hose couplings are as follows:

- Avoid dropping and/or dragging couplings.

- Prevent couplings from being run over by mobile equipment.

- Examine couplings when fire hose is washed and dried.

- Clean threads with a suitable brush.

- Inspect the swivel gasket, and replace if cracked or creased.

Fire Hose Testing

Two types of tests for fire hose are conducted: *acceptance testing* and *service testing.* Coupled fire hose is factory tested by the manufacturer before the hose is shipped to the user (acceptance testing). Acceptance testing is relatively rigorous, and the hose is subjected to extremely high pressures to ensure that it can withstand the most extreme conditions in actual use. Acceptance testing is not performed by crew members. Service testing, however, is performed periodically by the user to ensure that the hose is in optimum condition. This testing of fire hose confirms that it can still function under maximum pressure during fire fighting operations. Suggested guidelines for both types of tests are found in National Fire Protection Association (NFPA) 1962, *Standard for the Care, Use, and Service Testing of Fire Hose Including Couplings and Nozzles.* Because fire hose is required to be service tested periodically, crew members often assist in the process. Service testing of fire hose is important, not only to test the hose for reliability but to test the rest of the fire main system: fire pumps, hydrants, and nozzles. The procedure for service testing lined fire hose is given in Skill Sheet 6-1 at the end of the chapter.

Fire Hose Safety

Exercise care when working with fire hose, especially when it is under pressure. Pressurized hose is potentially dangerous because of its tendency to whip back and forth if a break occurs or a coupling pulls loose. Stand or walk near the pressurized hose only as necessary. Wear the appropriate safety gear when operating in the area of pressurized fire hoses. Other safety points are as follows:

- Open and close all valves and nozzles slowly to prevent water hammer in the fire hose, fire main system, and fire pump.

- Stand away from the discharge valve connection when charging fire hose because of its tendency to twist when filled with water and pressurized; the connection could twist loose.

Fire Hose Appliances and Tools

A complete hose layout for fire fighting purposes includes one end of a fire hose attached to a water supply source and the other to a nozzle or similar discharge device. Various other devices complete the arrangement. These devices are usually grouped into two categories: *hose appliances* and *hose tools*. A simple way to remember the difference between hose appliances and hose tools is that appliances have water flowing through them, and tools do not. Appliances include valves, valve devices (wyes and siameses), and fittings (adapters and reducers). Examples of hose tools include wheel keys, spanner wrenches, and hose strap (rope or chain) tools. The following sections highlight some of the more common hose appliances and hose tools found on board.

Figure 6.4 A ball valve helps to control water flow. A partially open (or closed) ball valve is shown.

Valves

The flow of water is controlled by various valve devices in fire hose, at hydrant outlets, and at portable pumping equipment. Regularly inspect valves and connections. Hydrant outlets located on weather decks may become corroded or encrusted with salt, causing valves to freeze in position and become inoperable. Exercise the valves and keep them lubricated to ensure their reliability. These devices include the following valve types:

- *Ball valve* — Used in fire pump piping systems and in wye appliances (see following section) to control water flow; valve opens when the handle is in line with the hose and closes when it is at a right angle to the hose; has either a handwheel or handle (Figure 6.4)

- *Gate valve* — Used to control the flow from a hydrant; has a baffle that is moved by a handle-and-screw arrangement (Figure 6.5)

- *Butterfly valve* — Used on large pump intakes; uses a flat baffle operated by a quarter-turn handle;

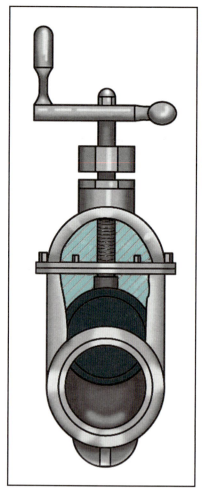

Figure 6.5 A gate valve controls the flow from a hydrant. A cutaway of a gate valve is shown.

baffle is in the center of the waterway when the valve is open (Figure 6.6)

- *Clapper valve* — Used in siamese appliances (see following section) to allow the connection and charging of only one intake fire hose before the addition of more hose; consists of a flat disk that is hinged on one side and swings in a door-like manner

Valve Devices

Valve devices (wyes or siameses) increase or decrease the number of fire hoses operating at an incident. Depending on the vessel layout and location of hydrant outlets, valve devices can maximize the efficiency of fire hose lays by connecting hoses of different diameters or enabling a large-diameter supply hose to divide into two small-diameter attack hoses. The large-diameter fire hoses give great volumes of water, but they are more difficult to maneuver. Small fire hoses are easier to move but give reduced pressure on long lays. Plan and practice the best use of fire hoses and appliances in advance of an emergency incident.

The siamese and wye adapters are often confused because of their close resemblance. *Wye appliances* have one inlet and two discharges while *siamese appliances* have two inlets and one discharge. Wye and siamese appliances are described in the following paragraphs.

Wye appliance. Certain situations make it desirable to divide a line of fire hose into two or more. Various types of wye connections are used for this purpose. The most common wye has a 2½-inch (64

mm) inlet and two 1½-inch (38 mm) outlets, although other combinations exist (Figure 6.7). The 2½-inch (64 mm) wye also divides one 2½-inch (64 mm) or larger line of fire hose into two 2½-inch (64 mm) hoses. Wye appliances are often gated with ball valves so that water fed into the fire hose is controlled at the wye.

Siamese appliance. Siamese fire hose layouts consist of two or more hoses that are brought into one hose or device. The typical siamese has two or three connections coming into the appliance and one discharge exiting the appliance (Figure 6.8). Siamese appliances may be equipped with or without clapper valves. Siamese appliances are seldom found or used in vessel fire fighting operations.

Fittings

A number of simple hardware accessories, called *fittings*, are available for connecting fire hoses of different sizes and thread types. A variety of special

Figure 6.7 Wye appliances divide fire hoses. This common wye has a 2½-inch (64 mm) inlet and two 1½-inch (38 mm) outlets.

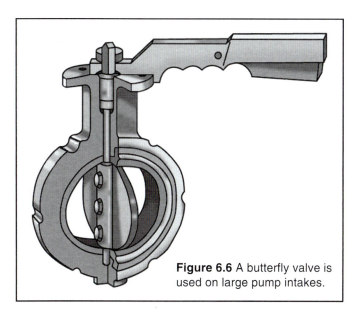

Figure 6.6 A butterfly valve is used on large pump intakes.

Figure 6.8 Siamese appliances combine two or more fire hoses into one.

hose appliances are sometimes used in various situations. Adapters and reducers are commonly used. Other fittings include elbows (change the direction of flow), hose caps (close off male couplings), and hose plugs (close off female couplings).

An *adapter* is a fitting for connecting fire hose couplings with dissimilar threads but with the same inside diameter. The *double male* and *double female adapters* are probably used more than any other special fire hose appliance (Figures 6.9 a and b). These appliances connect fire hoses when both couplings are of the same sex. This need most frequently arises when fire hose has been laid backwards, and it is more efficient to use adapters than to rearrange the hose.

A *reducer* is another common hose fitting that extends a large fire hose by connecting a smaller one to the end (Figure 6.10). Reducers are also commonly found on pump and fire main discharges so that smaller fire hoses may hook directly to the connections. Extending a fire hose with a reducer in this way limits options to just that hose, whereas using a gated wye at that point allows the option of adding another fire hose if needed.

Hose Tools

Although hose tools other than wrenches are not common on board a vessel, hose straps (or ropes and chains), spanner and hydrant wrenches, and wheel keys are found. Place hose tools where they are needed. Become familiar with their uses. Descriptions of these tools are given in the following paragraphs.

Hose strap (hose rope and hose chain). One of the most useful tools to aid in handling a charged fire hose is a hose strap. The hose rope and hose chain are similar tools. These devices can carry and pull fire hose, but their primary value is to provide a secure means to handle pressurized hose when applying water. Another important use of these tools is to secure fire hose to ladders and other fixed objects (Figure 6.11). Often fire hoses are tied or lashed by heaving lines or lashings when hose is advanced vertically.

Spanner and hydrant wrenches. The primary purpose of a *spanner wrench* (or *spanner*) is to tighten and loosen fire hose couplings. *Hydrant wrenches* are primarily used to remove caps from fire hydrant outlets and to open fire hydrant valves on shoreside or port

Figures 6.9 a and b Adapters connect hoses with different threads together. (a) A double female adapter. (b) A double male adapter.

Figure 6.10 A reducer fitting can extend a large fire hose to a smaller hose.

hydrants. A hydrant wrench is usually equipped with a pentagon opening in its head that fits most standard fire hydrant operating nuts. The lever handle may be threaded into the operating head to make it adjustable, or the head and handle may be a ratchet type. The head may also be equipped with a spanner.

Wheel key. Hydrant outlets on the fire main systems of vessels usually do not require wrenches to open them because they have valve wheels, but they may require wheel keys or wrenches if wheels are stiff. The *wheel key* operates the hydrant valve when it cannot be opened or closed otherwise (Figure 6.12).

 ## Water Fire Steams and Nozzles

Several different types of nozzles (also known as *branches*) are used for shipboard fire fighting. Nozzles are usually made of brass to resist salt air and water, although other lighter alloys that resist corrosion are available. Some nozzles in use today resemble the modern models used by shoreside firefighters, although most vessels remain equipped with the specialty-type nozzles traditionally found in maritime use.

Turn off all nozzles with a handle (bail) by pushing the handle forward. This procedure gives a better

chance of turning off the nozzle if the nozzle operator loses control of the hose. When water flows from a nozzle, a reaction is equally strong in the opposite direction, thus a force pushes back on the nozzle operator. As the nozzle reaction pushes backward, the bail can be caught with the hand, pushing it forward as the nozzle is slipping away.

Other shutoff features on nozzles include the twist type and the quarter-turn type. Turning the forward portion of the nozzle in one direction or the other to the closed position turns off the twist type. The quarter-turn type is like any other quarter-turn valve (such as a gas valve); the valve handle is turned a quarter turn to the closed position, typically perpendicular to the flow. The key is to know the equipment and be familiar with its operation.

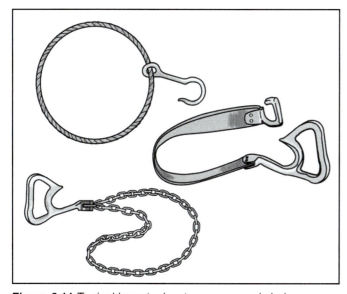

Figure 6.11 Typical hose tools: strap, rope, and chain.

Figure 6.12 A wheel key (also known as a portable reach rod) turns a wheel in the deckhead. *Courtesy of R. Wright/ Maryland Fire and Rescue Institute/United States Coast Guard.*

Three major types of streams are produced by different nozzles: *solid, fog,* and *broken* (Figure 6.13). The smaller the droplets of water, the greater the cooling power of the water but the less reach or penetration it has. Full fog shielding gives maximum protection but little depth, whereas solid streams give maximum penetration but less effective cooling. Descriptions of the types of fire streams and common marine nozzles used to produce the streams are given in the sections that follow. Nozzle care and maintenance are also discussed.

Solid Stream

The solid stream is the oldest form of water stream used for fire suppression. A *solid stream* is a stream of water that is as compact as possible with little shower or spray. The solid stream is formed by a smoothbore nozzle that is specially designed for that purpose

(Figures 6.14 a–c). The discharge end of the nozzle is tapered to less than one-half the diameter of the hose end. The tapering increases both the *velocity* of the water at the discharge end (nozzle pressure) and the *reach* (the distance that a solid stream travels before breaking or dropping). Reach is important when it is difficult to approach close to a fire. The maximum horizontal reach is attained with the nozzle held at an upward angle of 35 to 40 degrees from the deck. The maximum vertical reach is attained at an angle of 75 degrees.

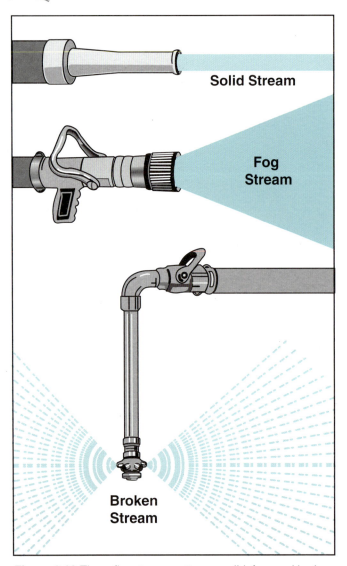

Figure 6.13 Three fire stream patterns: solid, fog, and broken.

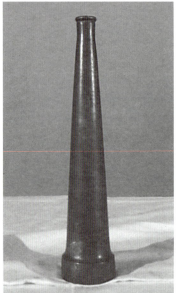

Figures 6.14 a–c
(a) Solid stream nozzle
(b) Another version of a solid stream nozzle. *Courtesy of Captain Mark Turner, Syndicated Management Services Ltd.*
(c) The basic design of a solid stream nozzle.

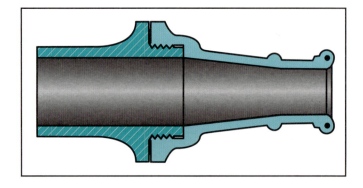

Although solid stream nozzles differ in shape and size, they all conform to the required physical characteristics for solid stream nozzles. A solid stream nozzle is equipped with a shutoff valve that matches the hose couplings used on the vessel. The shutoff valve with the tip(s) removed can extend a fire hose and provide an alternate means of shutting down the hose. A smooth-finished waterway contributes to both the shape and reach of the stream. Alteration or damage to the nozzle can significantly alter stream shape and performance.

Probably less than 10 percent of the water from a solid stream actually absorbs heat from a fire because only a small portion of the water surface actually contacts the fire — and only water that contacts the fire absorbs heat. The rest runs off — sometimes over the side — but more often, the runoff becomes free surface water and a vessel stability problem. The main use of solid streams is to break up the burning material and penetrate to the seat of a Class A fire.

It is often difficult to hit the seat of a fire even with the reach of a solid stream. Bulkheads with small openings can keep crew members from getting into the proper positions to aim the stream into a fire. If a stream is used before the nozzle is properly positioned, the water may hit a bulkhead and cascade onto the deck without reaching the fire. In some instances, an obstruction is between the fire and the nozzle operator. When this happens, deflect the stream off a bulkhead or the deckhead to get around the obstacle (Figure 6.15). This method can also break a solid stream into a spray-type (broken) stream that absorbs more heat. For example, a spray-type stream can cool an extremely hot passageway that is keeping fire crews from advancing toward a fire.

Fog Stream

A *fog stream* is composed of very fine water droplets that have a much larger total surface area than a solid stream. Thus, a given volume of water in fog form absorbs much more heat than the same volume of water in a solid stream. A fog pattern is produced by a high-velocity nozzle, has much more finely divided water particles, and is usually shaped into a uniform cone discharge. Most fog nozzles in use today are multipurpose types that are adjustable so that they may produce different stream patterns, including a straight stream.

The greater heat absorption of fog streams is important where the use of water is limited. With fog streams, less water is applied to remove the same amount of heat from a fire. In addition, more of the fog stream turns to steam when it contacts a fire. Consequently, there is less runoff, less free surface water, and fewer stability problems for the vessel. Improperly used, fog streams can cause injury to crew members if too much steam is produced in a confined area.

Fog streams do not have the accuracy or reach of solid streams. While they can be effectively used on the surface of a deep-seated fire, fog patterns are not as effective as solid streams in soaking through and reaching the heart of a fire. Fog patterns may combat high flash point Class B fires. However, it is important that crew members have actual experience in directing these streams during drills.

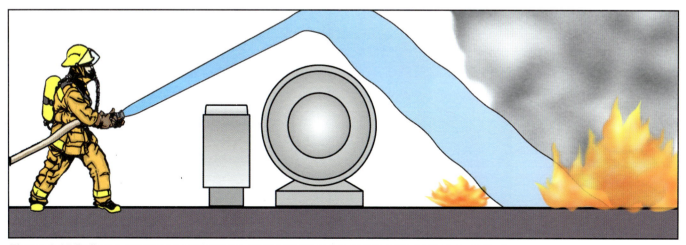

Figure 6.15 Deflect a straight or solid stream off the deckhead to reach a fire behind an obstacle.

Broken Stream

A *broken stream* is a stream of water that has been broken into coarsely divided drops. While a solid stream may become a broken stream when it begins to lose its forward velocity (breakover point), a true broken stream takes on that form as it leaves the nozzle. The coarse drops of a broken stream absorb more heat than a solid stream, and the broken stream has a greater reach and penetration than a fog stream, so it can be an effective stream in certain situations (confined spaces).

Low-velocity fog nozzles or applicators produce broken stream spray patterns. Spray patterns are very useful and have a much greater heat absorbing capability than solid streams. They are commonly used as a protective pattern for fire teams advancing into a fire area and for application into concealed spaces that cannot be entered.

Common Marine Nozzles

Basically two types of nozzles are used for shipboard fire fighting: *adjustable nozzles* (widely used and very popular in the shoreside fire service) and *marine all-purpose nozzles* (used for decades on board vessels). An *applicator* is often used with the marine all-purpose nozzle. Other nozzles used are piercing nozzles, distributor nozzles, and water curtain nozzles. The various nozzles are discussed in the sections that follow.

Adjustable Nozzle

Adjustable nozzles come in several varieties. On vessels, they are the set flow-rate type that are used for only one flow rate at a given discharge pressure. These types of fog nozzles are now used more often in maritime fire fighting, and in many ways they are superior to the traditional marine all-purpose nozzle. Figures 6.16 a and b show some examples of the more modern adjustable nozzles that have been approved by some agencies and are finding their place in marine fire fighting operations. These nozzles are extremely versatile and produce an excellent protective fog pattern. The pattern adjusts to a wide fog, a narrow fog, a straight stream, and all points in between (Figure 6.17).

Figures 6.16 a and b (a) Adjustable fog nozzles. (b) Another version of an adjustable fog nozzle. *Courtesy of Captain Mark Turner, Syndicated Management Services Ltd.*

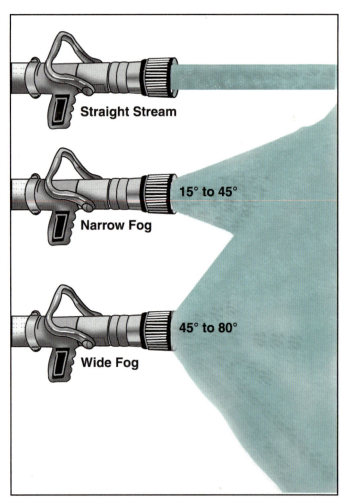

Figure 6.17 Adjustable fog nozzles are most commonly set to straight stream, narrow fog, or wide fog patterns.

An adjustable nozzle produces a straight stream that is very similar to a solid stream, but it is not truly solid. The straight stream has voids in the discharge pattern and typically has less reach than a solid stream of the same size. The straight stream is still very effective when reach and penetration are required.

The pattern adjustment on adjustable nozzles is typically always made in the same direction. Turn the adjustment left for wider fog and right for narrower fog to straight stream. Some nozzles may not follow this convention. Become familiar with how adjustable nozzles operate. If they follow the norm, remember this simple rule: *Left for LIFE — Right for REACH*. A full, wide fog can save a *LIFE* when faced with a flash fire situation. An automatic, instantaneous reaction to a flash fire is to drop to the deck and adjust the nozzle to full fog. An adjustment to the right only increases *REACH* and achieves little protection from the fire.

Marine All-Purpose Nozzle

The *marine all-purpose nozzle* produces a solid stream or high-velocity fog pattern. The pattern selection is made with the position of the bail. As with other nozzles, the closed position is with the bail all the way forward. A fog pattern is obtained with the bail in the center position (perpendicular to the body of the nozzle), and a solid stream is obtained with the bail all the way back toward the nozzle operator (Figures 6.18 a–c). Marine all-purpose nozzles are available to use with 1½- and 2½-inch (38 mm and 64 mm) hose couplings. In addition to solid stream and fog patterns, an applicator is also used with this nozzle to obtain spray patterns. When using an applicator (see Applicator section), the tip is changed and the bail placed in the center (fog) position.

Applicator

Applicators (sometimes referred to as *low-velocity fog applicators*) are tubes, or pipes, that are angled at either 60 degrees or 90 degrees at the water outlet end. A spray pattern is obtained by using an applicator along with a marine all-purpose nozzle. Applicators are stowed with a low-velocity head already in place on the pipe. Some heads are shaped somewhat like a pineapple, with tiny holes angled to cause minute streams to bounce off one another and create a mist. Some heads resemble a cage with a fluted arrow inside. The point of the arrow faces the opening in the applicator tubing. Water strikes the fluted arrow and then bounces in all directions, creating a fine mist.

For 1½-inch (38 mm) marine all-purpose nozzles, 4-foot (1.2 m) 60-degree angle and 10-foot (3 m) 90-degree angle applicators are approved for shipboard use. For 2½-inch (64 mm) nozzles, 12-foot (3.7 m) 90-degree angle applicators are approved (Figures 6.19 a and b). Other lengths with different angles are sometimes found. The 4-foot (1.2 m) applicator on the 1½-inch (38 mm) marine all-purpose nozzle is used in propulsion machinery spaces containing oil-fired boilers, internal combustion machinery, or fuel units.

Figures 6.18 a–c Marine all-purpose nozzle. *Courtesy of Maritime Institute of Technology and Graduate Studies.* (a) Closed position. (b) Fog position. (c) *(right)* Solid stream position.

Figures 6.19 a and b All-purpose fog applicators. (a) A 2½-inch (64 mm), 12-foot (1.2 m), 90-degree-angle applicator. *Courtesy of R. Wright/Maryland Fire and Rescue Institute/ United States Coast Guard.* (b) A 1½-inch (38 mm), 4-foot (1.2 m) 60-degree-angle applicator. *Courtesy of Maritime Institute of Technology and Graduate Studies.*

Figures 6.20 a–c (a) The high-velocity tip is removed from the marine all-purpose nozzle to attach an applicator. *Courtesy of Maritime Institute of Technology and Graduate Studies.* (b) Applicator pins. *Courtesy of Maritime Institute of Technology and Graduate Studies.* (c) Applicator discharge tip. *Courtesy of Maritime Institute of Technology and Graduate Studies.*

Spray patterns from applicators are effective in combating high flash point Class B fires in spaces where entry is difficult or impossible. Applicators can be poked into areas that cannot be reached with other nozzles. They are also used to provide a heat shield for crews advancing with high-velocity fog, adjustable streams, or foam. Low-velocity fog can be used to extinguish small tank fires, especially where the mist from the applicator can cover the entire surface of the tank. However, other extinguishing agents such as

foam or carbon dioxide are usually more effective in these cases. Use applicators during drills so that crew members can practice using them in concert with other nozzles.

The applicator is attached to the marine all-purpose nozzle. To attach the applicator, first remove the high-velocity tip (Figures 6.20 a–c). Then, snap the straight end of the applicator into the fog outlet and lock with a quarter turn. The desired stream is then obtained with the nozzle bail in the fog position (half-way back). The tip is replaced before operating the nozzle without the applicator.

Another applicator type is a lance-type attachment that can penetrate into partitions and other difficult-to-reach areas. It is used in conjunction with the marine all-purpose nozzle just like other applicators and is similar to the piercing nozzle (see next section).

Piercing Nozzle

A *piercing nozzle* is generally a 3- to 6-foot (1 m to 1.8 m), 1½-inch (38 mm) diameter hollow steel rod that produces a broken stream. This nozzle is used most commonly to extinguish fires in concealed spaces or areas that are otherwise inaccessible. The discharge end of the piercing nozzle has a hardened steel point that is suitable for driving through partitions and some bulkheads. Opposite the pointed nozzle end of the piercing nozzle is the driving end, which is struck with a sledgehammer to force the point through an obstruction. This feature makes the nozzle very useful in combating automobile fires on ro/ros and ferries. Built into the point is an impinging-jet nozzle that is generally capable of delivering about 100 gpm (379 L/min) of water. These nozzles may also deliver aqueous film forming foam (AFFF) to a confined area. See Aqueous Film Forming Foam section. Air-driven penetrating nozzles that bore through a vessel to reach the seat of the fire are used for the same purpose.

Water Curtain Nozzle

Water curtain nozzles were commonly used for exposure protection by creating a wall of water between the fire and exposure. However, research indicates that these nozzles are only effective if the water is sprayed directly against the exposure.

Distributor Nozzle

The *distributor nozzle* is a broken stream nozzle sometimes found on board vessels; it has a rotating head that sprays water in all directions. This nozzle is raised or lowered through holes in decks to attack fires on a level above or below that of the fire fighting crews.

Nozzle Care and Maintenance

Inspect nozzles periodically to make sure that they are in proper working condition. Inspect the following components:

- Condition of gasket
- External damage, for example, loose or missing parts or chains
- Internal damage (or lodged debris)

- Ease of operation
- Security of pistol grip (if applicable)

When necessary, clean nozzles thoroughly with soap and water using a soft bristle brush. Replace the gasket if it is worn or missing. Clean and lubricate any moving parts that appear to be sticking according to manufacturer's recommendations.

◆ Foam Equipment and Streams

Water alone is not always effective as an extinguishing agent. Under certain circumstances, fire fighting foam is needed. Fire fighting foam is used for deep-seated Class A fires and is especially effective on Class B flammable liquid fires and hazardous materials incidents (unignited spills). Even fires that are fought successfully using plain water are often more effectively extinguished if a foam concentrate is added.

In general, foam works by forming a blanket of bubbles on the burning fuel. The foam blanket excludes oxygen and stops the burning process. The water in the foam is slowly released as the foam breaks down, which provides a cooling effect on the fuel (Figure 6.21). Foam extinguishes fire in the following ways:

- *Separating* — Creates a barrier between the fuel and the fire
- *Cooling* — Lowers the temperature of the fuel and adjacent surfaces
- *Suppressing (sometimes called smothering)* — Prevents the release of flammable vapors and reduces the possibility of ignition or reignition

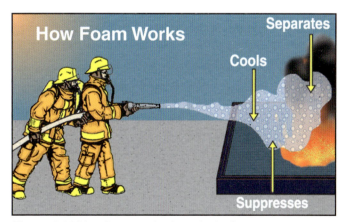

Figure 6.21 How foam works: It cools the fuel, separates the fuel and the fire, and suppresses (smothers) vapors.

A foam blanket needs time to work after its application to effectively extinguish fires. The life of a foam blanket depends on a tremendous number of variables. Among the most important are the following:

- Type of fuel (whether water miscible or not)
- Type of foam concentrate
- Heat of fuel surfaces
- Resistance of foam to reignition
- Degree of rolling of vessel, causing foam blanket to mix with fuel and edges of blanket to contact hot surfaces
- Whether foam is free to flow and cover all burning surfaces

Foams in use today are of the mechanical type; that is, they are *proportioned* (mixed with water) and *aerated* (mixed with air) before use. Along with how foam is generated and types of foam concentrates, foam proportioning, foam equipment (proportioners and nozzles), and foam application techniques are addressed in the following sections. Before discussing types of foams and the foam-making process, it is important to understand the following terms:

- *Foam concentrate* — Raw foam liquid as it rests in its storage container before the introduction of water and air; usually shipped in 5-gallon (19 L) pails or 55-gallon (208 L) drums; fixed foam extinguishing system storage tanks hold 3,000 gallons (11 356 L) or more

- *Foam proportioner* — Device that introduces the correct amount of foam concentrate into the water stream to make the foam solution

- *Foam solution* — Homogeneous mixture of foam concentrate and water before the introduction of air

- *Foam (finished foam)* — Completed product after air is introduced into the foam solution

How Foam Is Generated

Four elements are necessary to produce high-quality fire fighting foam: foam concentrate, water, air, and mechanical agitation. These elements must be blended in the correct ratios. Removing any element results in either no foam or poor-quality foam (Figure 6.22). Finished foam is produced in two stages. First, water is mixed with foam liquid concentrate to form a foam solution (proportioning stage). Second, the foam solution passes through the piping or hose to a foam nozzle or sprinkler that aerates the foam solution to form finished foam (aeration stage). Aeration produces an adequate amount of foam bubbles to form an effective foam blanket. Proper aeration also produces uniform-sized bubbles that form a long lasting blanket. A good foam blanket maintains an effective cover over the fuel.

Proportioners and foam nozzles or sprinklers are engineered to work together. Using a foam proportioner that is not hydraulically matched to the foam nozzle or sprinkler (even if the two are made by the same manufacturer) can result in either unsatisfactory foam or no foam at all. Numerous appliances are available for making and applying foam. A number of these are discussed later in this section.

Foam Concentrates

Foam is available in many different types for different applications—see Table 6.1. Foam expands (increases in volume) when it is aerated. Consider this expansion characteristic when choosing a foam concentrate for specific applications. Depending on its purpose, foam is described as low-, medium-, or high-expansion. Several foam types are discussed in the sections that follow: protein, fluoroprotein, film forming

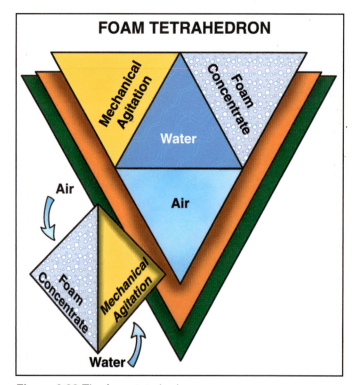

Figure 6.22 The foam tetrahedron.

Table 6.1
Foam Concentrate Characteristics/Application Techniques

Type	Characteristics	Storage Range	Application Rate	Application Techniques	Primary Uses
Protein Foam (3% and 6%)	• Protein based • Low expansion • Good reignition (burnback) resistance • Excellent water retention • High heat resistance and stability • Performance can be affected by freezing and thawing • Can freeze protect with antifreeze • Not as mobile or fluid on fuel surface as other low-expansion foams	35–120°F (2°C to 49°C)	0.16 gpm/ft² (6.5 L/min/m²)	• Indirect foam stream; do not mix fuel with foam • Avoid agitating fuel during application; static spark ignition of volatile hydrocarbons can result from plunging and turbulence • Use alcohol-resistant type within seconds of proportioning • Not compatible with dry chemical extinguishing agents	• Class B fires involving hydrocarbons • Protecting flammable and combustible liquids where they are stored, transported, and processed
Fluoroprotein Foam (3% and 6%)	• Protein and synthetic based; derived from protein foam • Fuel shedding • Long-term vapor suppression • Good water retention • Excellent, long-lasting heat resistance • Performance not affected by freezing and thawing • Maintains low viscosity at low temperatures • Can freeze protect with antifreeze • Use either freshwater or saltwater • Nontoxic and biodegradable after dilution • Good mobility and fluidity on fuel surface • Premixable for short periods of time	35–120°F (2°C to 49°C)	0.16 gpm/ft² (6.5 L/min/m²)	• Direct plunge technique • Subsurface injection • Compatible with simultaneous application of dry chemical extinguishing agents • Deliver through air-aspirating equipment	• Hydrocarbon vapor suppression • Subsurface application to hydrocarbon fuel storage tanks • Extinguishing in-depth crude petroleum or other hydrocarbon fuel fires
Film Forming Fluoroprotein Foam (FFFP) (3% and 6%)	• Protein based; fortified with additional surfactants that reduce the burnback characteristics of other protein-based foams • Fuel shedding • Develops a fast-healing, continuous-floating film on hydrocarbon fuel surfaces • Excellent, long-lasting heat resistance	35–120°F (2°C to 49°C)	*Ignited Hydrocarbon Fuel:* 0.10 gpm/ft² (4.1 L/min/m²) *Polar Solvent Fuel:* 0.24 gpm/ft² (9.8 L/min/m²)	• Cover entire fuel surface • May apply with dry chemical agents • May apply with spray nozzles • Subsurface injection • Can plunge into fuel during application	• Suppressing vapors in unignited spills of hazardous liquids • Extinguishing fires in hydrocarbon fuels

Continued on next page.

Type	Characteristics	Storage Range	Application Rate	Application Techniques	Primary Uses
FFFP *(continued)*	• Good low-temperature viscosity • Fast fire knockdown • Affected by freezing and thawing • Use either freshwater or saltwater • Can store premixed • Can freeze protect with antifreeze • Use alcohol-resistant type on polar solvents at 6% solution and on hydrocarbon fuels at 3% solution • Nontoxic and biodegradable after dilution				
Aqueous Film Forming Foam (AFFF) (1%, 3%, and 6%)	• Synthetic based • Good penetrating capabilities • Spreads vapor-sealing film over and floats on hydrocarbon fuels • Can use nonaerating nozzles • Performance may be adversely affected by freezing and storing • Has good low-temperature viscosity • Can freeze protect with antifreeze • Use either freshwater or saltwater • Can premix	25–120°F (-4°F to 49°C)	0.10 gpm/ft² (4.1 L/min/m²)	• May apply directly onto fuel surface • May apply indirectly by bouncing it off a wall and allowing it to float onto fuel surface • Subsurface injection • May apply with dry chemical agents	• Controlling and extinguishing Class B fires • Handling land or sea crash rescues involving spills • Extinguishing most transportation-related fires • Wetting and penetrating Class A fuels • Securing unignited hydrocarbon spills
Alcohol-Resistant AFFF (3% and 6%)	• Polymer has been added to AFFF concentrate • Multipurpose: Use on both polar solvents and hydrocarbon fuels (use on polar solvents at 6% solution and on hydrocarbon fuels at 3% solution) • Forms a membrane on polar solvent fuels that prevents destruction of the foam blanket • Forms same aqueous film on hydrocarbon fuels as AFFF • Fast flame knockdown	25–120°F (-4°C to 49°C) (May become viscous at temperatures under 50°F [10°C])	**Ignited Hydrocarbon Fuel:** 0.10 gpm/ft² (4.1 L/min/m²) **Polar Solvent Fuel:** 0.24 gpm/ft² (9.8 L/min/m²)	• Apply directly but gently onto fuel surface • May apply indirectly by bouncing it off a wall and allowing it to float onto fuel surface • Subsurface injection	Fires or spills of both hydrocarbon and polar solvent fuels

Continued on next page.

Type	Characteristics	Storage Range	Application Rate	Application Techniques	Primary Uses
Alc-Res AFFF *(continued)*	• Good burnback resistance on both fuels • Not easily premixed				
High-Expansion Foam	• Synthetic detergent based • Special-purpose, low water content • High air-to-solution ratios: 200:1 to 1,000:1 • Performance not affected by freezing and thawing • Poor heat resistance • Prolonged contact with galvanized or raw steel may attack these surfaces	27–110°F (-3°C to 43°C)	Sufficient to quickly cover the fuel or fill the space	• Gentle application; do not mix foam with fuel • Cover entire fuel surface • Usually fills entire space in confined space incidents	• Extinguishing Class A and some Class B fires • Flooding confined spaces • Volumetrically displacing vapor, heat, and smoke • Reducing vaporization from liquefied natural gas spills • Extinguishing pesticide fires • Suppressing fuming acid vapors • Suppressing vapors in coal mines and other subterranean spaces and concealed spaces in basements • Extinguishing agent in fixed extinguishing systems • Not recommended for outdoor use
Class A Foam	• Synthetic • Wetting agent that reduces surface tension of water and allows it to soak into combustible materials • Rapid extinguishment with less water use than other foams • Use regular water stream equipment • Can premix with water • Mildly corrosive • Requires lower percentage of concentration (0.2 to 1.0) than other foams • Outstanding insulating qualities • Good penetrating capabilities	25–120°F (-4°C to 49°C) (Concentrate is subject to freezing but can be thawed and used if freezing occurs)	Same as the minimum critical flow rate for plain water on similar Class A Fuels; flow rates are not reduced when using Class A foam	• Can propel with compressed-air systems • Can apply with conventional nozzles	Extinguishing Class A combustibles only

fluoroprotein, aqueous film forming, alcohol-resistant aqueous film forming, and medium- or high-expansion.

Protein

Protein foams are chemically broken down (hydrolyzed) animal and vegetable protein solids. The end product of this chemical digestion is protein liquid concentrate. Apply protein foam gently so it does not plunge into the fuel. Specific facts about protein foam include the following:

- First mechanical foam
- Use only on hydrocarbon fires
- Requires air-aspirating equipment
- Shorter shelf life than synthetic foams
- May deteriorate in storage
- Not compatible with some dry chemical agents

WARNING

Take great care when applying protein foam because application requires a close approach. Do not plunge protein foam into liquid fuel. Burning may continue and the fuel may splash, increasing the fire hazard.

Fluoroprotein

Like regular protein foams, fluoroprotein foams are based on hydrolyzed protein solids. Fluoroprotein foams are fortified with fluorinated surfactants that enable the foam to shed or separate from hydrocarbon fuels. This ability makes them ideally suited for direct foam application using a plunge technique, which means that fire crews can apply fluoroprotein foam streams from a distance. Fluoroprotein foams also provide a strong "security blanket" for long-term vapor suppression, which is especially critical with unignited spills. Some characteristics of fluoroprotein foam are as follows:

- Shorter shelf life than synthetic foam
- Requires air-aspirating equipment
- Maintains low viscosity at low temperatures during storage and use
- Can store premixed (proportioned with water) for a short period of time

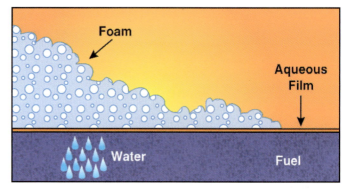

Figure 6.23 The AFFF film floats ahead of the foam blanket.

Aqueous Film Forming Foam (AFFF)

Several years after synthetic foam's entry into the fire service, the U.S. Navy discovered that when a fluorinated surfactant was added to detergent foam, the water that drained from the foam blanket actually floated on hydrocarbon fuel spills, creating an *aqueous film*. When AFFF is applied to a hydrocarbon fire, the following three things occur:

- A foam blanket immediately begins to drain water from its solution, creating an air-excluding film ahead of the blanket (Figure 6.23).
- The rather fast-moving foam blanket then moves across the fuel, adding further insulation.
- As the foam blanket continues to drain, more film is released, giving AFFF the ability to "heal" or reseal over areas where the foam blanket was disturbed.

Some facts about AFFF are as follows:

- Premixable in portable fire extinguishers and water tanks
- Uses either aerating or nonaerating nozzles
- Has penetrating capabilities in baled storage fuels or high surface tension fuels such as treated wood

Film Forming Fluoroprotein (FFFP)

Film forming fluoroprotein foam concentrate is based on fluoroprotein foam technology with aqueous film forming foam capabilities. Film forming fluoroprotein foam incorporates the benefits of AFFF for fast fire knockdown and the benefits of fluoroprotein foam for long-lasting heat resistance. Some specific characteristics of FFFP are as follows:

- Can store premixed (proportioned with water) for use in portable fire extinguishers and other portable foam equipment

- Uses either aerating or nonaerating nozzles (gives penetrating capabilities)

Alcohol-Resistant AFFF

Alcohol-resistant AFFF is available from most foam manufacturers. On most polar solvents (liquids that mix with water), alcohol-resistant AFFF is used at 3 or 6 percent concentrations, depending on the particular brand. Stronger polar solvents may require greater flow rates. Alcohol-resistant AFFF can also be used on hydrocarbon fires at 3 percent proportions, which makes this foam a true multipurpose hydrocarbon/polar solvent foam.

When an alcohol-type AFFF is applied to polar solvent fuels, it creates a membrane (rather than a film) over the fuel, which separates the water in the foam blanket from the solvent. The blanket then acts in much the same way as regular AFFF. Alcohol-resistant AFFF is applied gently to the fuel so that the membrane can form first. Some characteristics of alcohol-resistant AFFF are as follows:

- Not premixable

- Adversely affected if stored at temperatures below 50°F (10°C) (concentrate becomes viscous and may not be usable with some proportioning equipment)

- Multipurpose (use at 3 percent for hydrocarbon fires and at 3 or 6 percent for polar solvent fires, depending on manufacturer)

- Can store at temperatures from 35°F to 120°F (2°C to 49°C)

Medium- and High-Expansion

Medium- and high-expansion foams are special-purpose foams. Because they have a low water content, they minimize water damage and have fewer adverse effects on vessel stability than low-expansion foams. The major use for these foams is in fixed extinguishing systems for engine rooms and machinery spaces (Figures 6.24 a and b).

Unlike conventional foam, which provides a blanket a few inches (millimeters) thick over the burning surface, medium- and high-expansion foams are truly three-dimensional. Their area of coverage is measured in length, width, height, and cubic feet (cubic meters). These foams are designed for fires in confined spaces. Heavier than air but lighter than water or oil, they flow down openings and fill compartments,

Figures 6.24 a and b (a) A fixed high-expansion foam system for an engine room. The foam concentrate is in the tank (right) and the machinery is in the vent trunking (left). *Courtesy of Captain John F. Lewis.* (b) High-expansion foam nozzle. *Courtesy of R. Wright/Maryland Fire and Rescue Institute/ United States Coast Guard.*

spaces, and crevices, replacing the air in these spaces. In this manner they deprive the fire of oxygen. Because of their water content, they absorb heat from the fire and cool the burning material. When the foam absorbs sufficient heat to turn its water content to steam, it has absorbed as much heat as possible. The steam then continues to replace oxygen and thus combat the fire.

Medium-expansion foam provides a heavier, thicker foam blanket than the types mentioned previously; expansion rates range from 20:1 to 200:1. High-expansion foam completely fills large areas such as machinery spaces; expansion rates range from 200:1 to 1,000:1. High-expansion foam displaces the oxygen in the space, but because it is light, it may float on thermal currents, allowing hidden pockets of hot gas and fire under the foam blanket. The advantages and disadvantages of medium- and high-expansion foam concentrates are given.

Advantages of Medium- and High-Expansion Foams

♦ Highly visible white bubbles; act as warning to those entering a foam-filled environment

♦ Nontoxic foams, even when exposed to fire

♦ Excellent postfire cooling effect on hot metal; prevent reignition

♦ Do not require high-pressure storage cylinders and small-bore distribution pipework; save space and weight

♦ An effective cover for burning fuels; work best in a vented space, so ventilation to release excess air pressure, heat, and steam desirable; sealing the space not required

♦ Use either freshwater or saltwater to make finished foams

♦ Quick testing of fixed extinguishing systems at minimal expense and no damage to equipment or the environment

♦ Regular shipboard fire pumps pressurize fixed systems

♦ Quick discharging of fixed extinguishing systems; fills spaces and extinguishes fires quickly

♦ Quick and easy cleanup; requires only water and compressed air

♦ Cost effective; high ratio of water to foam concentrate; high ratio of foam expansion to water

 ### Disadvantages of Medium- and High-Expansion Foams

♦ Poor heat resistance; air-to-water ratio is very high

♦ May chemically attack galvanized or raw steel surfaces; avoid prolonged contact

♦ Foam-filled environment blocks vision, impairs hearing, and makes breathing difficult; difficult to escape from large foam-filled space; exit the area once foam begins to flow

♦ Require SCBA or line mask and lifeline to enter foam-filled environment if rescue necessary

♦ High-expansion foam not recommended on weather decks; slightest breeze may remove the foam blanket in sheets, reexposing the fuel to ignition sources

♦ Electrically conductive; do not use on live electrical equipment

♦ Require sufficient supplies of concentrate to ensure coverage of entire surface areas or filling of spaces

CAUTION: Although some regulating authorities require walking through a high-expansion foam-filled space as a part of a drill, those directing the publication of this manual believe that walking blind in a foam-filled space is a dangerous practice and do not recommend it.

Foam Proportioning

Proportioning is the mixing of water with foam concentrate to form a foam solution (Figure 6.25). To ensure maximum effectiveness, foam concentrates are proportioned only at the specific percentage for which they are designed. This percentage rate is set by the approving standards authority and is clearly marked on the outside of every foam container. Fire fighting foam is 94 to 99.9 percent water. Foams are used at 0.2-, 1-, 3-, or 6-percent concentrations. For example, when using a 3-percent concentrate, 97 parts

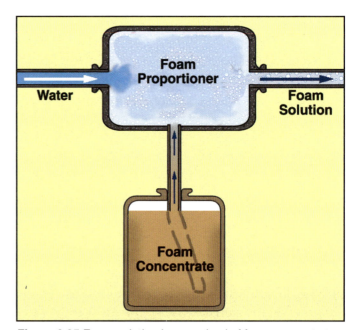

Figure 6.25 Foam solution is comprised of foam concentrate and water.

water mixed with 3 parts foam concentrate equals 100 parts foam solution. Class A foams use 0.2- to 1-percent concentrations. In general, foams designed for hydrocarbon fires use 1- to 6-percent concentrations. Polar solvent fuels require 3- or 6-percent concentrates, depending on the particular brand. Medium- and high-expansion foams typically use 1.5-, 2-, or 3-percent concentrations.

To be effective, foam concentrates must also match the fuel to which they are applied. It is extremely important to identify the type of fuel involved before applying foam. Foams designed for hydrocarbon fires do not extinguish polar solvent fires, regardless of the concentration at which they are used. However, foams that are designed for polar solvent fires may be used on hydrocarbon fires.

Foam Equipment

The seafarer is likely to encounter a number of different types of foam equipment. Along with proportioning equipment, a foam delivery device (nozzle or generating system) is needed. The nozzle or generating system adds air to the foam solution to produce finished foam. The equipment must be compatible to produce usable foam (Figure 6.26). The following sections describe the portable types used on board

vessels. Foam equipment used in fixed systems is discussed in Chapter 8, Fixed Fire Suppression Systems.

Foam Proportioners

Several types of foam proportioners are in common use, including *line eductors* and *combination nozzle eductors* (also known as *foam branch pipes*). Others are also available. Eductors are relatively tolerant of different operating conditions, but they observe several operation and maintenance principles to maximize their performance. These principles include the following:

- An eductor controls the flow through the system; otherwise, foam concentrate does not induct into the water.

- The pressure at the outlet of the eductor (also called *back pressure*) does not exceed 65 to 70 percent of the eductor inlet pressure. If this pressure is excessive, foam concentrate does not induct into the water.

- Foam solution concentration is only correct at the rated inlet pressure of the eductor, usually 150 to 200 psi (1 034 kPa to 1 379 kPa) {10 bar to 14 bar}. Using eductor inlet pressures lower than the rated pressure for the eductor results in rich foam concentrations. Conversely, using inlet pressures

Figure 6.26 AFFF foam hose station. *Courtesy of R. Wright/Maryland Fire and Rescue Institute/United States Coast Guard.*

greater than the rated inlet pressure produces lean foam concentrations. Neither too-rich nor too-lean concentrations work properly.

- Properly flush and maintain eductors after each use. Flush an eductor by submerging the foam pickup tube in a pail of clear water and inducting water through it for at least one minute. Inspect the strainer after each use, and clean if necessary.

- Metering valves match the foam concentrate percentage and the burning fuel, otherwise poor-quality foam results.

- The foam concentrate inlet to the eductor is not more than 6 feet (1.8 m) above the liquid surface of the foam concentrate. If the inlet is too high, the foam concentration will be very thin, or foam may not induct at all.

Line eductor. The line eductor is the simplest and least expensive proportioning device. It has no moving parts in the waterway, which makes it durable and dependable. The line eductor may attach to the hose or be part of the nozzle. Line eductors are not used on fixed extinguishing systems. The two different types of line eductors are the *in-line eductor* and the *self-educting master stream nozzle.*

Both types of eductors use the Venturi principle to draft foam concentrate into the water stream. As water at high pressure passes over a reduced opening, it creates a low-pressure area near the outlet side of the eductor, which creates a suction effect. The eductor pickup tube connects to the eductor at this low-pressure point. The pickup tube submerged in the foam concentrate draws concentrate into the water stream, creating a foam water solution.

The in-line eductor is the most common type of portable foam proportioner. It either attaches directly to the hydrant discharge or connects in the middle of a hose lay (Figures 6.27 a and b). Follow the manufacturer's instructions about inlet pressure, maximum hose lay from the eductor, and the appropriate nozzle.

Combination nozzle eductor. Self-educting foam nozzles have been available for many years and are still used. Their primary disadvantage is their lack of mobility: They have to operate next to the foam concentrate containers. The operation of these devices is primarily the same as the line eductor, except that a matched nozzle is added to the appliance (Figure 6.28).

Foam Nozzles

After a foam concentrate and water are mixed together, the resulting foam solution is next mixed with air (aerated) and delivered to the surface of the fuel by nozzles or other generating systems. Most foam nozzles or applicators have air-aspirating vents. The Venturi effect of the foam solution passing through causes it to mix with air thus giving bubbles to the finished foam. These bubbles aid in extinguishment by insulating the surface of the burning liquid while a film seals in the vapors.

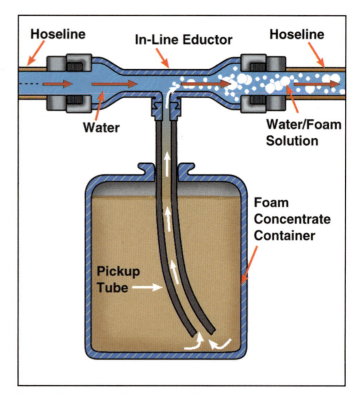

Figures 6.27 a and b (a) In-line eductor. *Courtesy of Captain Mark Turner, Syndicated Management Services Ltd.* (b) Operating principles of an in-line eductor.

The most effective appliance for the generation of low-expansion foam is the air-aspirating foam nozzle. Medium- and high-expansion foam is generated from water-aspirating nozzles and mechanical blowers. Water-aspirating nozzles are similar to the other foam-producing nozzles except they are larger and longer.

In order for the nozzle and eductor to operate properly, both must have the same rating in gallons per minute or liters per minute (gpm or L/min). As mentioned earlier, the eductor — not the nozzle — controls the flow. If the nozzle has a lower flow rating than the eductor, the eductor cannot pick up concentrate. An example of this situation would be a 60 gpm (227 L/min) nozzle with a 95 gpm (360 L/min) eductor. Using a nozzle with a higher rating than the eductor also gives poor results. A 125 gpm (473 L/min) nozzle used with a 95 gpm (360 L/min) eductor results in improper eduction of the foam concentrate. Low nozzle inlet pressure, however, results in poor-quality foam because of the lack of proper aeration.

Foam Application Techniques

It is important to use correct techniques when manually applying foam from nozzles. If incorrect techniques are used, the effectiveness of the foam is reduced. The following are standard techniques for applying foam:

- Have an individual (*spotter*) to one side direct the application because the nozzle person may not see where the stream is going.

- Do not plunge the stream directly into the fuel; it may splash and cause the fire to spread. Plunging can also disturb an existing foam blanket and allow vapors to escape, causing reignition.

- Apply sufficient foam to cover the surface of the fire.

- Allow time for fuels to cool.

- Precool with fog spray to reduce flame radiant heat before applying foam if a fire is very hot. Do not apply fog and foam together or where a foam blanket already exists.

- Do not walk through a foam blanket because the surface may release vapors, and reignition may occur.

Some effective applications methods include the following:

- ***Bank-down or bounce-off method.*** Bank or deflect the foam off a bulkhead or other obstruction. Allow the foam to run down onto the surface of the fuel. It may be necessary to direct the stream off various points around the fuel area to achieve total coverage and extinguishment. Take particular care to apply the foam as gently as possible (Figure 6.29).

Figure 6.28 A foam nozzle-eductor combination in an engine room. *Courtesy of R. Wright/Maryland Fire and Rescue Institute/United States Coast Guard.*

Figure 6.29 Bank-down foam application method.

- **_Roll-on method._** Direct the foam stream on the deck near the front edge of a burning fuel, and allow the foam to accumulate and roll across the fuel's surface. Continue applying foam until it spreads across the entire surface of the fuel and extinguishes the fire (Figure 6.30).

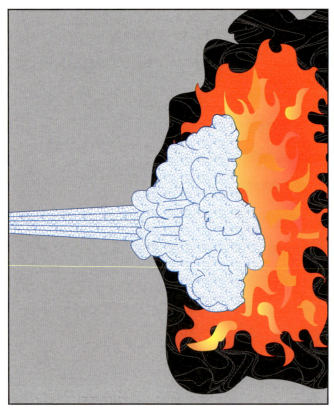

Figure 6.30 Roll-on foam application method.

- **_Rain-down method._** Direct the foam nozzle into the air above the fire, and allow the stream to reach its maximum height and break into small droplets. The foam floats gently down onto the surface of the fuel. Adjust the altitude of the nozzle so that the fallout pattern matches that of the fire or spill area. This method may not be practical or effective in high winds (Figure 6.31).

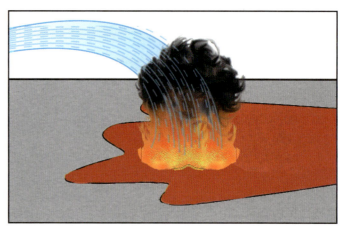

Figure 6.31 Rain-down foam application method.

Step 1: Connect a number of hose sections (check the gaskets before connecting) into test lengths.

Step 2: Connect an open test valve to a deck hydrant. Tighten each connection.

Step 3: Connect a test length to each test valve. Tighten each connection.

Step 4: Tie a rope, hose rope tool, or hose strap to each test length of hose 10 to 15 inches (254 mm to 381 mm) from the test valve connections.

Step 5: Secure the other end to a nearby strong support.

Step 6: Attach a shutoff nozzle (or any device that permits water and air to drain from the hose) to the open end of each test length.

Step 7: Fill each fire hose with water with a pump pressure of 50 psi (345 kPa) {3 bar}.

Step 8: Open the nozzles as the hoses are filling.

Step 9: Hold nozzles above the level of the pump discharge to permit all the air in the hose to discharge.

Step 10: Discharge the water away from the test area.

Step 11: Close the nozzles after all air has been purged from each test length.

Courtesy of R. Wright/Maryland Fire and Rescue Institute/United States Coast Guard.

Step 12: Make a chalk or pencil mark on the hose jackets against each coupling.

Step 13: Check that all hose is free of kinks and twists and that no couplings are leaking.

NOTE: Take any length that is leaking from BEHIND the coupling out of service; repair before testing.

Step 14: Retighten any couplings that are leaking at the connections.

NOTE: If the leak does not stop, depressurize, disconnect the couplings, replace the gasket, and start over at Step 7.

Step 15: Close each hose test gate valve.

Step 16: Increase the pump pressure to a test pressure equivalent to the maximum expected working pressure (but not less than 100 psi [689.5 kPa] {7 bar}).

Step 17: Closely monitor the connections for leaks as the pressure increases.

Step 18: Maintain the test pressure for 5 minutes.

Step 19: Inspect all couplings to check for leaks at the point of attachment.

Step 20: Slowly reduce the pressure after 5 minutes.

Step 21: Close each discharge valve.

Step 22: Turn off the testing machine or pump.

Step 23: Open each nozzle slowly to bleed off pressure in the test lengths.

Step 24: Break all hose connections and drain water from the test area.

Step 25: Inspect the marks placed on the hose at the couplings. If a coupling has moved during the test, tag the hose section for recoupling. Tag all hose that has leaked or failed in any other way.

NOTE: Expect a ¹⁄₁₆- to ⅛-inch (1.6 mm to 3.2 mm) uniform movement of the coupling on newly coupled hose. This slippage is normal during initial testing but should not occur during subsequent tests.

Step 26: Record the test results for each section of hose.

Chapter 7

Fire Detection Systems

The most desirable way to stop fire losses is to prevent fires from occurring in the first place. In a perfect world this would be the case. Unfortunately, we do not live in a perfect world, and despite our best efforts, fires still occur. What we can do in this imperfect world is to detect those fires that do occur early enough to prevent them from becoming uncontrollable. The most efficient means of providing early detection is a well-designed and properly maintained automatic fire detection system. However, manual fire alarm mechanisms also allow crew members to activate alarm systems when fires are discovered. All systems and equipment are designed to withstand temperature changes, vibration, humidity, and other conditions normally encountered in vessels according to regulations from the *International Convention for the Safety of Life at Sea (SOLAS)*. The intent of all fire detection systems is to give early warning of a fire so that enough time is available for persons to safely escape and emergency teams to conduct an effective fire attack.

Many different devices and types of systems that detect fires on board vessels are available such as smoke detectors (ionization, photoelectric, and air-sampling types), heat detectors (rate-of-rise and fixed-temperature types), and flame detectors. Depending on the design of the systems, various combinations of these detection devices may be used in a single device (for example, fixed-temperature/rate-of-rise detectors, heat/smoke detectors, and ionization/photoelectric detectors). These combinations give the benefit of both services and increase the detector's responsiveness to fire conditions.

This chapter describes the automatic fire detection devices and systems most widely used in the marine industry. How alarms are controlled and powered to ensure the devices operate properly is also explained. Other means of achieving early detection such as manual fire alarm mechanisms, fire watch systems, and supervised patrols are discussed. The automatic system operation that some alarm systems provide (such as closing fire doors and smoke dampers) are also explained. Fixed fire suppression systems often have fire detection capabilities. Inspecting, testing, and maintaining fire detection systems and alarm system control units are also important functions.

Smoke Detection

The fact that virtually every fire produces smoke is what makes a smoke detector one of the most dependable warning devices available. It takes only a small amount of the products of combustion to enter the device and trigger an alarm. It does not have to wait for heat generation. Many types of smoke detectors are available. The most common are the ionization and photoelectric detector types and the air-sampling detection system that brings the air sample to the detector from a remote location such as a cargo hold.

Ionization Smoke Detector

During a fire, molecules ionize (lose electrons) as they undergo combustion. The ionized molecules have an electron imbalance and tend to steal electrons from other molecules. These particles cannot be seen by the naked eye, but devices can detect them as they enter a chamber within the device. The *ionization smoke detector*, sometimes referred to as the *invisible products-of-combustion smoke detector*, uses this phenomenon (Figure 7.1). The detector has a sensing chamber that samples the air in the protected space. Two electrical plates are inside the chamber; one is positively charged, and one is negatively charged. A small amount of radioactive material (usually americium) adjacent to the opening of the chamber ionizes the negative plate, and the electrons travel to the positive plate. Thus, a minute current normally flows between the two plates. When ionized products of combustion enter the chamber, they pick up the electrons freed by the radioactive ionization. The current between the plates ceases, and an alarm signal is initiated. An ionization detector works satisfactorily on all types of fires; however, it generally responds more quickly to flaming fires than to smoldering ones. The ionization detector automatically resets when the atmosphere is clear.

Photoelectric Smoke Detector

A *photoelectric smoke detector,* sometimes called a *visible products-of-combustion smoke detector,* uses a photoelectric cell coupled with a specific light source. The photoelectric cell detects smoke in one of two ways: beam application or refractory application. A photoelectric smoke detector works satisfactorily on all types of fires; however, it generally responds more quickly to smoldering fires than an ionization smoke detector. The photoelectric smoke detector automatically resets.

The *beam-application* method uses a beam of light focused across the area being monitored onto a photoelectric cell. The cell constantly converts the beam into electrical current, which keeps a switch open. When smoke blocks the path of the light beam, the current is no longer produced, the switch closes, and an alarm signal is initiated (Figure 7.2).

The *refractory* photocell uses a light beam that passes through a small chamber at a point away from the light source. Normally, the light does not strike the photocell, and no current is produced. When no current is produced, a switch in the current remains open. When smoke enters the chamber, it causes the light beam to refract (scatter) in random directions. A portion of the scattered light strikes the photocell, causing current to flow. This current closes the switch and activates the alarm signal (Figure 7.3).

Air-Sampling Detection System

In some applications on board a vessel, smoke detectors are not located directly in the protected space. An air-sampling system is employed instead. In these instances, air is continuously sampled and channeled from the protected space to a cabinet located on the bridge. During this process, a portion of the sample is routed through tubing to one of the heat detectors described earlier, typically the photoelectric smoke detector. The function of the particular device itself, however, is the same as if it were located in the pro-

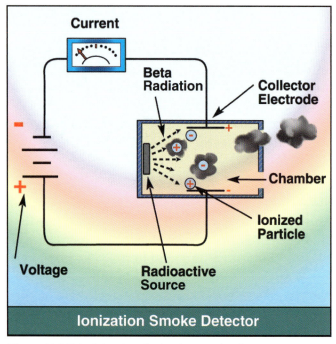

Figure 7.1 Principle of an ionization smoke detector.

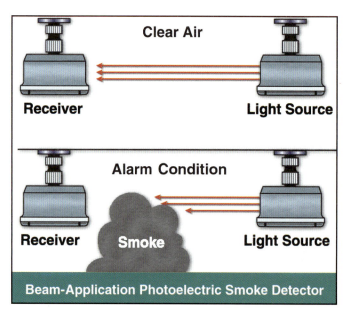

Figure 7.2 Principle of a beam-application photoelectric smoke detector.

tected space. The air-sampling method of protection is usually used in cargo holds, rather than accommodation spaces or other areas where crew members are usually stationed. Similar use of smoke detectors is achieved by placing detectors or sampling tubes that lead to detectors inside air-handling ducts. This method provides an extra degree of protection because the duct detector is usually equipped to provide shutdown of the air-handling equipment to prevent the spread of smoke.

◆ Heat Detection

Three primary principles of physics explain how heat is detectable by certain devices: (1) Heat causes an expansion action of material, (2) Heat causes a melting action of material, and (3) Heated materials have

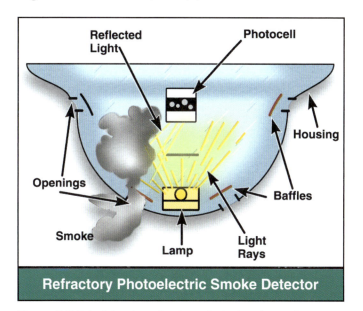

Refractory Photoelectric Smoke Detector

Figure 7.3 Principle of a refractory photoelectric smoke detector.

thermoelectric properties that are detectable. All heat-detection devices thus operate on one or more of these principles. For maximum effectiveness, place heat detectors near the highest point of the space being protected. To prevent false activations, select a detector at an activation temperature rating that gives a reasonable margin of cushion above the normal deckhead temperatures that are expected in that area. Heat detection is available in two basic types of detectors: *fixed-temperature* (responds when temperature reaches a predetermined level) and *rate-of-rise* (responds when temperature increases quickly).

Fixed-Temperature Heat Detectors

Fixed-temperature heat detectors initiate an alarm when the temperature of the protected space reaches a predetermined level. Most heat detectors are the *spot* detector type; that is, they detect the conditions only in a relatively small area surrounding the specific place where they are located. Some types detect heat over larger areas. Several means of activating these devices are available: *fusible devices, frangible/quartzoid bulbs, bimetallic,* and *continuous-line.*

Fixed-temperature devices are rated at various temperature ranges, depending on the normal ambient temperature range of a given space. Keep a supply of spare detectors of each temperature range so that replacements have the same temperature-range rating. These ranges are the expected normal temperatures of the space. The actual activation temperature range of the device for each classification of detector is given in Table 7.1. Detectors are classified as ordinary, intermediate, or hard, depending on the normal temperatures at the device.

Table 7.1
Limits of Rated Temperature for Fixed-Temperature Detector Operation

Normal Ambient Temperature Range (Degrees)		Detector Rating	Temperature Limits for Detector (Degrees)	
Exceeds	**Does Not Exceed**		**Maximum**	**Minimum**
—	100°F (38°C)	Ordinary	165°F (74°C)	135°F (57°C)
100°F (38°C)	150°F (66°C)	Intermediate	225°F (107°C)	175°F (79°C)
150°F (66°C)	225°F (107°C)	Hard	300°F (149°C)	250°F (121°C)

See NFPA 72, *National Fire Alarm Code®,* and UL 521, *Standard for Heat Detectors for Fire Protective Signaling Systems,* for more details. Information may also be found in Title 46 CFR 161 Electrical Equipment.

Fusible Devices and Frangible/Quartzoid Bulbs

A fusible device is normally held in place by a solder with a known melting (fusing) temperature. The fusible device holds a spring-operated contact device inside the detector in the open position (Figure 7.4). When the melting point of the fusible device is reached, the solder melts, causing the spring to release and touch an electrical contact that completes the circuit and sends an alarm signal. Replacing the fusible device restores some of these detectors; others require replacing the entire heat detector.

Frangible/quartzoid bulbs contain liquids that expand when heated. A bulb is inserted into the detection device to hold two electrical contacts apart, much like the fusible device. Once the liquid reaches a vapor pressure that exceeds the strength of the bulb, the bulb breaks. The contacts complete the circuit to send the alarm. A frangible/quartzoid bulb fails at a specific temperature. To restore the system, the entire detector unit must be replaced. Replacements must have the same temperature ratings.

Bimetallic Detector

A *bimetallic detector* uses two types of metal that have different heat-expansion ratios. In one type of detector, these metals are each formed into thin strips that are then bonded together. The fact that one metal expands faster than the other causes the combined strip to distort when it is subjected to heat, resulting in

an arched shape. The size of the arch depends on the characteristics of the metals and the amount of heat to which they are subjected. These factors are calculated into the design of the detector. The bimetallic strip is positioned with either one or both ends secured in the device. In either case, the detector uses the bowing action of the strip to either open or close a set of electrical contacts, initiating the alarm.

Another type of bimetallic device is the *snap-action disk* that works on the same principle as the bimetallic strip but employs a distinct snapping action of the disk to put pressure on the electrical contacts to initiate the alarm (Figure 7.5). When installed in the detector, the disk's natural shape is concave on the top. As the temperature increases, the top half of the disk expands faster than the bottom half. At the activation temperature, this distortion of the disk causes it to snap upward into a convex shape and strike the electrical contacts.

Most bimetallic detectors reset automatically when cool. However, they require testing after activation to ensure that they function properly and have not been damaged.

Continuous-Line Detector

The *continuous-line* detection device detects heat over a linear area parallel to the detection device. It is

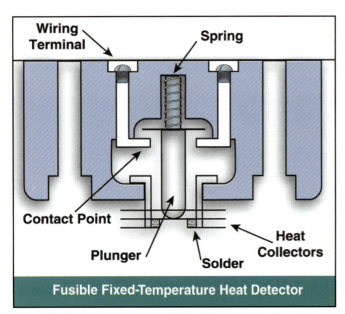

Figure 7.4 Cutaway of a fusible fixed-temperature heat detector.

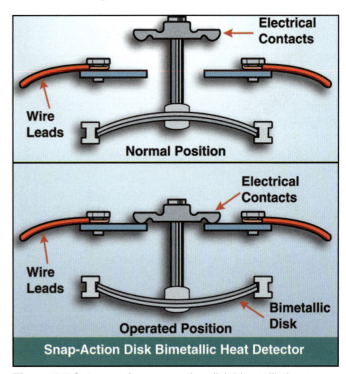

Figure 7.5 Cutaway of a snap-action disk bimetallic heat detector.

useful and cost-effective in large open areas that would require a large number of spot detectors to adequately protect the area. A disadvantage of continuous-line detectors is that they do not indicate the precise location of a fire as do spot detectors.

One type of continuous-line detection device, known as a *resistance detector,* consists of a conductive metal inner-core cable that is sheathed within stainless steel tubing (Figure 7.6). These cables can be strung over extremely large areas. An electrically insulating semiconductor material separates the inner core and the sheath and keeps them from touching, but it allows a small amount of current to flow between the two. This insulation loses some of its electrical resistance capabilities at a predetermined temperature. When the heat at any given point anywhere along the line reaches the resistance reduction point of the insulation, the current between the two components increases. This increase in current sends an alarm signal to the alarm system control unit. This detection device restores itself when the heat level lowers.

The *thermostatic cable* type uses two wires that are each insulated and bundled within an outer covering. When the melting temperature of each wire's insulation is reached, the insulation melts and allows the two wires to touch. This completes the circuit to send an alarm signal to the alarm system control unit (Figure 7.7). To restore this type of line detector, remove the fused portion of the wires and replace with new wire.

Still another type of continuous-line detector, the *fire detection loop system,* often found on offshore drilling platforms, is made with a flexible, plastic tube running throughout the platform. The tube is connected to pressure switches and

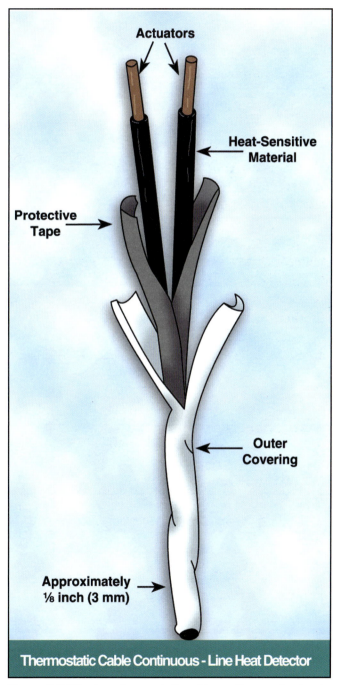

Figure 7.7 Wire-type thermostatic cable continuous-line heat detector.

Resistance Continuous-Line Heat Detector

- Center Conductor
- Insulating Semiconductor Material
- Stainless Steel Tubing

Figure 7.6 Tubing-type resistance continuous-line heat detector.

electrical circuits and is pressurized with air. When heat from a fire melts the plastic, the air pressure escapes, thus releasing the pressure switches and initiating the alarm (Figure 7.8). A variation on this type of detection device consists of stainless steel tubing with fusible caps installed at intervals. When high temperatures melt the caps, pressure releases from the tubing, allowing pressure switches to initiate the alarm. In both of these cases, replace the segment of tubing damaged by a fire to restore operation of the entire length of tubing.

Rate-of-Rise Heat Detectors

Fire rapidly increases the temperature in a given area, and the *rate-of-rise heat detector* senses these quick increases in temperature. These detectors respond at substantially lower temperatures than the fixed-temperature heat detectors. Typically, the rate-of-rise heat detectors send a signal when the rise in temperature exceeds 12°F to 15°F (7°C to 8°C)

per minute because temperature changes of this magnitude are not expected under normal, nonfire circumstances.

Most rate-of-rise heat detectors are reliable and not subject to false activations. However, they can occasionally activate under nonfire conditions. An example would be when a rate-of-rise heat detector is placed near the discharge of a heating unit. When the heating system begins to discharge warm air, the temperature in the immediate area can rise dramatically, causing a nearby detector to activate. Avoid this situation by placing the detector in a more appropriate location. Installing a rate-compensated detector in an area where regular temperature changes occur is another option (see Rate-Compensated Detector section).

Several different types of rate-of-rise heat detectors are available such as the *pneumatic rate-of-rise spot detector, pneumatic rate-of-rise line detector,*

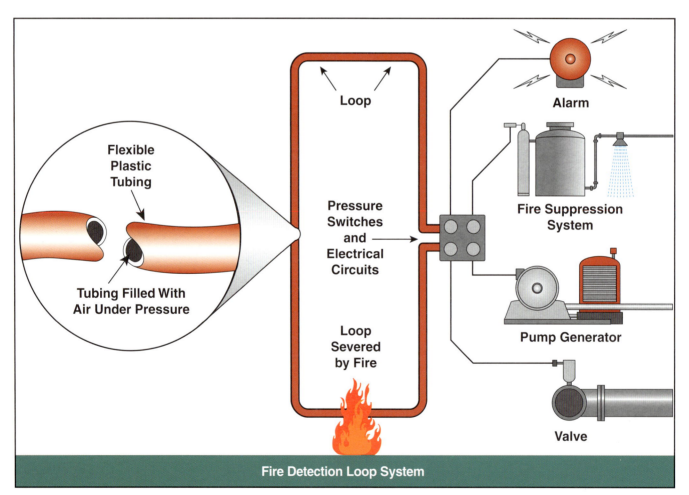

Figure 7.8 Fire detection loop system. The loop is flexible plastic tubing whose hollow core is filled with air under pressure. When the tubing is severed, the pressure is lost. The reduced pressure allows switches to activate emergency equipment, including alarms, suppression systems, generators, and control valves.

rate-compensated detector, and *thermoelectric detector.* All rate-of-rise detectors reset automatically. These detectors are discussed in more detail in the following sections.

Pneumatic Rate-of-Rise Spot Detector

A pneumatic spot detector monitors an exact area and is the most common type of rate-of-rise heat detector. It contains a small chamber filled with air and has a flexible metal diaphragm in the base (Figures 7.9 a and b). As the temperature rises, air inside the chamber expands and the diaphragm is forced out to a predetermined level. Depending on the design of the particular detector, this action causes a set of electrical contacts to either open or close, thus sending an alarm signal to the alarm system control unit. The air chambers in these detectors contain vents to prevent activation caused by normal changes in ambient temperature or changes in barometric pressure. The vents allow air to slowly enter or exit the chamber in response to slow changes in temperature or barometric pressure. In a fire situation, however, pressure builds due to rapid heating, and the air cannot escape as quickly, thus initiating the alarm.

Pneumatic Rate-of-Rise Line Detector

The pneumatic rate-of-rise line detector is similar to the pneumatic rate-of-rise spot detector except it monitors large areas with a system of tubing arranged over the area of coverage (Figure 7.10). The space inside the tubing acts like the chamber in the pneumatic rate-of-rise spot detector. The pneumatic rate-of-rise line detector also contains a diaphragm and air vents. When any area being served by the tubing experiences a temperature increase, the detector functions in the same manner as the pneumatic rate-of-rise spot detector. The tubing in these systems is limited to about 1,000 feet (305 m) in length and is arranged in rows that are not more than 30 feet (9 m) apart and 15 feet (4.6 m) from bulkheads.

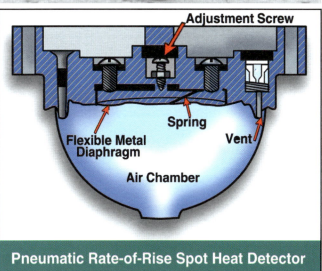

Pneumatic Rate-of-Rise Spot Heat Detector

Figures 7.9 a and b (a) A pneumatic rate-of-rise spot heat detector. (b) A pneumatic rate-of-rise spot heat detector uses a pocket of expanding air to detect fire. The pocket is vented to allow the detector to adjust to moderate temperature and barometric changes.

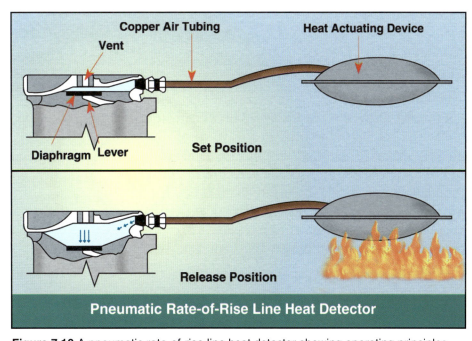

Pneumatic Rate-of-Rise Line Heat Detector

Figure 7.10 A pneumatic rate-of-rise line heat detector showing operating principles. The system is equipped with large air chambers that increase the sensitivity of the detection system.

Rate-Compensated Detector

The *rate-compensated detector* is used in areas that are subject to regular temperature changes under normal conditions (slower than fire conditions). It contains an outer bimetallic sleeve (shell) with a moderate expansion rate. This outer sleeve contains two bowed struts that have a slower expansion rate than the sleeve (Figure 7.11). The bowed struts have electrical contacts on them. In the normal position, these contacts do not come together. When the detector is heated rapidly, the outer sleeve expands in length. This expansion reduces the tension on the inner strips and allows the contacts to come together, thus sending an alarm signal to the alarm system control unit. If the rate of temperature rise is fairly slow, such as 5°F (2°C to 3°C) per minute, the sleeve expands at a slow rate that maintains tension on the inner strips. This tension prevents unnecessary system activations.

Thermoelectric Detector

The *thermoelectric detector* contains two wires made of dissimilar metals that are twisted together. When the wires are heated at one end, an electrical current is generated at the other. The rate at which the wires are heated determines the amount of current generated. These detectors dissipate small amounts of current, which reduce the chance of a small temperature change activating an alarm. Large changes in temperature result in larger amounts of current flowing and activation of the alarm system.

 ## Flame Detection

A *flame detector* is sometimes called a *light detector*. Three basic types exist: UV detectors (which detect light in the ultraviolet wave spectrum), IR detectors (which detect light in the infrared wave spectrum), and those that detect light in both ultraviolet *and* infrared wave spectrums. Flame detectors are often found on board vessels and in aircraft hangars. They are also used as negative detectors in boilers to prevent explosion in case of flameout. Flame detectors are used in addition to smoke and heat detectors.

While these types of detectors are among the fastest to respond to fires, they are also easily tripped by nonfire conditions such as welding and other bright light sources. Place them in areas where these possibilities are avoided. Position these detectors so that they have an unobstructed view of the protected area.

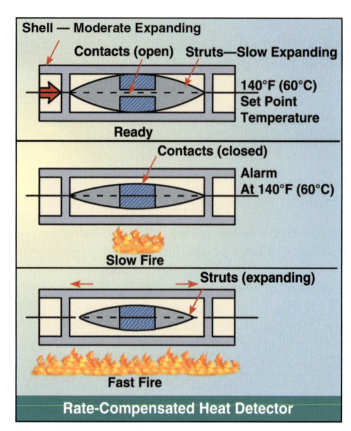

Figure 7.11 Cutaway of a rate-compensated heat detector.

If their lines of sight are blocked, they cannot activate. An IR detector is effective in monitoring large areas. To prevent activation from infrared light sources other than fires, an IR detector requires the flickering action of a flame before it sends an alarm.

 ## Other Fire Detection Methods

All the detection devices discussed so far are the automatic type; that is, they work independently of human actions. Most fixed fire suppression systems also have automatic detection devices that can detect a fire and then activate the system. Other devices are manually operated so that a person can quickly activate an alarm system when a fire is discovered. Watch and patrol systems place trained individuals in various locations throughout a vessel to look for signs of fire or other dangerous conditions. These various methods are described in the sections that follow.

Detection Capability of Fixed Fire Suppression Systems

Fixed fire suppression systems such as galley systems and sprinkler systems are often designed to provide a means of detecting a fire and then activating the system. For example, once a sprinkler opens in response to a fire, water begins to flow through a system of piping. At least one point in the system is equipped with water flow detecting devices that are electronically wired into a complete alarm system. All the features of fixed fire suppression systems are discussed in more detail in Chapter 8, Fixed Fire Suppression Systems.

Manual Fire Alarm Mechanisms

An important component of fire alarm systems involves providing a means for crew members or passengers to quickly activate an alarm when they discover a fire. These alarm devices, *manual pull stations* (also known as *manual alarm boxes* or *fire alarm push buttons*), are located in passageways, stairway enclosures, and public spaces. A manual pull station can often give the earliest report of a fire before any of the automatic detection devices have time to operate.

While pull station boxes come in a variety of shapes and sizes, they are distinctively red in color and usually have white lettering that clearly specifies their purpose and how to use them. A pull station box is used for fire signaling purposes only. The pull station box is mounted on either bulkheads or vertical supports where it is plainly visible so that its bottom is no less than 3½ feet (1.1 m) and no more than 5 feet (1.5 m) from the deck. At least one pull station box is located on each deck. In all cases, a person travels no more then 200 feet (61 m) to reach a pull station box. The ideal location for pull stations is at exits so that people can activate them as they escape from a fire area.

Watch and Supervised Patrol Systems

The watch and supervised patrol systems provide different guards to look for fires on a vessel. Under the *watch system,* a *sentry* person is assigned to a designated portion of the vessel. The *supervised patrol,* conversely, provides for people who move about, watching for signs of fire or other dangerous conditions along a predetermined route. A patrol makes a *round* within a certain time frame and visits all checkpoints or stations during each round. A fire watch or patrol is also posted to spaces when automatic detection systems are disabled and after fires are extinguished to watch for possible reignition.

Passenger vessels are required to provide supervised patrols between the hours of 10 p.m. and 6 a.m. when passengers are aboard. The patrol tours every part of the vessel accessible to passengers and crew members with the exception of occupied sleeping accommodations. The engine room or other machinery spaces require a full-time watch and are exempt from the patrol requirement. Patrols are required to check in with the deck officer of the watch (OOW) at regular intervals, at least once per hour.

Even on nonpassenger vessels, it is essential that the deckhand on watch or rating (unlicensed person) on duty perform patrols during the night watches (8 p.m. to 6 a.m.). The OOW performs a fire patrol after each watch by turning over responsibilities to a relief person, touring the accommodation areas, and then phoning the bridge to report conditions.

A uniform or special badge easily identifies a patrol. The patrol carries a flashlight, keys to any secured areas that may need entering, and a notebook for recording observations (Figure 7.12). Other equipment includes a two-way radio for direct communication with the bridge, a whistle to alert passengers and/or crew members, and a folding utility tool that serves

Figure 7.12 Fire patrol checking an air-sampling system cabinet on the bridge. *Courtesy of R. Wright/ Maryland Fire and Rescue Institute/United States Coast Guard.*

as pliers and screwdriver. Patrol recording methods (mechanical and electronic systems) verify that patrols made their rounds. Watch and patrol duties and recording methods are described in the sections that follow.

Duties

Those on watches and patrols are held to very rigid standards when performing their duties. First and foremost, their responsibility is to activate the alarm system when discovering a fire or even just suspecting a fire because of smoke or an odor of possible smoke. Before taking any other actions, their first objective is to report the fire by pulling the nearest manual fire alarm or calling the bridge. Taking time to investigate a fire before reporting it could let a fire grow from one that is easily controlled to one that is fully developed. After transmitting the alarm, a patrol can perform any of the following duties if possible:

- Wake passengers and crew members in the area.

- Close doors or other barriers to prevent spread of the fire.

- Use portable fire extinguishers if practical.

- Report conditions/observations to the emergency response team.

- Report to assigned muster station for further instructions.

For the purposes of investigating a fire and/or establishing the cause, the watch or patrol who first discovers a fire is a very valuable asset. A watch or patrol is trained to have keen observation skills. The person records the following information as soon as practical (while the information is fresh):

- Time the fire was discovered

- Exact location where fire or smoke was seen

- What doors were open or closed

- Who (if anyone) was in the area prior to the discovery

- Condition of any fire extinguisher used

- Other conditions or circumstances

Patrol Recording

Using a recording device (either mechanical or electronic) ensures that patrols make their assigned rounds. In cases where a vessel is not equipped with recording mechanisms, on-duty patrols report to the bridge in person at regular intervals not to exceed one hour apart. When there are multiple patrol routes, one patrol may report on behalf of others after first establishing direct contact with them.

With the *mechanical system,* a patrol carries a special portable clock. A special key at each station along the patrol route is inserted into the clock to register the patrol's presence. The time and location of the visit is recorded. The clock is sealed to keep the permanent record intact. This record is examined regularly to ensure that the rounds are conducted within the established time frame.

The *electronic system* works somewhat the opposite of the mechanical system. The patrol carries a special key that is inserted into a registering mechanism at each key station along the patrol route. Inserting the key into the device sends a signal to a central station where it is recorded. If no signal is received within an allowable time period, an officer investigates. This method ensures that a patrol is making rounds by maintaining contact with the bridge electronically.

Alarm System Control Units

Detection hardware and methods are useless if they are not connected to some sort of "brain" or control system that processes signals and makes the devices operate properly. The *detection and alarm system control unit,* also called the *fire alarm panel* or *annunciator panel,* is located on the bridge (Figure 7.13). All controls for the system are located in the main system control unit. A secondary or remote repeater panel is located in the engine room on most vessels and in selected officers' quarters, offices, or the fire control room. Power supplies, alarm-indicating devices, and visual indicators are parts of the alarm control system. The alarm system control unit is responsible for performing the following actions:

- Processing alarm signals from actuating devices

- Indicating a particular zone or space in which the activation occurred

- Indicating any problems or trouble with the system (or a portion of the system)

- Sounding the appropriate warning devices

Figure 7.13 Detection and alarm system control unit (annunciator panel). *Courtesy of R. Wright/ Maryland Fire and Rescue Institute/United States Coast Guard.*

Power Supplies

A separate branch circuit of the vessel's main electrical system usually provides the primary electrical power supply for the alarm system. A secondary or emergency power supply also ensures that the system is operational even if the main power supply fails. The secondary system must make the detection and signaling system fully operational within 30 seconds after the failure of the main power supply. The secondary power source comes from either a separate branch circuit of the vessel's emergency power and lighting switchboard or from a separate set of storage batteries. When power is switched from the main power supply to the secondary one, both audible and visual alarms are activated.

Alarm-Indicating Devices

When an alarm-initiating device (detector) activates and sends a signal to the alarm system control unit, the control unit processes that signal and takes the appropriate action (Figure 7.14). This action usually includes the sounding and lighting of alarm indicators on the bridge, in the engine room, and at local alarms in the area of the suspected fire. Local alarm devices include bells, buzzers, horns, strobe lights, and other warning lights. The watch officer on the bridge (or the senior officer in the engine room if the vessel is undergoing repair in port) is responsible for investigating the cause of an alarm and/or sounding the general alarm as appropriate (Figure 7.15).

Figure 7.14 Smoke detection system. *Courtesy of R. Wright/ Maryland Fire and Rescue Institute/United States Coast Guard.*

Figure 7.15 General alarm bell. *Courtesy of Maritime Institute of Technology and Graduate Studies.*

Visual Indicators

Visual indicators fulfill two purposes: (1) alert the hearing impaired or those in a noisy environment of a fire alarm and (2) identify the location of the alarm on the alarm panel. For alerting purposes, strobe indicators or some other type of flashing light usually emanate from a device marked FIRE. An alarm panel, on the other hand, has a multitude of small lamps to indicate different zones, whether the problem is a fire alarm or a trouble alarm, whether the system has an adequate source of power, and the like.

Automatic System Operations

Another tremendous benefit of fire alarm systems is that they can be designed to perform other important operations automatically—operations that someone would ordinarily have to do manually in order to gain control of a fire. Do not confuse these automatic fire-emergency features with other automatic features that control other situations such as engine overspeed or engine emergencies. Automatic fire-emergency system operations can perform the following actions:

- Shut down air-handling equipment or reverse the flow of air.

- Close fire doors and/or smoke or fire dampers.

- Operate air vents and mechanical space ventilation systems.

- Pressurize stairwells for evacuation purposes.

- Override control of elevators to prevent them from opening at the deck where fire is suspected.

- Return elevators automatically to the main deck.

- Operate heat and smoke vents.

- Prepare a fixed fire suppression system for activation.

- Activate fixed fire suppression systems such as automatic sprinkler systems (See Chapter 8, Fixed Fire Suppression Systems).

- Shut down engines or other systems.

 ## Inspection, Testing, and Maintenance

In order to ensure operational readiness and proper performance, fire detection and signaling systems are tested both when they are installed *(acceptance testing)* and then on a continuing basis *(service testing)*. In many cases, crew members are assigned the responsibility of inspecting systems, performing system tests, and performing maintenance. Crew members who routinely conduct inspections need a working knowledge of these systems. Ideally, crew members are trained by representatives of the system's manufacturer. The engineering department has the technical manuals and parts lists that assist in inspecting, testing, and maintaining the systems. Maintaining complete records of these procedures is also necessary.

Regulatory agencies may board a vessel in ports under their control and either conduct random testing of systems or ask to see the inspection/testing/maintenance records. The following components are inspected/tested regardless of who is conducting the inspection/test:

- *Wiring* — Look for proper support, wear, damage, or any other defects that may render the insulation ineffective.

- *Conduit* — Look for solid connections and proper support where circuits are enclosed in conduit.

- *Batteries* — Check for clean contacts and proper charge, particularly when they are designated as emergency power sources. Many batteries have floating-ball indicators that show whether the battery is properly charged.

All equipment, especially initiating and indicating devices, must be kept free of dust, dirt, paint, and other foreign materials. When either dust or dirt is found, clean with a vacuum cleaner rather than by wiping. Wiping tends to spread the debris around, causing it to settle on electrical contacts, which may inhibit the future operation of the system.

Obviously, detection systems (whether manual or automatic) will not work unless the alarm-initiation devices are in proper working order and send the appropriate signal to the alarm system control unit. The following sections highlight the procedures for inspecting and maintaining these devices and alarm system control units.

Alarm-Initiating Devices

Without functional detection devices, the most elaborate wiring and signaling systems are virtually useless. The reliability of the entire system is, in fact, based mostly on the reliability of the detection devices regardless of whether they are manual or automatic.

Numerous items are checked when testing and inspecting a manual alarm-initiating device. Access to the device must be unobstructed, and each unit must be easy to operate. The housing must be tightly closed to prevent dust and moisture from entering the unit and disrupting service. If the device is equipped with a cover or door, make sure that it opens easily and that all the components behind the door are in place and ready for service. Inspectors may wish to activate selected devices to ensure that the device and the system are operational.

Test automatic alarm-initiating devices after installation, after a fire, and periodically based on regulations of the authority having jurisdiction. Maintain a record of all tests. The minimum information included is the date, alarm-initiating type, location, type of test, and results of the test. Replace the following detectors or send them to a recognized testing laboratory for testing in the following situations:

- When systems are being restored to service after a period of disuse

- When obviously corroded, mechanically damaged, or abused

- When painted, even if attempts were made to clean them

- Where circuits were subjected to current surges, overvoltages, or other electrical disturbances

- When subjected to foreign substances that might affect their operation

Various detectors are tested in various ways. Some of these detector tests are described as follows:

- *Nonrestorable fixed-temperature detector* — Obviously, this detector cannot be tested periodically. Fifteen years after installation, remove 2 percent of the detectors and replace with new detectors. Send the old detectors to an approved testing laboratory for testing. If one detector fails, remove several other detectors and send them for testing.

- *Line detector* — Perform resistance testing semiannually.

- *Restorable heat detection device* — Test with heat from a hair dryer, heat lamp, or specialized testing appliance. Never use an open flame. Test one detector on each signal circuit semiannually. Select a different detector each time.

- *Fusible-device detector with replaceable link* — Test semiannually. Remove the link, and observe whether or not the contacts close. Replace the fusible link after the test. Replace fusible links with new ones at 5-year intervals.

- *Pneumatic detector* — Test semiannually with a heating device or a pressure pump. Follow the manufacturer's instructions closely if a pressure pump is used.

- *Smoke detector* — Test semiannually in accordance with manufacturer's recommendations. The instruments required to perform performance and sensitivity testing are usually provided by the manufacturer. Perform sensitivity testing after the detector's first year of service and every 2 years after that. Blowing cigarette smoke into the detector is not an acceptable method of testing.

Alarm System Control Units

Do not store objects in or around alarm system control panels, recording instruments, and other devices. Many alarm system control panels have locking doors with storage areas for extra relays, lightbulbs, and test equipment. If this space is not designed into the unit, store these devices elsewhere. Otherwise, they may foul moving parts or cause electrical shorts that can result in system failure.

Inspect and test the alarm system control unit to ensure that all parts are operating properly: All switches perform their intended functions, and all indicators light or sound on demand. Trigger individual detectors to see whether the alarm system control unit sends the appropriate signal and the warning lamps light properly. Report any failure during the tests to the persons responsible for repairing the system. Check auxiliary devices (local evacuation alarms, air-handling-system shutdown controls, fire dampers, and the like) also at this time. Restore all devices to proper operation after testing.

Chapter 8

Fixed Fire Suppression Systems

The fire alarm panel on the bridge captures everyone's attention — lights are blinking and alarms are buzzing. An automatic detection system indicates fire and smoke in an engineering space. If a fire cannot be easily extinguished with portable or semiportable extinguishers, a fixed fire extinguishment system may be the next step to saving the vessel. Fixed fire suppression systems, like fire detection systems, come in a large variety of types and configurations for different uses and to protect various areas of the vessel.

When to use a fixed fire suppression system is an important decision that the designated officer in charge of fire control must make after becoming well informed of the situation and its surrounding circumstances. For instance, it would be poor judgment to discharge the vessel's carbon dioxide (CO_2) fire suppression system for a machinery space fire before the space is fully evacuated and all ventilation systems to the space are secure. Besides endangering crew members, the vessel's entire supply of CO_2 agent could be wasted, leaving fire fighting crews without an important major weapon to use in controlling the fire. Likewise, an officer would not want to activate that same CO_2 system into the generator room, only to discover that the fire location was not confirmed and the fire was actually in the engine room.

This chapter discusses the various types of fixed fire suppression system found on board vessels: carbon dioxide, halogenated agent, steam smothering, chemical agent (dry and wet), water-based (automatic sprinkler, water spray, and water mist), fire main, and foam (machinery space and deck). Inert gas systems

on tank vessels, which are designed to *prevent* fires, are also discussed. A special system described is the galley ventilator washdown system that also prevents fires by keeping ductwork clean. In the event of a galley fire, it can also prevent fire from spreading through the ductwork to another location. The following sections describe the components, features, indicated uses, operating instructions, cautions, and other pertinent information about these shipboard fixed fire suppression systems.

◆ Carbon Dioxide Suppression Systems

Carbon dioxide systems are widely used for shipboard application in controlling fires. The two main types of CO_2 systems are the *local application type* (protects particular pieces of equipment and machinery) and the *total flooding type* (protects an entire space). The total flooding type is most commonly used to protect engine rooms, other machinery spaces, and cargo holds. Local application systems or independent (freestanding) systems are also used for small specific hazard spaces such as paint lockers and lamp lockers. Cargo-hold systems are provided on some types of cargo vessels. Some CO_2 systems are large enough to require bulk storage.

Carbon dioxide is colorless, odorless, and inert. It discharges as a liquid under pressure. The basic components of all CO_2 systems are piping, storage cylinders, discharge heads, and control mechanisms. The high-pressure storage cylinders are connected to the system through a manifold. All systems are capable of operating independently of electricity or other external power source. The agent itself powers the required discharge and predischarge alarm devices and in some cases operates pressure switches that ensure ventilation of the space. Total flooding systems used in large spaces on board vessels require manual activation. It is permissible and recommended that the freestanding systems protecting smaller spaces be capable of automatic operation — controlled by heat-sensing devices within the space — in addition to the manual capability.

Although CO_2 is most effective against flammable liquids, it is also useful for controlling fires involving Class A fuels. A minimum *soaking time* (time required for a space to cool) allows for complete extinguishment of smoldering Class A materials. Spaces that involve excessively heated surfaces must also be kept tightly closed for an extended period to allow possible reignition sources to cool. Because leaks may occur in these situations, intermittent supplementation of CO_2 may be required to maintain the desired concentration within the space. Unsecured openings are the most common reasons for CO_2 system failure.

Carbon dioxide concentrations as low as 9 percent can cause unconsciousness in humans after a few minutes of exposure (see *Fire Protection Handbook*, 18th edition, National Fire Protection Association, for details). Prolonged exposure to a 9-percent concen-

tration results in death. Even in smaller concentrations, inhalation of CO_2 affects the capability of blood to receive oxygen from the lungs. Thus it is a life-threatening situation *anytime* CO_2 is discharged into a space. Any situation/space involving an elevated CO_2 concentration is unsafe to enter without wearing self-contained breathing apparatus (SCBA). Many fatalities occur when individuals believe they can safely enter an oxygen-depleted or otherwise hazardous atmosphere by holding their breaths long enough to rescue another person. A hazardous atmosphere to one person is hazardous to another!

The specific characteristics of the different types of systems are described in the sections that follow. Along with descriptions of the different system types, discussions are given on warning alarms, maintaining adequate carbon dioxide concentrations, reentering spaces, and emergency actions. Crew members also need to know proper inspection and maintenance procedures for CO_2 systems and the advantages and disadvantages of CO_2.

 Advantages of Carbon Dioxide

♦ Clean agent; leaves no residue that could damage equipment or cargo

♦ Noncorrosive

♦ Nonconductive; acceptable for use on electrical fires

♦ Stores easily as a liquid under pressure

♦ Maintains quality or extinguishing capability during long storage

 Disadvantages of Carbon Dioxide

♦ Extremely cold discharge; little cooling effect on fire

♦ Frost-producing nature of discharge potentially damaging to sensitive electronic equipment

♦ Limited amount carried on vessels

♦ Hazardous to humans; even fatal in concentrations

♦ Ineffective on materials that generate their own oxygen; for example, reactive metals or metal hydrides

♦ Reignition possible; smothering atmosphere dissipates, leaving smoldering embers or hot surfaces

♦ Initial cloud of agent hinders visibility

Total Flooding System

In a total flooding fixed system, CO_2 expands to a gas as it contacts the atmosphere, filling the space or area protected by the system. Because CO_2 is heavier than air, it settles to the lowest areas of the space and builds upward until the space is filled. Carbon dioxide does not substantially cool a space that has been heated by a fire; therefore, a minimum concentration of CO_2 must be reached in order to exclude the oxygen and extinguish the fire. An oxygen concentration of 15 percent or lower stops flaming combustion. Most situations require at least a 34 percent atmospheric CO_2 concentration to adequately displace oxygen. The percentage is higher in many cases, depending on the major fuel source of the fire.

Reserve the use of a total flooding system until other means of combating a fire have proven unsuccessful or too hazardous to attempt. However, do not unduly delay using a CO_2 system. The chances of successfully extinguishing a fire with CO_2 diminish in proportion to the time the fire is allowed to burn. Make the decision to employ CO_2 in a timely manner, but not hastily. Once the decision to discharge CO_2 is made, ensure that the space is evacuated, all ventilation is secure, and exhaust dampers are closed before activating the CO_2 system.

A carbon dioxide system for machinery spaces must discharge 85 percent of the total volume needed for the space within two minutes to be effective. This is particularly true in engine rooms and other spaces where flammable liquids are involved. The vessel's main system or a freestanding (independent) system that is dedicated to a particular space may provide CO_2 total flooding for smaller machinery spaces and in paint lockers. See Independent System section.

Total flooding systems on board vessels require a two-step procedure to discharge the system. This procedure is typically achieved by pulling two separate cables that are housed in clearly marked pull boxes that identify their purpose and provide instructions for proper operation (Figure 8.1). Simply opening the box *may* take care of the first step in some installations. To prevent tampering or accidental discharge, the cables require either a double action to pull each one such as releasing a latching device before pulling or breaking a glass to gain access to the pull lever. In the latter case, a small mallet is provided at the pull box to break the glass. In addition, the cables must be pulled out straight in the appropriate sequence in order for the system to discharge. Most regulatory bodies require locating the pull boxes outside the protected area, usually next to a doorway that provides the normal means of escape from the area.

The cables physically operate valves through a series of pulleys and guides between the pull box and the discharge heads of the cylinders. One cable operates the discharge heads on pilot cylinders, while the other operates the control head on the pilot port valve. Once the system activates, carbon dioxide flows to a stop valve that provides a 20-second (minimum) discharge delay. During the delay period, CO_2 pressure in the manifold acts upon switches to turn on discharge alarms and turn off ventilation systems. Discharging the system automatically activates the alarm. After the delay period, CO_2 pressure acts upon the stop valve, causing it to open and allowing discharge of the agent into the protected space.

Large total flooding systems are also activated from the CO_2 room. Whether the system is activated locally — near the protected space — or from the CO_2 room, many variations exist in the activation procedures from one system to another (Figures 8.2 a–d). It is important that crew members know their protected

Figure 8.1 Total flooding carbon dioxide system pull boxes. The pull cables activate the carbon dioxide system. Pull the cables in the proper order as noted in the posted instructions. *Courtesy of Captain John F. Lewis.*

Figures 8.2 a–d (a) *(top left)* Carbon dioxide room. (b) *(lower left)* Carbon dioxide control valve console that directs CO_2 to desired location. (c) *(top right)* Carbon dioxide time delay release. (d) *(lower right)* Key box typically found at emergency lockers and carbon dioxide rooms. *All photos courtesy of Maritime Institute of Technology and Graduate Studies.*

areas, their systems, and the activation procedures on board their vessels.

CAUTION: Before discharging CO_2 into a space, evacuate the space, stop all fans, and close all openings (vents, doors, fiddleys, hatches, etc.). Even if vents are remotely closed, check their actual operation because heat can warp vents and prevent full closure.

Cargo-Hold System

Cargo-hold fixed fire suppression systems are provided on a number of cargo vessels, including container, ro/ro, and break bulk vessels. Vessels carrying both passengers and cargo may be equipped with cargo-hold fixed systems. Tank vessels built before 1962 may also employ cargo-hold fixed systems to protect their cargo tanks. Tank vessels built after 1970, however, are required to have deck foam systems and may also have a water spray or inert gas system protecting the tanks. See Foam Suppression Systems, Water-Based Suppression Systems, and Inert Gas Systems on Tank Vessels sections.

Like a total flooding system, a cargo-hold fixed system is not activated in haste. Take steps first to ensure that the use of the cargo-hold system is necessary. Confirm the fire's presence and location. Then manually seal the hold before activating the system. The cargo-hold system is activated in much the same manner as a machinery-space total flooding system. After sealing the hold, introduce carbon dioxide at the prescribed flow rate, which is determined by the volume of the hold and the fuel burning (with allowances for leakage and other factors). The percentage of CO_2 required to reduce the oxygen content within the space to a level that does not support combustion is engineered into the system and is usually in the range of 30 to 50 percent, although some fuels require up to 75 percent (see Tables 8.1 and 8.2). With some systems, CO_2 continues to flow in order to maintain the desired concentration. Some systems have a table of percentages for each hold giving both the number of supply cylinders needed for initial discharge and the number of cylinders to replenish agent lost through leakage or heat dissipation. In some cases, agent is added every half hour until extinguishment is complete.

Some lighter aboard ship (LASH) vessels and barge-carrying vessels are equipped with special cargo-hold CO_2 systems that work in conjunction with independent smoke detection systems. When a fire is detected on a LASH vessel, CO_2 is released into the entire hold. On barge-carrying vessels, the smoke-monitoring lines are connected to each individual barge. When fire is detected in a particular barge, acknowledgment of the alarm and activation of the manual release sends CO_2 to that barge only. The amount of CO_2 released is given in the system control instructions.

WARNING

Do not enter a hold filled with CO_2 without SCBA; this atmosphere does not support life.

Independent System

As mentioned earlier, small confined spaces may be equipped with an independent or freestanding carbon dioxide system that protects that space only. Paint lockers, small machinery spaces, and the like are the most common examples of these types of

Table 8.1
Minimum Percentages of Carbon Dioxide Needed to Extinguish Fires in Various Materials

Material	Carbon Dioxide (Percent)
Most flammable liquids	34
Most combustible materials	65
Dry electrical wiring insulation	50
Small electrical machines Wire enclosures (under 2,000 cubic feet [57 m²])	50
Record (bulk paper) storage Ducts	65
Fur storage vaults Dust collectors	75
Acetylene	66
Coal or Natural Gas Benzene	37
Gasoline Butane Kerosene	34
Quench and lube oils	34
Hydrogen	78

For more information, see *Fire Protection Handbook,* 18th edition, National Fire Protection Association, 1997.

Table 8.2
Carbon Dioxide Needed to Achieve a 34-Percent Concentration

Volume (cubic feet)	Carbon Dioxide Needed (pounds per cubic foot)	Conversion Factor Formula*
0 to 140	1 per 14	0.072 x total cubic feet = pounds needed
141 to 500	1 per 15	0.067 x total cubic feet = pounds needed
501 to 1,600	1 per 16	0.063 x total cubic feet = pounds needed
1,601 to 4,500	1 per 18	0.056 x total cubic feet = pounds needed
4,501 to 50,000	1 per 20	0.050 x total cubic feet = pounds needed
Over 50,000	1 per 22	0.046 x total cubic feet = pounds needed

* Examples: *40,000 cubic feet x 0.050 (conversion factor) = 2,000 pounds of CO_2*
40,000 cubic feet ÷ 20 = 2,000 pounds of CO_2

These examples mean that 2,000 pounds of CO_2 are required to achieve a 34% concentration in a 40,000 cubic foot compartment. To convert to metric units, use the following equations: 1 ft^3 = 0.0283 m^3 and l lb = 0.454 kg.

Other conversion factors for a 40,000 cubic foot compartment: 50% = 1.60, 65% = 2.46, 66% = 2.53, 75% = 3.22, and 78% = 3.50. For example, to achieve a 65% concentration, multiply the conversion factor for 65% by the 2,000 pounds required for a 34% concentration. Example: *2.46 x 2,000 = 4,920 pounds of CO_2*

For more information, see *Fire Protection Handbook,* 18th edition, National Fire Protection Association, 1997.

spaces (Figures 8.3 a and b). Each independent system has its own CO_2 supply. For spaces that require less than 300 pounds (136 kg) of agent, the supply cylinders may actually be located within the protected space as long as the system is designed to operate automatically. Automatic systems are also equipped so that they can be activated manually. The independent automatic CO_2 system operates like a pneumatic smoke detector (see Chapter 7, Fire Detection Systems). The sudden pressure buildup within the detection device operates a diaphragm lever on the cylinder control head, initiating release of the agent from all cylinders. Whether the system operates manually or automatically, it must provide a discharge alarm to alert crew members of the emergency.

Local Application System

Local application suppression systems are used to protect fixed objects or equipment such as large electric motors (Figure 8.4). In the marine industry, local application systems most often protect generators and paint lockers. Local application protection is achieved by discharging CO_2 through nozzles directed at the machinery. The discharge of the system creates a cloud of agent that surrounds the object, excluding oxygen from a localized fire. Most local application

Manually Operated Independent CO_2 System Activation

Step 1: *Make certain that crew members have evacuated the space.*

Step 2: *Turn off the power to fans,*

Step 3: *Close all openings to the space: doors, hatches, vents, etc.*

Step 4: *Go to the pull box, and pull the cable out straight. Depending on the system manufacturer, a second cable may need to be pulled.*

systems are manually operated upon discovery of a fire. Some systems are equipped with heat-sensing devices that activate the system in response to the heat of a fire. These devices are installed on the object. The manually activated local application system is operated in the same manner as the independent system.

A carbon dioxide *galley system* is a special type of local application system. Although not as common as the dry chemical galley system (discussed later in this

Figures 8.3 a and b (a) *(right)* Carbon dioxide system protecting flammable liquids storage room. (b) *(below)* Carbon dioxide system protecting paint locker. *Both photos courtesy of R. Wright/Maryland Fire and Rescue Institute/United States Coast Guard.*

Figure 8.4 *(below)* Local application carbon dioxide system located at the electrical panel. It has two 50-pound (23 kg) cylinders, 30 feet (9 m) of hose, and a diffuser nozzle. *Courtesy of Captain John F. Lewis.*

chapter), the carbon dioxide system protects ranges, deep fryers, and ducting. The system has the same components as other local application systems with nozzles arranged to flood the ductwork and range top with a cloud of agent. The heat detector for a CO_2 galley system is usually a fusible-link type that activates the system at a set temperature. A manual pull station is also provided so that the system can be activated upon discovery of a fire before the fusible link melts.

Extinguishment by a CO_2 galley system can take several forms: Any fire in the ducting is extinguished by excluding oxygen because the duct is a confined space. Fire on the range top is also extinguished by smothering because the cloud of agent saturates the protected target, again excluding oxygen. Supplementing the smothering effect is the force of the gas being expelled, which serves to sweep the flames from the fuel area.

Of particular concern in ensuring the continued effectiveness of a CO_2 galley system is the positioning of the nozzles. If the nozzles are moved to any degree from their designed positions, gaps in the smothering cloud can occur, allowing a fire to continue to burn in that location. Once the system has completely discharged, a small fire left unextinguished can reignite the rest of the fuel area. The fire could become the same size as before, only now a fixed system is not available to attack it.

Most CO_2 galley systems are also designed to turn off the fuel supply or other power source when activated (Figure 8.5). Make certain all burners and elements are in fact turned off. If heat from the power source is still being supplied to the area, the extinguished fire could easily reignite once the smothering effects of the CO_2 system are exhausted. CO_2 galley systems provide no lasting extinguishing effects on the fuel — their biggest disadvantage.

Bulk Storage System

Although rare, some CO_2 systems are large enough to require bulk storage. It is then often referred to as a low-pressure system. Bulk storage is used when the amount of agent required to operate a large system is so great that it is not practical to store it in high-pressure cylinders. The low-pressure system is supplied by liquefied carbon dioxide that is stored in a large, specially insulated refrigerated tank at a temperature of 0°F (-18°C) and at a pressure of only 300 psi (2 068 kPa) {21 bar}. The low-pressure bulk storage container is connected to the piping and nozzle system through remotely operated discharge valves in the same way as the high-pressure cylinders.

Carbon Dioxide Warning Alarm

Every space protected by a CO_2 system must have an audible warning device if that space is normally occupied by crew members when the vessel is underway. Visual warning devices such as strobe lights or rotating beacons are also very useful in noisy machinery spaces, and they can be incorporated into the audible alarm. The warning devices must operate for a minimum of 20 seconds before the agent is released into the space and continue to operate during the discharge phase. The CO_2 agent is the sole power source for operating these audible alarms. The warning devices are located so that they are easily heard throughout the protected space, and they are clearly marked with signs that contain the following message: WHEN ALARM SOUNDS, VACATE AT ONCE. CARBON DIOXIDE IS BEING RELEASED (Figure 8.6).

> ## WARNING
>
> Provide emergency escape breathing devices to all personnel in a space when testing the operation of warning devices on CO_2 system alarms. Notify all personnel that the procedure is being performed. Some deaths have occurred when the system inadvertently discharged.

Emergency Actions

In the event of fire in the engine room or other large machinery space, take emergency actions swiftly. Ideally, crew members are aware of a fire before the CO_2

Figure 8.5 Galley fuel shutoff. *Courtesy of Don Merkle, Maritime Institute of Technology and Graduate Studies.*

Figure 8.6 Carbon Dioxide total flooding system warning sign. *Courtesy of David Ward.*

alarm sounds. But if the first indication of fire is the sounding of the CO_2 discharge alarm, a crew member may have *as little as 20 seconds* to accomplish the following tasks:

- Warn others of the danger, and implement evacuation procedures.

- Secure all machinery.

- Close all openings to the space when evacuating (doors, hatches, ventilation and exhaust ducts, etc.).

- Ensure that the CO_2 system is actually discharging when the space is clear.

- Follow the procedures described at the CO_2 pull station if the system has not activated.

Maintaining Adequate Carbon Dioxide Concentration

It may be necessary to keep a space sealed for a period of time to achieve complete extinguishment. The soaking time may be measured in hours or even days. Although every effort is made to thoroughly seal the space before activating the CO_2 system, leakage can occur. For this reason, it may be necessary to introduce additional CO_2 agent into the space after the initial discharge. The amount of additional agent needed depends on the following variables:

- Volume of the space

- Amount of leakage

- Type of fuel burning

- Amount of heat remaining in the space

In most cases, a profile chart is available that indicates how many supply cylinders to discharge in a space to maintain the desired concentration and at what intervals. The fire control officer considers several factors in determining the best use of the CO_2 supply and other resources. Some important factors related to releasing additional CO_2 are as follows:

- Quantity of agent available

- Distance to a port where an additional supply of CO_2 is available

- Temperature of the fire (evaluated by increasing heat and blistering of bulkheads, decks, deckheads, and plates)

- Quantity of smoke production

Reentering the Space

Regardless of what type of space or hold has been filled with CO_2, the fire space is kept sealed for several hours (even days) if possible. If heat, smoke, and other conditions indicate that the fire is extinguished, crew members (in full personal protective clothing and SCBA) can prepare to reenter the space by advancing a charged fire hose to the door or opening. When crew members are ready to attack, partially open the door or opening. If fire erupts, quickly close the door or opening and seal it again. Fire control officers may wish to reevaluate the situation, introduce additional agent, and allow a longer soaking time.

Inspection and Maintenance

With proper care and maintenance, a carbon dioxide system is one of the most reliable means of safeguarding vessels against fire. When poorly maintained or neglected, CO_2 systems are notorious for failure. Almost all documented failures involving CO_2 systems have been from neglect or operator error. When crew members are highly trained in the use of the onboard CO_2 systems, they are more likely to concern themselves with proper care and maintenance of the equipment as well.

Carbon dioxide systems do not require a tremendous amount of in-depth maintenance. The best thing that can be done to ensure proper operation is regular inspection. Observe operating controls and mechanisms for obstructions or other conditions that may interfere with their operation. Perform a careful examination of CO_2 equipment every month. Pay particular attention to the stowage of materials that may obstruct access to controls or interfere with moving parts. Ensure that nozzles and piping are not clogged with debris, paint, lubricants, or other foreign substances. Report any deficiencies that are not readily correctable to the engineering department for immediate repair.

Have a qualified fire protection technician or engineer perform an annual inspection of carbon dioxide extinguishing systems to ensure that they will continue to perform as designed. During the annual inspection, weigh each cylinder to determine how much agent it contains. In some cases, a weigh bar is installed above the cylinder to allow for easy measurement without completely removing the cylinder from the rack. Otherwise, remove the cylinder and place on a scale. The gross (full) and tare (empty) weights of each cylinder are stamped into the cylinder itself. If a cylinder is within 10 percent of its full weight, it is considered full. This convention is taken into account during the design of the systems.

◆ Halogenated Agent Suppression Systems

The properties of halogenated extinguishing agents (halon type) were discussed in Chapter 5, Portable and Semiportable Fire Extinguishers. Two of these agents, Halon 1211 and Halon 1301, are also widely used in fixed suppression system applications. Halon 1211 is primarily used in local application systems; however, it is used in some total flooding systems. Halon 1301 is used almost exclusively in total flooding systems. The majority of halogenated agent suppression systems in use today are the total flooding variety. Halogenated agent suppression systems are similar in design and application to the carbon dioxide suppression systems (Figure 8.7).

A halogenated agent suppression system primarily protects against flammable liquid and electrical fires, but it has limited effectiveness on fires involving ordinary combustibles as well. A halogenated agent system is particularly useful in settings that require a clean agent. These settings usually contain high-value processes or equipment that would be damaged by other extinguishing agents such as water or dry chemical. Halogenated agents are not effective on most reactive metals or in locations containing self-oxidizing fuels.

As mentioned in Chapter 5, Portable and Semiportable Fire Extinguishers, halogenated agents have fallen from favor due to environmental damage. Due to the harmful effects on the earth's ozone layer, most regulatory agencies prohibit the installation of new halogenated agent systems. Those that are already in service are typically allowed, but modifications and extensive repairs to existing systems are not permitted. Alternative replacements for halogenated agents are under development. An alternative type of suppression system is a high-pressure water mist system (discussed later in this chapter). Descriptions of halogenated agent suppression systems and their inspection, maintenance, and testing requirements are given in the sections that follow.

System Description

Most halogenated agent suppression systems are custom designed or engineered. However, some smaller systems may be pre-engineered, standard models — particularly the local applications system. Components common to all halogenated agent systems are containers, activators, nozzles, detectors, manual releases, and control panels. Some similarities exist between a halogenated agent system installation and a CO_2 system, but do not confuse the two nor make any attempt to interchange parts.

A variety of container sizes, colors, or shapes exists in halogenated agent systems. Container capacities range from 5 to 600 pounds (2.3 kg to 272 kg). The amount of agent in a container and the number of containers depend on the system design. Pressure gauges indicate the pressure in the container. However, due to super pressurization with nitrogen, the pressure gauge reading may not truly indicate whether or not the container is completely full.

All containers have one or more valves to permit release of the agent into the hazard area. Although valves may be activated mechanically,

Figure 8.7 A halogenated agent fixed suppression system protecting the room containing the cryogenic liquid oxygen manufacturing plant and the air compressor used to refill self-contained breathing apparatus cylinders. *Courtesy of R. Wright/Maryland Fire and Rescue Institute/United States Coast Guard.*

hydraulically, or pneumatically, many are activated electrically in response to smoke or heat detection devices. Most systems also have a manual activation control. Although they are not recommended, many systems have one or more abort switches to stop system activation. Activation of the system is accompanied by visual and audible warning signals for occupants of the area.

Containers are connected to a system of fixed piping that terminates at the nozzles. No standard designs exist for piping or nozzles; each manufacturer has its own favored style. However, piping has some minimum requirements. Piping must withstand pressures of 620 psi (4 275 kPa) {43 bar} in low-pressure systems and 1,000 psi (6 895 kPa) {69 bar} in high-pressure systems. The piping must not be a cast iron or nonmetallic variety.

A series of system controls and status indicators are contained on the system control panel. These controls and indicators are not standard from system to system. This control panel is located in a highly accessible area.

Inspection, Maintenance, and Testing

Only qualified technicians or engineers perform inspections, maintenance, and testing of halogenated agent suppression systems. These systems, along with their detection and control systems, are typically too complex for the average crew member to service without special training. Also only properly trained individuals inspect, maintain, or test these systems because of the expense of the agent and the environmental impact of accidental discharge.

Ensure that a halogenated agent system is inspected semiannually. Thoroughly review all components. Check agent cylinders for loss of agent. Weigh the cylinder or inspect pressure gauges. Replace any cylinder whose weight has decreased by 5 percent or whose pressure gauge has decreased by 10 percent (adjusted for temperature differences). The temperature at which the cylinder is stored affects the pressure on the gauge. The pressure goes down as the temperature goes down and vice versa; therefore, this factor is taken into account. Inspect all hoses on the system for damage, and test them if damage is suspected.

 ## Steam Smothering Suppression Systems

Under most flags, it has not been permissible to install steam smothering suppression systems on vessels since 1962. Such systems consist of permanent connections to spaces or compartments so that steam may be introduced as an extinguishing medium much like CO_2. Some boiler protection systems use steam lances or semiportable appliances to direct steam on a fire. The microscopic droplets provide some cooling and extinguishment. These systems were banned due to concerns about electrostatic hazards. However, many of those in service before the ban are still in use. Repair of existing systems is permitted, provided that no changes are made to the original design and that they are properly maintained.

Either the main or auxiliary boilers provide the steam used by these systems. The steam pressure is required to be at least 100 psi (689.5 kPa) {7 bar}. Steam is delivered at a rate of 1 pound of steam per hour per 12 cubic feet (1.14 kg of steam/hour/m³) of the largest space or compartment.

The size of piping in steam smothering suppression systems is determined largely by the size of the largest space or compartment. The piping from the boiler to all manifolds must be large enough to supply all branch lines in the largest compartment and also the compartments on all six sides. Similarly, from each manifold to the branch lines, the cross-section of the distribution piping must be equal in area to the sum of the branch line cross-sectional areas. It is forbidden to route steam piping through accommodation spaces or other areas that passengers or crew members may occupy while the vessel is underway. Machinery spaces and corridors are exceptions. Steam piping for dry cargo spaces, machinery spaces, and lockers is separate from the piping for bulk cargo tanks to prevent accidental spread of petroleum fires. Drain cocks are required in all piping to prevent freezing.

The inlet side of each manifold is equipped with a master valve to close that particular manifold. Likewise, each outlet on the manifold is equipped with a valve that clearly indicates which space it serves. Valves leading to liquid cargo tanks are open at all times to ensure that steam flows to each tank when the master valve is opened in response to a fire. Once it is confirmed which tanks are not burning, those corresponding valves are turned off. Steam smothering systems on dry cargo vessels work just the opposite: The valves leading to individual compartments or holds are kept closed, and the master valve is always open.

The location of all controls and valves pertaining to the operation of a steam smothering suppression system are outside the protected space (Figure 8.8). Furthermore, steam control valves are not permitted in any area that could be blocked or made inaccessible by fire conditions. Control valves for a pump room steam smothering system are located outside the pump room, next to the exit. In pump rooms, the steam outlets are located just above the floor plates. The steam outlets in cargo holds are in the lower portion of each hold or tween deck.

Figure 8.8 A steam smothering system control station. *Courtesy of Maritime Institute of Technology and Graduate Studies.*

 ## Chemical Agent Suppression Systems

A chemical agent suppression system encompasses two types of systems: *dry chemical* and *wet chemical*. The most common dry chemical systems are *galley systems* and *deck systems* where fires can spread quickly if they are not extinguised. The wet chemical system is effective on flammable liquids and cooking hazards (grease, oil, etc.). Each system is discussed in the sections that follow.

Dry Chemical System

A dry chemical suppression system is used in locations where a rapid fire knockdown is required. The agents used in this system are the same as those described for portable dry chemical fire extinguishers in Chapter 5, Portable and Semiportable Fire Extinguishers. In the marine industry, the most common types of dry chemical systems are galley systems and deck systems.

Galley System

A local application, pre-engineered dry chemical galley system is the most common system used to protect against galley fires (Figure 8.9). This system discharges a dry chemical agent onto a specific surface such as the cooking surface of a grill or deep fryer. Galley systems are designed and calculated to protect the total area of potential involvement and the plenum and duct system that collects and exhausts heat and gases from the cooking area. Basically, all galley systems have the following components:

- Storage cylinder for expellant gas and agent
- Piping to carry the gas and agent
- Nozzles to disperse the agent
- Activating mechanism

All the dry chemical agents discharge a cloud that leaves a powder residue that creates a cleanup problem after system operation. The powder can usually be vacuumed; however, the very fine particle size makes it somewhat difficult to contain in an ordinary vacuum cleaner bag. Normal mopping is difficult because the chemicals are water-repellent. Due to these cleanup problems, dry chemicals are not considered clean agents. A dry chemical system is not recommended in areas where sensitive electronic equipment

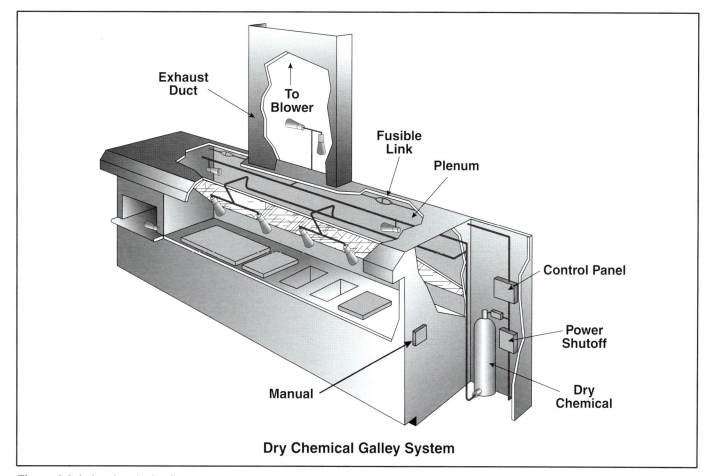

Figure 8.9 A dry chemical galley system.

is located. The chemical residue has insulating characteristics that hinder the operation of the equipment unless extensive restorative cleanup is performed.

Dry chemical agent containers come in a variety of sizes. The storage container may contain both the agent and the pressurized expellant gas (stored pressure) or the agent and the gas may be stored separately.

These systems are presently pre-engineered; very specific components are designed to work together for a particular installation. Proper operation depends on pipe length, pipe size, and many other variables. Do not interchange parts with those that are merely similar or modify the system from its original design.

Detailed maintenance of dry chemical galley systems is left to reputable contractors. A detailed annual inspection by a reputable fire protection firm is recommended. However, crew members can inspect these systems for mechanical damage, obstructions in nozzles, proper positioning of nozzles, and proper pressures on stored-pressure containers. Keep a log indicating that each inspection was made. Correct

any defects immediately. In some cases, engineering department personnel can make minor repairs, but others require notification of the system service provider. Perform the following inspection procedures monthly to ensure system readiness:

- Check all parts of the galley system to ensure that they are connected and in their proper location.

- Inspect manual activators for obstructions.

- Inspect the tamper indicators and seals to ensure that they are intact.

- Check the maintenance tag or certificate for proper placement and expiration date.

- Examine the system for any obvious damage.

- Check the pressure gauges to ensure that they read within their operable ranges.

Deck System

Dry chemical deck systems are required on many types of vessels carrying liquefied gases in bulk. These systems are also very effective fire suppression weapons on vessels that are subject to flammable liquid spills and aircraft crashes. Twin-agent deck systems (typically but not always a combination of dry chemical and foam systems) are used on spill fires. Dry chemical is used to quickly knock down a spill fire, and then a foam blanket to suppress vapor production covers it. Both agents are applied from the same nozzle that is fed by two separate hoses (Figure 8.10).

A dry chemical deck system protects the cargo deck and loading station manifolds. To provide complete coverage, several deck systems may be necessary to provide overlapping reach of the protected area. Each unit is self-contained and independent of an outside power source. Most systems are skid-mounted and contain the following components:

- Agent storage tank with a capacity of up to 3,000 pounds (1 361 kg)

- Nitrogen (propellant) storage cylinder(s) with a capacity of 400 cubic feet (11.3 m³)

Figure 8.10 Twin agent skid-mounted deck system. *Courtesy of Captain Mark Turner, Syndicated Management Services Ltd.*

- Hose reels equipped with 100 to 150 feet (30.5 m to 46 m) of rubber hose

- Turret nozzle (optional)

A single deck system may have up to six hoses, depending upon the size of the system and whether or not it is equipped with a turret nozzle. Minimum standards require that turret monitors discharge at a rate of not less than 22 pounds per second (10 kg/sec) and that hose nozzles have a minimum discharge rate of not less than 7.7 pounds per second (3.5 kg/sec). Maximum discharge rates for hoses are also established so that a hose is easily controlled by a single crew member. The required stream reach for turret nozzles is established based on the discharge rate given in Table 8.3.

The reach of a dry chemical hose stream is considered equal to its length; it has no actual stream reach. The length of the hose limits its effective coverage range. If a fire is located significantly higher than the point of discharge from the nozzle, the ability to provide proper reach and coverage is affected. Coverage is also affected by the wind. The storage capacity of each unit must provide 45 seconds of continuous discharge from all nozzles.

After resolving a fire emergency, restore the deck system to a state of readiness. Exhaust any remaining dry chemical from the hoses by operating them with only nitrogen gas flowing. If agent is allowed to settle in the hoses, it can cake and cause a clog that renders the hose useless. Refill the dry chemical agent tank with identical agent. Some agents are not compatible with each other and can cause serious problems. Reset any activating devices, and provide recharged propellant cylinders as required. Follow the manufacturer's procedures closely to ensure safety in the recharging and system readiness processes.

Table 8.3 Stream Reach of Turret Nozzles in a Deck Dry Chemical System	
Maximum Flow Pounds/Second (kg/sec)	**Maximum Reach Feet (m)**
22 (10)	33 (10)
55.4 (25)	99 (30.2)
99 (45)	132 (40)

Perform weekly inspections of the deck systems to ensure operational readiness. Refer to the manufacturer's operating manual for specific inspection and maintenance instructions. Crew members can make some minor repairs provided they are done in strict accordance with the manufacturer's directions. Perform each of the following checks during an inspection:

- Visually observe all components of the system exposed to the weather for physical damage or corrosion.

- Ensure that operating instructions are provided on the unit and that they are legible and easily understood.

- Check pressure or level gauges. Consult the manufacturer's operating manual if needed. Be sure the dry chemical agent cylinder is filled to the required level. Only hand-tighten the fill cap when replacing it after inspection.

- Ensure that all hose reels are in the unlocked position and that they roll freely on their bearings when the hose is pulled.

- Observe the nozzle discharge openings for foreign matter, insect nests, or other obstructions. Exercise the nozzle bails by operating them several times. Make sure that nozzles are closed when stored.

- Ensure that all operating valves are properly secured by ring pins and seals.

Wet Chemical System

A wet chemical system is best suited for applications in galley cooking hoods, plenums, ducts, and associated cooking appliances. A wet chemical system operates and is designed similarly to a dry chemical system. The wet chemical agent is typically a solution of water and either potassium carbonate or potassium acetate that is delivered to the hazard area in a spray. It is an excellent extinguishing agent for fires involving flammable liquids or gases, grease, or ordinary combustibles such as paper and wood. It is not recommended for electrical fires because the spray may act as a conductor.

A wet chemical system is most effective on fires caused by cooking hazards. The wet chemical agent reacts with animal or vegetable oils and forms soap. A wet chemical agent extinguishes grease or oil fires by fuel removal, cooling, smothering, and flame inhibition. Wet chemical systems are messy, however, particularly when food greases are involved. A *thorough* clean-up operation is required to restore the area.

For the most part, the components and activation procedures for wet chemical systems are the same as those for dry chemical systems. The inspection and testing procedures are also the same.

CAUTION: Wear personal protective equipment for protection against wet chemical agents; they are toxic, electrically conductive, and corrosive.

 ## Water-Based Suppression Systems

Several types of fixed systems fall into the category of water-based suppression systems. These include automatic sprinkler systems, water spray systems, high-pressure water mist systems, and deluge systems. Each of these systems is discussed in the following sections. Fire main systems are described separately following these discussions.

Automatic Sprinkler System

In a perfect world, the perfect fire suppression system would be simple, reliable, and automatic. It would discharge a readily available and inexpensive extinguishing agent directly on a fire while it was still small. While the world is not perfect, the system that best satisfies these requirements is the automatic sprinkler system. All new cruise vessels are required to have automatic sprinkler systems, and older vessels are being retrofitted.

The fundamental components of the automatic sprinkler system are very simple: piping, control valves, sprinklers (sometimes called *sprinkler heads*), and a reliable water supply. In the case of an automatic sprinkler system on board vessels, the water supply consists of a pressure tank and a pump. The sprinkler system components described in the following sections are usually found on vessels built and registered in the United States (U.S.). Sprinkler systems on vessels built in other countries are similar but may not have the exact counterparts. Information on supplementing, restoring, inspecting, testing, and maintaining automatic sprinkler systems is also given.

Despite a high degree of reliability, automatic sprinkler systems can fail to control or extinguish a fire because of closed valves, a frozen water supply or an inadequate water supply, corroded pipes, obstructed sprinkler discharges, and impaired sprinklers. For automatic sprinkler systems to function properly, they must be designed, installed, and maintained properly.

In general, automatic sprinkler systems control shipboard fires with less water than fire hoses. However, an operating sprinkler system can introduce tremendous amounts of water into an area in a relatively short time. Large quantities of water trapped in compartments above the waterline can rapidly affect the stability of a vessel. The ability of liquids to flow unobtrusively from one side of the vessel to the other (free-surface effect) has an enormous impact. Finding a way to drain this water overboard or to a lower level is a priority. Instability and free surface effect are very serious situations; they are discussed in more detail in Chapter 13, Shipboard Damage Control.

Fire fighting water can be supplied to both the shipboard fire main and automatic sprinkler system through the international shore connection should a fire occur in port (see Fire Main Systems section). See Chapter 6, Water and Foam Fire Fighting Equipment, and Chapter 15, In-Port Fire Fighting and Interface with Shore-Based Firefighters, for more information.

Sprinklers

A *sprinkler* is that portion of an automatic sprinkler system that senses the fire, reacts to that sense, and delivers water to the fire area. Think of it as a small, fixed-spray nozzle that turns on automatically but only turns off manually to stop the water flow. Each sprinkler is equipped with a heat-sensing device that allows only that sprinkler to open in response to a fire. If a fire spreads beyond the area covered by the first sprinkler, additional sprinklers open, discharging water onto those portions of the fire. As sprinklers open, water under pressure begins to flow through the piping from a pressurized tank. The loss of pressure in that tank causes the sprinkler pump to start. In many cases, a fire is extinguished or controlled by the opening of only one or two sprinklers.

The designer of each system takes into account the pressure available and many other factors in determining the appropriate sprinkler for each application.

Three main components of a sprinkler are of interest: *activation (heat-sensing) device, deflector,* and *discharge orifice.* Any change in the orifice size, position, or temperature rating of a sprinkler can severely interfere with the performance of the system and its ability to control a fire. The following paragraphs highlight each of these components.

Activation device. A sprinkler is a nozzle that is under pressure; a simple heat-sensing device controls its operation. The most common heat-sensing device is the *fusible link.* In its simplest form, the fusible link is a solder link with a low, precisely established melting point (Figure 8.11a). Various links and levers connect the solder link to a cap that restrains the water at the nozzle orifice. As illustrated in Figure 8.11b, the solder in the fusible link melts at a predetermined melting point. The lever arms release and spring clear of the sprinkler frame. As the lever arms drop, the seated cap releases, which permits water to flow (Figure 8.11c).

Other types of heat-sensing devices include *frangible/quartzoid (glass) bulbs, fusible alloy pellets,* and *chemical pellets.* Fusible alloy pellets and chemical pellets work in much the same way as fusible links. Frangible/quartzoid bulbs are inserted between the sprinkler frame and the discharge orifice in a similar way. Frangible/quartzoid bulbs contain liquids that expand when heated. The expansion of the liquid within the bulb increases vapor pressure, and the bulb breaks once the vapor pressure exceeds the bulb's strength.

Just like the heat-sensing devices used in fire detectors (Chapter 7, Fire Detection Systems), sprinklers are rated at various temperature ranges, depending on the normal ambient temperature range of a given space. The temperature rating is stamped on the link in soldered link sprinklers. On other types of sprinklers, the temperature rating is stamped into the frame or some other part of the device, the sprinkler frame arms are color-coded, or different colored liquids are put in the bulks. By varying the composition of the fusible link solder or the liquid used in frangible/quartzoid bulbs, the operating temperature of a sprinkler is changed. Table 8.4 lists the actual activation temperature range of sprinklers for each temperature classification along with color codings. Maintain a supply of spare sprinklers of each temperature range (along with sprinkler wrenches). Only replace an

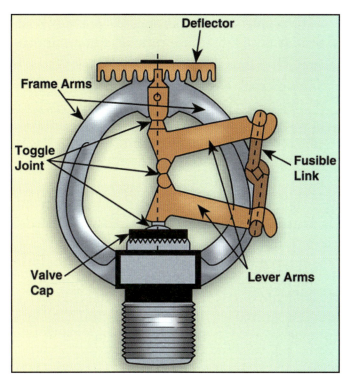

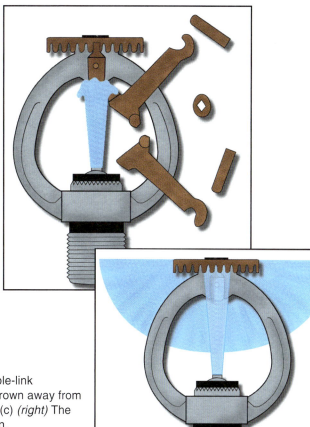

Figures 8.11 a–c (a) *(above left)* Components of an upright fusible-link sprinkler. (b) *(above right)* As the fusible link melts, the plug is thrown away from the discharge orifice by the water pressure in the piping system. (c) *(right)* The water bounces off the deflector into an effective discharge pattern.

Table 8.4						
Temperature Ratings, Classifications, and Color Codings						
Maximum Deckhead Temperature		**Temperature Rating**		**Temperature Classification**	**Color Code**	**Glass Bulb Colors**
°F	°C	°F	°C			
100	38	135–170	57–77	Ordinary	Uncolored or Black	Orange or Red
150	66	175–225	79–107	Intermediate	White	Yellow or Green
225	107	250–300	121–149	High	Blue	Blue
300	149	325–375	163–191	Extra High	Red	Purple
375	191	400–475	204–246	Very Extra High	Green	Black
475	246	500–575	260–302	Ultra High	Orange	Black
625	329	650	343	Ultra High	Orange	Black

Reprinted with permission from NFPA 13-1999, *Installation of Sprinkler Systems,* Copyright © 1999, National Fire Protection Association, Quincy, MA 00269.

opened or damaged sprinkler with one of the proper temperature range rating.

Provide a margin of safety between the normal temperature of a space and the operating temperature of a sprinkler. The use of sprinklers with appropriate temperature ratings is important. Temperature ratings that are too low for a given location may result in the opening of sprinklers at normal temperatures. Sprinklers with operating temperature ratings that are too high operate more slowly in a fire situation and delay sprinkler operation, permitting fire growth.

Deflector component. A deflector is attached to the sprinkler frame and forms the discharge pattern of the water. Discharging water is directed against the deflector to convert it into a spray pattern. The deflector configuration is fundamental to the effectiveness of the sprinkler. Three basic types of sprinklers are available, depending on how they are used: *upright, pendent,* and *sidewall* (Figures 8.12 a–c).

- ***Upright sprinkler*** — Sits on top of the piping; the solid deflector deflects the spray of water into a hemispherical pattern that is redirected toward the deck. Do not invert upright sprinklers in the hanging or pendant position; the spray would deflect toward the deckhead.

- ***Pendant sprinkler*** — Extends down from the underside of the piping; the deflector breaks the pattern of water into a circular pattern of small water droplets. Use in locations where it is impractical or unsightly to use sprinklers in an upright position.

- ***Sidewall sprinkler*** — Extends from the side of the piping; the deflector creates a fan-shaped pattern of water. By modifying the deflector, the sprinkler can discharge most of its water to one side. Use in accommodation spaces and passageways.

Sprinklers produced before 1955 were designed with deflectors that discharged a portion of the water upward toward the deckhead. This design did not produce a good downward distribution of water. Modern sprinklers produce a more uniform discharge pattern. Because discharge patterns are different, do not use old-style sprinklers to replace modern sprinklers. However, substitute modern sprinklers for old-style sprinklers in an existing system when changing sprinklers or upgrading a system.

Special purpose sprinklers are available to use in specific applications and environments. Special wax-coated sprinklers are available to place in a corrosive atmosphere. A variety of chrome-plated and flush-type sprinklers are available where appearance is important. If a sprinkler is located where mechanical damage is possible, cover it with a protective cage (Figure 8.13).

Discharge orifice. The actual amount of water discharged by a sprinkler is determined by the size of the discharge orifice and the pressure available at the sprinkler. The most common orifice size in automatic sprinklers is ½ inch (13 mm). As with other parts of a sprinkler, the size of the orifice can vary for different applications.

Figures 8.12 a–c Three sprinkler designs
(a) *(top left)* A pendent sprinkler.
(b) *(top right)* An upright sprinkler.
(c) *(right)* A sidewall sprinkler.

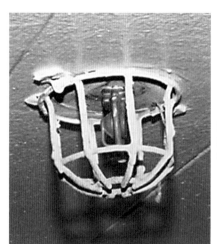

Figure 8.13 Some sprinklers are covered with protective cages.

System Piping

Sprinkler system piping is usually made of steel, although copper is acceptable for use in some situations. The cost of copper, however, usually makes steel piping the economical choice. Hangers and clamps support the piping. Sprinkler piping that penetrates deckheads or bulkheads must preserve their fire integrity. Pipe supports are usually welded to the vessel's primary structural members so that they cannot loosen during vessel motion or vibration. Pipe supports are braced so piping does not shift with vessel movement. The piping of a sprinkler system consists of the following components:

- *Feed main* — Piping that delivers the total volume of water to the system before any branching of pipes.

- *Riser* — Vertical piping that carries water from the source of supply (feed main) through the main shutoff valve, check valves, and flow detection and alarm devices. Risers connect to the feed main.

- *Cross main* — Portion of a main that has branch lines attached to it. The risers may supply one or more cross mains, but none of these pipes have sprinklers installed on them.

- *Branch lines* — Portion of the piping that actually has the sprinklers attached to them. Cross mains serve branch lines.

How much water is needed to flow from a sprinkler is determined during the design of the system. It is based on the expected fuel load and other considerations of the protected area. With hydraulic calculations, the designer determines what pipe sizes are necessary to discharge that minimum amount of water from any sprinkler in the system. However, the system is not capable of discharging that amount of water from every sprinkler in the system at the same time. Only a portion of the sprinklers in the design area needs to operate. This reasoning is based on the historical performance of sprinkler systems — that a few sprinklers can control most fires. Thus, the pipe sizes gradually get smaller and smaller between the riser and the last sprinkler on each branch line.

Control Valves

Control valves in an automatic sprinkler system are found at the riser (or pump manifold) and outside each protected area or zone of the system (Figure 8.14). These valves are necessary for turning off the system or a portion of the system when making repairs, replacing sprinklers, or resetting the system after a fire. All valves are readily accessible at all times and clearly marked as to their purpose (sprinkler control) and specific function (main control, main drain, zone control, etc.). After work is complete, return the control valves to the open position. Numerous losses have occurred because the water supply to a sprinkler system was turned off before the start of a fire.

On vessels built in the U.S., most valves that control water to sprinkler systems are of the indicating type to ensure that they are not closed inadvertently. Vessels built elsewhere may have different components. An *indicating valve* shows at a glance the position of the valve — open or closed. The most common type of indicating valve used in sprinkler systems is the *outside screw and yoke (OS&Y)* (also known as the *rising stem valve*), which has a yoke on the outside with a threaded stem that controls the opening and closing of the gate. The threaded portion of the stem is outside the yoke when the valve is open and inside the yoke when the valve is closed.

Figure 8.14 A cruise vessel sprinkler system valve box. *Courtesy of Robert E. "Smokey" Rumens.*

Another type of indicating control valve on vessels built in the U.S. is the *butterfly,* which has a paddle indicator or a pointer arrow that shows the position of the valve in relation to the pipe. If the indicator or arrow is in line with the pipe, the valve is open. If it is crosswise with the pipe, the valve is closed.

To further ensure that sprinkler control valves are kept open, secure (chain or lock) the valves in the open position or electronically supervise or do both. A valve that is electronically supervised has a switch attached to it. Closing of the valve causes an electrical circuit to open and transmit a signal to the fire alarm control panel.

Pressure Tank

A pressure tank keeps the automatic sprinkler system on a vessel ready to respond to a fire. The pressure tank is usually filled to two-thirds of its capacity with freshwater and the other third with pressurized air. When a sprinkler opens, the air pressure expels the water from the tank and forces it through the piping to the open sprinkler. The water capacity of the pressure tank must equal the amount of water required to fill all pipes in the largest area or zone of protection plus 200 gallons (757 L). This initial supply of water allows the sprinkler pump time to start. The air pressure in the pressure tank must deliver water to the open sprinkler (regardless of its location within the system) at a discharge pressure of at least 15 psi (103 kPa) {1 bar}.

Sprinkler Pump and Alarm Feature

The sprinkler pump automatically starts in response to a pressure drop in the pressure tank. If needed, water is also pumped from the sea chest until the sprinkler system is turned off. The drop in pressure that starts the sprinkler pump can also activate a water flow alarm on the vessel's fire alarm control panel. Another device that can warn the crew that the sprinkler system is responding to a fire is a water flow switch: a paddle-type device placed in the riser or on cross mains to particular areas or zones. The flow of water in the pipe causes the paddle to bend in one direction, closing the switch contacts and initiating the alarm. Depending on the location of the flow switch, the alarm may be designed to indicate a particular area or zone.

Supplementing the System

Even with an automatic sprinkler system, crew members must prepare for a fire attack just like they would for any other fire situation. Do not rely on an automatic sprinkler system to take care of a fire emergency on its own. Some component of the sprinkler system could fail, or it might only hold a fire in check and not completely extinguish it. Keep the sprinkler system operating until a fire is completely extinguished and the danger of reignition has passed. Prematurely turning off the system could allow a fire to grow beyond control before an effective attack is made or before the system is placed back in operation.

Restoring the System

Restore the sprinkler system to ready condition immediately after a fire incident is resolved. Replace any sprinklers that were heated by the fire (even if they did not open) with spares of the same size, orifice diameter, deflector type, and temperature rating. If seawater was pumped into the system, thoroughly flush with freshwater to prevent electrolysis from destroying the system components. Reset all valves to their designated position, switch the water supply back to freshwater, and refill and recharge the pressure tank. Refer to system specification markings for the designed volume and pressure.

Zoning of Sprinkler Systems

When a large portion of a vessel is protected by automatic sprinkler systems, the protected areas are usually divided into smaller fire zones by Class A bulkheads. This division serves to confine the fire into smaller areas and prevents overtaxing the sprinkler system. The zone control valves for a sprinkler system prove very useful in this type of arrangement.

Vessels equipped with sprinkler systems must have a sprinkler system *zoning chart* that depicts the arrangement of fire zones. The chart identifies each zone and clearly illustrates the sprinkler system piping, valves, and pumps related to each zone on a schematic drawing (Figure 8.15). Post this chart on the bridge, in the engine room, at the fire control station, and in any other location that has an annunciator panel.

Inspecting, Testing, and Maintaining Sprinkler Systems

Automatic sprinkler systems require periodic inspections and maintenance in order to perform properly during a fire situation. Crew members can perform routine maintenance, but more extensive mainte-

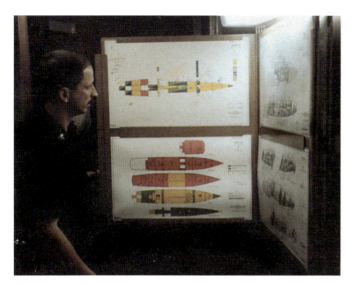

Figure 8.15 Chart showing arrangement of fire zones and system components for an automatic sprinkler system. *Courtesy of R. Wright/Maryland Fire and Rescue Institute/ United States Coast Guard.*

nance and repairs are left to a contracted sprinkler system company. Crew members charged with the responsibility of inspecting and maintaining sprinkler systems can also conduct system tests for the purposes of prefire familiarity and to verify system readiness. Before proceeding with an inspection or test, take the following steps:

- Review the records of prior inspections, and identify the make, model, and type of equipment and what spaces or zones are protected by the system.

- Coordinate the time of the inspection or test with the master and the chief engineer to avoid conflicting with other operations.

- Notify crew members on the bridge and in the engine room just before beginning a test. Tell them which system is being tested if equipment is electronically supervised to prevent confusion as to what indications or alarms are results of the test and which are potential indications of an actual fire or water flow.

Notify the bridge and engine room crew members when testing is complete. Make note of which alarms sounded at specific times. Compare notes with the log kept on the bridge to ensure the alarm features worked properly. Guidelines for inspecting, testing, and maintaining the various system components are given in the following paragraphs.

Sprinklers. Examine sprinklers to make sure that they are clean, undamaged, and free of corrosion.

Note if guards are needed to protect them from mechanical injury. Carefully examine sprinklers in spaces subject to high temperatures for appropriate temperature ratings. Ensure that sprinklers exposed to a corrosive atmosphere have a special protective coating. Replace sprinklers that are corroded, painted, or filled with foreign material with a sprinkler of the proper temperature rating, size, and type.

Ensure that partitions or stores do not obstruct the distribution of water discharge from sprinklers and that the discharge area is free of obstructions from suspended storage. Maintain a clearance under sprinklers of at least 18 inches (457 mm) measured from the deflector.

Sprinkler system piping and pipe hangers. Inspect all sprinkler system piping and pipe hangers to determine that they are in good condition. Check for corrosion, physical damage, and leaks. Corrosion inside the pipes, often microbe-induced, can cause ruptures or blockages. Internal inspection and water treatment are necessary to prevent corrosion. Ensure that sprinkler system piping is not used as a support for ladders, stores, or other materials.

The flowing of water through sprinkler system pipes can produce very significant forces and stresses. Properly support sprinkler system piping because improperly supported piping can break and cause improper drainage. Repair loose sprinkler system hangers promptly.

Protect piping in sprinkler systems against freezing. Be alert for bends in the piping, low points, lodgment breaks, and horizontally fitted valves where water could be trapped when a system is drained. Freezing of sprinkler piping can stop the flow of water to sprinklers or cause the failure of control and alarm devices. The greatest danger from frozen piping, however, is the rupture of a pipe or fittings, resulting in severe water damage or expensive repairs and interruption of protection. Look for these potential problem areas and make recommendations to the master on ways to prevent them. If it is necessary to disable and drain a sprinkler system to prevent freezing, maintain a fire watch until the automatic system is restored.

Water supply. Pay particular attention to the electrical power for electrically driven sprinkler pumps. Visually check the pump control panel to ensure that the circuit breaker is closed and that the power indicating light is on. Check the water level and air pressure in pressure tanks weekly. Inspect the heating source for pressure tanks daily during cold weather.

Ensure that all valves controlling water supplies to the sprinkler system and within the system (zone valves) are open at all times. Examine each control valve to ensure the following conditions are met:

- Control valve is fully open, secured, and supervised in an approved manner.

- Operating wheel or crank is in good condition.

- Control valve is accessible. If a permanent ladder is provided to elevated valves, ensure that it is in good condition.

- Operating stem is not subject to mechanical damage. Provide guards if necessary.

- Main drain valve and any auxiliary drains are closed.

Functional tests. Perform functional tests at the required frequencies, and record results. Flow test sprinkler pumps annually. Test run diesel engines that power pumps weekly and electric motors monthly. When performing functional tests, notify the bridge and engine room crew members that alarm activations will be the result of tests. Notify all concerned persons again after the system is restored to readiness.

Water flow alarm test. Conduct a water flow alarm test on automatic sprinkler systems by using the inspector's test connection: a 1-inch (25 mm) pipe equipped with a shutoff valve and a discharge orifice that is equal in size to the smallest sprinkler in the system. This test connection simulates the operation of a single sprinkler and ensures that the alarm operates even if only one sprinkler is activated in a fire. The test connection is usually located at the section control valve. With an observer at the riser, open the inspector's test valve. The system should cause a sprinkler flow indication on the alarm panel within 15 seconds of opening the test valve and only a slight variation in pressure at the riser. After the alarm indication, close the inspector's valve, and restore the system to operational condition.

Deluge System

A *deluge system,* often referred to as a *manual sprinkler system,* is very similar to the automatic sprinkler system with two major exceptions: All sprinklers are

Figure 8.16 These deluge sprinkler controls are located on one side of the engine room on the vehicle deck of a ferry. The valve colors link to the various zones. *Courtesy of Captain John F. Lewis.*

open, and no water is in the pipes or a pressure tank. Crew members activate the deluge system manually (Figure 8.16). The spaces installed with a deluge system are fitted with an alarm system as well. When crew members discover a fire in a space protected by a deluge system, they open a control valve located near the vessel's fire pump manifold or sometimes just outside the protected space. The opening of this valve starts the fire pump, which then discharges a large volume of water from all sprinklers at one time. This large flow of water has tremendous knockdown capability, often more than enough to suppress a fire.

A deluge system is usually installed to protect the more hazardous spaces of a vessel such as the vehicle decks of ro/ros and ferries. Cargo spaces that are accessible to crew members while a vessel is underway are other common spaces where deluge sprinklers are installed. Spaces protected by a deluge system must drain well or dewater easily to prevent flooding. Due to the tremendous flow of a deluge system, the potential for creating free surface effect and instability is even greater than with most other systems.

The details relating to piping, control valves, and pumps for deluge systems are not significantly different from those for automatic sprinkler systems. Likewise, the needs to supplement and properly restore a deluge system are of equal concerns.

Water Spray System

A water spray system is sometimes designed to protect fire exposures, rather than suppress or control a fire itself. A common application of water spray systems is providing exposure protection for lifeboats, piping, loading stations and manifolds, and storage tanks on vessels. Water spray is also occasionally used to protect the superstructure of a vessel when it is exposed to large flammable liquid fires. The spray is applied directly to the surfaces of bulkheads and decks, which provides exceptional cooling. Some cargo pump rooms on tank vessels are equipped with water spray systems.

Water spray systems are typically manually operated, although some are designed to automatically operate in response to fire detection devices. When a fire is discovered, opening the main control valve activates the manually operated type. If water is provided by the vessel's fire main system, the fire pump may start automatically in response to a drop in pressure. Otherwise, the dedicated pump may require manual starting before the system can activate. One of the vessel's main fire pumps may provide the water supply for spray systems. However, the pump must provide an adequate supply of water to both the spray system and the vessel's fire main when both are used. Otherwise, a separate pump is dedicated for the spray system.

The spray heads for water spray systems are open nozzles that shape the discharge of water into a cone. They apply a maximum amount of water to the protected area. Because all the spray nozzles are open, the volume of water flow required by a spray system can be substantial (Figure 8.17).

Figure 8.17 Water spray system nozzle. *Courtesy of R. Wright/Maryland Fire and Rescue Institute/United States Coast Guard.*

Water spray systems are inspected and tested at least quarterly in much the same manner as automatic sprinkler systems. The following points are recommended for a thorough inspection and testing program:

- Inspect the system for damage, inappropriately closed valves, and improper nozzle aim.

- Flow the system, and inspect all nozzles for good spray patterns and adequate volume of water flow.

High-Pressure Water Mist System

As far back as the 1940s, Factory Mutual tested the use of fine water spray to extinguish flammable liquid fires. It was observed that droplets of water could be *"entrained into the fire, producing cooling and oxygen dilution in the combustion zone."* The only disadvantage at the time was the difficulty in producing the high pressures required to operate the system.

Because of high system failure rates for gaseous marine fixed fire suppression systems, the potential lethality of CO_2, and the dangers posed to the ozone layer by halogenated hydrocarbons, a water mist replacement system has come to the forefront for machinery-space protection. This system combines the extinguishing characteristics of water with the penetrative qualities of the previously deployed gases without any safety hazards for crew members or the environment. Compared to traditional sprinklers, high-pressure water mist sprinklers have faster fire extinguishment and reduced levels of fire and smoke damage with minimal danger of reignition. Due to the small amount of water used to suppress fires, minimal water damage is caused to facilities and equipment. In addition, international requirements mandating the installation of automatic sprinkler systems or their equivalent in oceangoing passenger vessels has led to the usage of water mist systems on some international cruise vessels.

A high-pressure water mist system uses the energy of a high-pressure pump or bank of compressed nitrogen to atomize sprays of water directed at a fire. This action gives the water droplets high momentum and great penetration throughout the protected areas. Water mist systems offer the advantage of extinguishing flammable liquid fires with water using flow rates one-tenth those of traditional deluge sprinkler systems. In addition, water mist can flow around obstructions more easily than the sprays from traditional sprinklers. High-pressure water mist is employed for both machinery-space protection and accommodation-space (including public space and service space) protection. The system flow rates, nozzle spacing, and water droplet sizes generally differ between systems extinguishing Class A fires in accommodation spaces and Class B fires in machinery spaces.

Water mist systems used to protect accommodation, public, and service spaces are arranged very similarly to traditional automatic sprinkler systems. The piping is filled with pressurized water, and each sprinkler individually operates when a glass bulb inside the sprinkler reaches its activation temperature. These systems require a filtration system to ensure that debris in the water or piping system does not block the very small orifices of the nozzles.

Because the flow rates for water mist systems are generally much smaller than those for traditional sprinkler systems, the associated weight of the water-filled piping is also much lower. The accommodation and public spaces are typically located high in a cruise vessel so that the weight of the sprinkler system piping impacts the center of gravity of the vessel. The lower weight of water mist systems allows vessel designers to reduce the impact of the fire suppression system on a vessel's stability. The amount of water on the deck from a water mist system is also less than that of a conventional sprinkler system. Thus a water mist system operating the same length of time as a conventional system has less impact on a vessel's center of gravity.

Water mist systems employed on vessels in international trade must meet fire suppression performance requirements set by the International Maritime Organization (IMO). In addition, such systems on U.S. vessels generally must meet the requirements of National Fire Protection Association (NFPA) 750, *Standard on Water Mist Fire Protection Systems.* The U.S. Environmental Protection Agency convened a medical panel of experts to consider water mist and found that water mist fire suppression systems pose no significant health concerns and that the use of water mist systems is appropriate without restrictions in areas occupied by the public.

Water mist systems in machinery spaces are generally arranged to either flood the entire space with mist or apply mist locally around high-hazard machinery (Figure 8.18). Because a water mist system breaks water into very fine droplets, the surface area of the water is maximized, allowing the water to quickly absorb large quantities of heat. The cooling effect of the mist slows the burning rate of a fire and also acts to cool the burning liquid fuel. Water vapor generated by the evaporating mist is cooled by the ongoing and surrounding mist and becomes entrained in the fire, reducing the combustion process. The fine mist droplets block heat radiated by a fire, and propagation of the fire is slowed. These combined effects allow water mist to extinguish flammable liquid fires in enclosed spaces and also in open spaces under limited circumstances. The cooling afforded by water mist can be so great that during controlled tests using very large flammable liquid fires in simulated machinery spaces, fire teams have reported that they were able to advance to the seat of fires during water mist discharge without discomfort.

Water mist is most effective when all the openings to the enclosed space are closed. However, tests show that water mist can be very effective even when doors, vents, or other openings are left open.

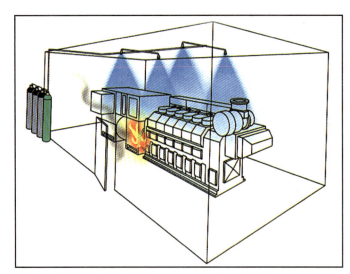

Figure 8.18 After the high-pressure water mist fire suppression system is activated by a detection device, a fine mist is discharged into the machinery space. The high-momentum water mist effectively penetrates the fire plume and extinguishes the fire. After the fire is extinguished, the high-density water mist is circulated around the space to cool the environment and scrub the smoke and hot gases, thus preventing reignition. *Courtesy of Marioff, Inc., Finland.*

Fire Main Systems

The fire main system is a vessel's primary mechanism for getting water to the location of a fire. It is required no matter what other fire extinguishing systems are installed. The fire main system supplies water to all areas of the vessel. Fortunately, the supply of water at sea is limitless. The movement of water to the fire location is restricted only by the system itself, the effect of the water on the stability of the vessel, and the capacity of the supply pumps.

Water may be moved or pumped by several different means through the fire main system when the vessel is at sea. Aside from fire pumps, other alternatives are general service pumps, ballast pumps, butterworth pumps, etc. (Figure 8.19). Fire main systems include fire pumps, hydrant outlets, piping (main and branch lines), control valves, fire hose, and nozzles. One or more international shore connections are required to access a shoreside water supply. Aside from the shore connections, other water enters the fire main system through the sea chest, which is frequently covered with marine growth. It is good practice to fit all hydrant outlets with self-cleaning marine strainers. These strainers remove matter that might clog nozzles, particularly the fine holes in all-purpose nozzles and applicators.

Every crew member needs knowledge of the use and operation of the vessel's fire main system and various pumps. Before discussing the many components and applications of the fire main system in the sections that follow, information on water characteristics and pressure theory is given first.

Water Characteristics/Pressures

The basic physical characteristics of water dictate the way in which it is used for fire suppression purposes. In order for crew members to understand and appreciate their fire fighting water supply, it is important that they fully understand the principles and theory concerning the water that is pumped as well as the mechanical nature of the pump itself. A crew member who knows these characteristics and principles can use water more efficiently and in a manner that protects the pump and fire main system from damage. Properties of water and types of pressure are discussed in the following sections to complete a crew member's understanding.

Properties of Water

Water is an effective coolant (absorbs heat) and thus very effective in fire extinguishment. Water also has a smothering effect by excluding oxy-

Figure 8.19 Two examples of main fire pumps. Fire pumps are usually dual service, so although most vessels have two main fire pumps, they may employ several other pumps on the fire main. *Courtesy of R. Wright/Maryland Fire and Rescue Institute/United States Coast Guard.*

gen. Smothering occurs to some extent when water converts to steam in a confined space. Steam expansion helps cool the fire area by driving heat and smoke from the area. The amount of expansion varies with the temperature of the area. At 212°F (100°C), water expands approximately 1,700 times its original volume.

Water does have some limitations. Water may react with combustible metals, and it readily conducts electricity. Between 32°F and 212°F (0°C and 100°C), water is usually in the liquid state, which is the form most useful for extinguishing fire. As a liquid, water handles easily. It pumps through the fire main system and fire hoses and applies very effectively to a fire. When water freezes and changes from a liquid to a solid, it expands. This action exerts strong pressures and can cause extensive damage to equipment.

Water in a standing state is *static* (at rest or without motion). Several characteristics of water at rest are important. First, water is virtually incompressible. In its fluid state, it would take a pressure of 33,000 psi (227 527 kPa) {2,275 bar} to reduce its volume by 1 percent. For this reason, water does not store pressure. As soon as pressure is removed, no appreciable expansion of water occurs, and the pressure is relieved. Water pumps easily because it cannot compress when pressure is applied; therefore, it moves rapidly into an area where pressure is lower (for example, from fire main to fire hose to nozzle).

Several characteristics of water in *motion* are also important. Water in motion tends to remain in motion, without changing direction. Water frequently moves through fire hose at a rate of 1,000 feet (305 m) per minute or more. If the flow is interrupted suddenly, the high velocity plus the weight of the water create momentum that causes *water hammer* (often heard as a distinct sharp clank). This force can damage fire hoses and equipment and can also be dangerous for fire hose handlers. Most fire pumps on vessels have a relief valve that protects them, but it is best to close valves slowly to prevent water hammer.

The size of the nozzle opening and the velocity of the water as it leaves a fire hose determine the amount of water that flows through a hose. The velocity of the water leaving an opening is measured by using a pressure gauge connected to a pitot tube inserted into the center of the stream. From this measurement, it is possible to calculate how much water is flowing from the opening.

When water moves through any type of waterway, the friction of the water against the surface of the waterway creates a resistance to its movement known as *friction loss*. The pressure loss due to this resistance applies to fire hose as well as to the pipe and fittings used in fire main systems. Friction loss increases with the amount of water flowing through a given size waterway, and it decreases with an increase in the size of the waterway with a constant flow.

Types of Pressure

Pressure is expressed in several ways and several different units of measurement. See Chapter 1, Fire Science and Chemistry, for more information. Pressure results when water is forced into a confined area faster than it can move out. Any time additional fire hoses are placed in service or water flow is increased, pressure can drop. Increasing the supply without a corresponding increase in discharge causes the pressure to rise.

Pressure exerted in all directions in any system with no water moving is *static pressure:* stored potential energy that is available to force water through pipes, fittings, fire hose, and adapters. The pressure normally found in a fire main before water flows from a hydrant is static.

Residual pressure is that part of the total available pressure not used to overcome friction or gravity while forcing water through the fire main system, fittings, fire hose, and adapters. *Residual* means a remainder or that which is left. If a pressure gauge is attached to a hydrant to which flowing fire hoses are connected, the pressure indicated is the residual pressure. However, residual pressure is relevant only at a particular point. A pressure gauge at another location on the fire main system would likely indicate an altogether different residual pressure.

Water moving through a system loses pressure while it is in motion. The residual pressure of a water system is the static (no movement) pressure of the system minus the friction loss between the source and the point of measurement. See the following example.

Water Pressure Example

A fire pump can create 150 psi (1 034 kPa) {10 bar} in a fire main system when no water is flowing (static pressure). When water flows, the pressure at the hydrant is reduced to 100 psi (689.5 kPa) {7 bar} (residual pressure). The reduction in pressure is due to the friction loss between the fire pump and the hydrant.

Static pressure - Residual pressure = Friction loss

OR

150 psi (1 034 kPa) {10 bar} - 100 psi (689.5 kPa) {7 bar} = 50 psi (345 kPa) {3 bar}

Water pressure is generated *naturally, mechanically,* or *chemically.* Naturally created pressure comes from the force of gravity and the mass of the water. Pressure differences due to gravity also occur as a result of change in elevation between different points in a water supply operation. When water is stored at a higher elevation than the point of discharge, 1 psi (6.89 kPa) {0.069 bar} is generated for each 2.3 feet (0.7 m) of height differential. Water pressure due to elevation is *head pressure.* Fire pumps on ships are designed to pump against a specified head pressure in order to deliver adequate pressure through the fire main to the uppermost deck. Fire pumps generate mechanically created pressure. To determine how much head pressure the pump can handle, divide the feet of head rating by 2.3 (or 0.7 for meters) and convert it to psi (kPa) {bar}.

Chemically created pressure is generated when two substances combine, resulting in a chemical reaction. This principle was used in early fire extinguishers such as soda acid chemical foam extinguishers. Chemical extinguishers were also used on early shoreside fire apparatus for rapid fire extinguishment before fire hoses were laid and fire pumps put into service.

Fire Pump Operations

The number of fire pumps required and their capacity, location, and power sources are specified by regulatory agencies. Because of the different requirements that are unique to the country where the vessel is registered, crew members may encounter various types of fire pumps. A review of the most common types, centrifugal and positive displacement, is given in the sections that follow. Pump performance, testing, safety, uses, and alternatives to the main fire

pump are also discussed. Discussions of the fire main piping and the international shore connection are included.

Number and Location

The number of fire pumps varies, but the most common requirement for most vessels is at least two. Refer to the regulating authority to determine the requirements for a particular vessel. Fire pumps are located in separate spaces on vessels that are required to have at least two. The fire pumps, sea suction, and power supply are arranged so that a fire in one space does not remove all the pumps from operation and leave the vessel unprotected. Any alternative to the two separate pump locations requires the prior approval of the regulatory agency and the installation of a CO_2 flooding system to protect at least one fire pump and its power source. This arrangement is permitted only in the most unusual circumstances. Generally, it is used only on special vessels where safety would not be improved by separating the pumps.

Performance and Testing Requirements

Each fire pump must be capable of delivering at least two powerful streams of water (having at least 40 feet [12 m] of reach using standard fire hoses and nozzles) from the two farthest or highest hydrant outlets from the pump. These outlets must also be at opposite ends of the vessel. According to regulations, each fire pump on cargo vessels must deliver water simultaneously from the two highest outlets at a pressure of approximately 50 psi (345 kPa) {3 bar}. Pumps on tankers may require from 71 to 75 psi (490 kPa {4.9 bar} to 517 kPa {5.17 bar}). These requirements are the same for the fire main piping. Requirements must be met when the system is tested. See Title 46 Code of Federal Regulations (CFR) 95.10-5 and Title 46 CFR 34.10-5 for details.

Other fire fighting systems, such as an automatic sprinkler system, may also be connected to the fire main pumps. However, the capacity of the fire pumps must then be increased sufficiently so that they can supply both the fire main system and the other system with adequate water volume and pressure at the same time.

Vessels are designed, built, and, when necessary, refitted to comply with regulations. Crew members are responsible for keeping pumps in good condition. All pumps require regular maintenance. Engineering personnel are usually charged with the responsibility of maintaining and testing the vessel's fire pumps periodically to ensure their performance and reliability during an emergency. Test conditions must prove that the pumps meet the minimum performance requirements at the highest possible draft of the vessel.

Safety

Every fire pump is equipped with a relief valve on its discharge side. The relief valve is set slightly above the pressure necessary to provide the required fire streams, but below the pressure at which the fire main system is designed to withstand. A pressure gauge is also located on the discharge side of the pump. Do not close valves while a pump is operating.

Do not use a pump that is connected to an oil line as a fire pump. The pump could possibly pump a flammable liquid rather than water through the fire main system. In addition, the pump could contaminate the system with oil, which would clog applicator and nozzle openings. Oil in the water would also ruin the linings in fire hoses.

Other Uses

Various pumps may be used for supplying water to fire mains. However, one of the required pumps must be available for use on the fire main system at all times. The reliability of fire pumps is probably improved when they are dual purpose, for example used for other services such as general service and ballast pumps.

When control valves for other services are located at a manifold adjacent to the pump, water is readily obtained by opening the valve to the fire main. Fire main piping that is already installed in most spaces is a tempting source of water. However, improper or careless use quickly reduces the reliability of the fire main system. If the fire main pumps are used for purposes other than fire fighting (such as deck washing or tank cleaning on tankers), make connections only to a discharge manifold near the pump. Use the fire main system only when specific exceptions are granted.

Connections to the fire main system for low-water demand services in the forward portion of the vessel (such as anchor washing, forepeak eductor, or chain-locker eductor) are frequently allowed. In such cases, each fire pump must meet its water flow requirements with the other service connection open. This requirement ensures that the effectiveness of the fire main system is maintained if the other service connection is accidentally opened. Water availability is maintained on deck at all times when such activities are conducted.

A good precaution when in port is to keep the fire main pressurized using the anchor-wash as an outlet. This procedure ensures that fire-fighting water is instantly available merely by opening a hydrant. In cold weather it also keeps the fire main from freezing, but be aware that exterior branch lines may freeze during severe weather conditions. If the pump stops for some reason, an alert deck watch will notice when the sound of water from the hawse pipe ceases.

Centrifugal Pumps

Nearly all fire pumps used on vessels at the present time are the centrifugal type. The *centrifugal pump* does not pump a definite amount of water with each revolution. Rather, it imparts velocity to the water and converts it to pressure within the pump itself, which gives the pump a flexibility and versatility that has made it popular for fire fighting purposes.

In theory, the operation of a centrifugal pump is based on the principle that a rapidly revolving disk *(impeller)* tends to throw water introduced at its center toward its outer edge *(centrifugal force)*. The faster the disk turns, the farther the water is thrown or the more velocity the water has. When the water is contained at the edge of the disk, the water at the center of the container *(casing)* begins to move outward. The velocity created by the spinning disk is converted to pressure by confining the water within the casing. Then, the water is directed to the discharge of the pump. Fundamentally, the centrifugal pump consists of these two parts: the impeller (which transmits energy in the form of velocity to the water) and the casing (which collects the water and confines it to convert the velocity to pressure) (Figure 8.20).

Water is confined in its travel by the shrouds of the impeller, which increases the velocity for a given speed of rotation. The height to which water rises or the extent to which it overcomes the force of gravity depends upon the speed of rotation. The impeller is mounted off-center in the casing, which creates a water passage *(volute)* that gradually increases in cross-sectional area as it nears the discharge outlet of the pump. The increasing size of the volute is necessary because the amount of water passing through the volute increases as it approaches the discharge outlet. The gradually increasing size of the waterway reduces the velocity of the water, thus enabling the pressure to build proportionately (Figure 8.21). The following factors greatly influence a centrifugal fire pump's discharge pressure:

- ***Amount of water discharged*** — With all other factors remaining constant, the amount of output pressure that a pump may develop is directly dependent upon the volume of water discharging. If the discharge outlet is large enough in diameter to allow the water to escape as it is thrown from the impeller and collected in the volute, pressure buildup will be very small. If the discharge outlet is closed, a very high pressure buildup results. In other words, the greater the volume of water flowing, the lower the discharge pressure.

- ***Speed of impeller*** — The greater the speed of the pump, the greater the pressure that develops because the discharge pressure results from the velocity and the amount of water that is moving. This pressure increase is approximately equal to the square of the change in impeller speed. For example, doubling the speed of the impeller results in four times as much pressure if all other factors remain constant.

- ***Pressure of water entering the pump (intake pressure)*** — Isolation of discharge pressure from intake pressure in the impeller occurs entirely by the velocity of the moving water. No positive mechanical blockage exists between the intake and the discharge outlet. Water flows through the pump even if the impeller is not turning. When water under

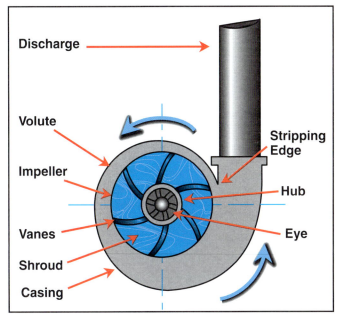

Figure 8.20 The major parts of a centrifugal fire pump.

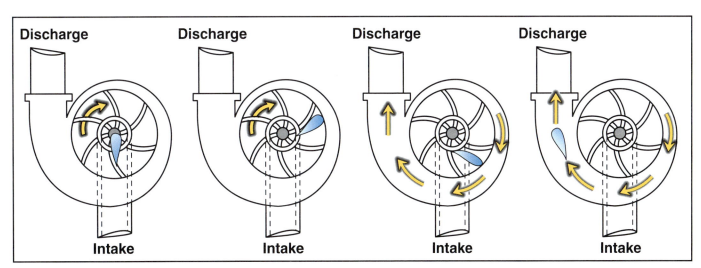

Figure 8.21 Schematics (left to right) tracing the path of water flow through a centrifugal fire pump.

pressure is supplied to the eye of the impeller, it moves through the impeller by itself. Any movement of the impeller increases both the velocity of the water and the corresponding pressure buildup in the volute. Since the incoming pressure adds directly to the pressure developed by the pump, incoming pressure changes are reflected in the discharge pressure.

Most centrifugal fire pumps are constructed with a single impeller (single-stage). Pumping capacities of main fire pumps can vary from 500 to 1,500 gallons per minute (gpm) (1 893 L/min to 5 678 L/min) or more.

Impurities, sediment, and dirt are present in seawater. As these pass through a centrifugal pump, they cause wear when they come in contact with the impeller, which is turning nearly 4,000 revolutions per minute (rpm) when the pump approaches its capacity. This wear greatly accelerates when pumping water with a high sand content. The sand particles passing between the impeller and the pump casing act like sandpaper in wearing down the metal surfaces. Normally a small opening (0.01 inch [0.25 mm] or less) is maintained between the pump casing and the hub of the impeller, which prevents water from escaping back into the intake. As wear increases the gap, greater amounts of water (water that is not available at the discharge) are allowed to slip back into the intake. Eventually, the pump is no longer able to supply its rated capacity. The first indication that wear is becoming a problem is when engine rpm has to be increased in order to pump the rated capacity in pump tests.

To restore the capacity of the pump without replacing the pump itself, place replaceable wear rings or clearance rings in the pump casing to maintain the desired spacing between the hub of the impeller and the casing. If the hub of the impeller is also worn down, it is possible to install smaller wear rings to compensate for the smaller size and maintain the proper clearance.

When discharges are closed, the energy supplied to the impeller is dissipated in the form of heat because the water within the pump is allowed to churn. No harm results from closing all discharges for short periods of time, but if this situation continues for extended periods, the water in the pump can become quite hot, and the metal parts may expand. If the wear rings and the impeller expand too much, they may contact each other. The friction of the two surfaces rubbing together may cause even more heat. In extreme cases, the wear rings may seize, causing serious pump damage. Make sure that some water is moving through the pump at all times.

The impellers are fastened to a shaft that connects to a gearbox that transfers the necessary energy to spin the impellers. A semitight seal must be maintained at the point where the shaft passes through the pump casing. Packing rings make this seal in most fire pumps. The most common type of packing is a material made of rope fibers impregnated with graphite. The material is pushed into a stuffing box by a packing gland driven by a packing adjustment mechanism. As packing rings wear with the operation of the shaft, the packing gland can be tightened.

Positive Displacement Pumps

Positive displacement pumps are necessary in situations where the pump intake is above sea level because they can pump air and centrifugal pumps cannot. For this reason, positive displacement pumps are used as priming pumps to get the water into the centrifugal pump in these situations. The basic principle of the positive displacement pump is a hydraulic law that stems from the near incompressibility of water: *When pressure is applied to a confined liquid, the same pressure is transmitted through the liquid, outward against the walls of the confining container and equally in all directions.* By removing the trapped air in the pump, water is forced into the pump casing by atmospheric pressure.

Two basic types of positive displacement pumps exist: *piston* and *rotary*. Rotary pumps are the simplest of all fire pumps from a design standpoint. In recent years, their use has been confined to small capacity booster-type pumps and priming pumps. Most of the rotary-type pumps in use today are either *rotary gear* or *rotary vane*.

Piston pumps. A piston pump contains a piston that moves back and forth inside a cylinder. The pressure developed by this action causes intake and discharge valves to operate automatically and moves the water through the pump. As the piston drives forward, the air within the cylinder compresses, creating a higher pressure inside the pump than the atmospheric pressure in the discharge manifold. This pressure opens the discharge valve, and the air escapes through the discharge hose. This action continues until the piston completes its travel on the forward stroke and stops. At that point, pressures equalize, and the discharge valve closes. As the piston begins the return stroke, the area in the cylinder behind the piston increases and the pressure decreases, creating a partial vacuum. The intake valve then opens, allowing some of the air from the intake hose to enter the pump.

As the air from the intake side of the pump evacuates and enters the cylinder, the pressure within the intake area of the pump reduces. Atmospheric pressure forces the water to rise within the intake until the piston completes its travel and the intake valve closes. As the forward stroke repeats, the air is again forced out of the discharge. On the return stroke, more of the air in the intake section is removed, and the column of water in the intake hose rises. This action repeats until all the air is removed, and the intake stroke introduces water into the cylinder. The pump is now primed, and further strokes force water instead of air into the discharge.

The forward stroke causes water to discharge, and the return stroke causes the pump to fill itself with water again *(single-acting piston pump)*. Obviously, this action does not produce a usable fire stream because no water flows during the return stroke. The discharge is a series of water surges followed by an equal length of time with no water flow. A more constant stream is produced with the addition of two more valves because then the pump both receives and discharges water on each stroke of the piston *(double-acting piston pump)* (Figure 8.22). Even with the double-acting pump, the output is a series of pressure surges with two periods of no flow when the piston ends its travel in each direction.

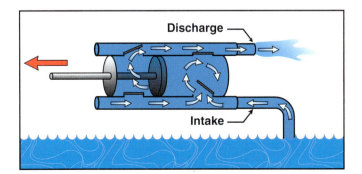

Figure 8.22 A double-action piston pump pushes water on both the forward and return strokes.

Since the pump cylinder contains a definite amount of water, just that amount is delivered by each stroke of the piston. The output capacity of the pump is determined by the size of the cylinder and the speed of the piston travel. In practice, it is more practical to build multicylinder pumps than one large single-cylinder pump. Multiple cylinders are more flexible and efficient because some of them can disengage when the pump's full capacity is not needed. Multicylinder pumps also provide a more uniform discharge because the cylinders are arranged to reach their peak flows at different parts of the cycle.

Rotary gear pumps. The rotary gear pump consists of two gears that rotate in a tightly meshed pattern inside a watertight case. The gears are constructed so that they contact each other in close proximity to the case (Figure 8.23). With this arrangement, the gears within the case form watertight and airtight pockets as they turn from the intake to the outlet. As each gear tooth reaches the discharge chamber, the air or water contained in that pocket is forced out of the pump. As the tooth returns to the intake side of the pump, the gears are meshed tightly enough to prevent the discharged water or air from returning to the intake.

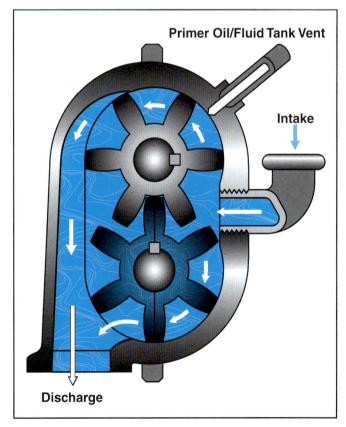

Figure 8.23 Basic design of a rotary gear pump.

The total amount of water that a rotary gear pump can pump depends upon the size of the pockets in the gears and the speed of rotation. Each pocket in the gears contains a definite amount of water, and this water is forced out of the pump each time the gears turn. If the pump is trying to move more water than the discharge hose can take away, pressure builds. Provide an adequate pressure relief device to handle excess pressure.

Rotary vane pumps. The rotary vane pump is constructed with movable elements that automatically compensate for wear and maintain a tight fit with close clearances as the pump is used (Figure 8.24). The rotor is mounted off-center inside the housing. The distance between the rotor and the housing is much greater at the intake than it is at the discharge. The vanes are free to move within the slot where they are mounted. As the rotor turns, the vanes are forced against the housing by centrifugal force. When the surface of the vane in contact with the casing becomes worn, centrifugal force causes it to extend farther, thus automatically maintaining a tight fit. This self-adjusting feature makes the rotary vane pump much more efficient at pumping air than a standard rotary gear pump.

As the rotor turns, air is trapped between the rotor and the casing in the pockets formed by adjacent vanes. As the vanes turn, this pocket becomes smaller, which compresses the air and causes pressure to build. This pocket becomes even smaller as the vanes progress toward the discharge opening. At this point, the pressure reaches its maximum level, forcing the trapped air out of the pump. The close spacing of the rotor prevents the air or water from returning to the intake. The air evacuated from the intake side causes a reduced pressure (similar to a vacuum), and water is forced into the pump by atmospheric pressure. When the pump fills completely with water, it is primed. It then forces water out of the discharge in the same manner as air was forced out.

Alternatives to Main Fire Pump

As mentioned earlier, pumps other than the main fire pump or pumps may supply the fire main with water at sufficient pressures. Such pumps, depending on the type of vessel, may include the following:

- Bilge pumps
- Ballast pumps
- General service pumps
- Sanitary pumps
- Tank cleaning pumps
- Cooling pumps
- Fixed and portable dewatering pumps with eductors

Regardless of what is available in the engine room, every vessel must have an emergency fire pump located outside the engine room. It may be located in the steering flat or forecastle or other suitable location. Take precautions with ventilation if the emergency fire pump is in an oxygen-deficient atmosphere or is difficult to access. If located forward, take care during cargo operations because excessive stern trim may leave the suction out of the water, leaving the vessel without protection. Likewise, smaller vessels that lie aground while loading may require an emergency pump that draws from a ballast or freshwater tank (Figure 8.25).

Which fire pump do you start first in an emergency? It would seem logical that it would be the emergency fire pump. However, that is *not* the case. The emergency fire pump is the *backup pump*. The primary pump is the *main fire pump* although it is prudent to start the emergency (backup) pump at the same time in case the primary pump fails during operation.

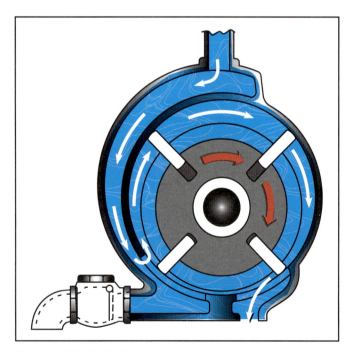

Figure 8.24 A typical rotary vane pump design.

Figure 8.25 Examples of emergency fire pumps located in different locations on different vessels. *Courtesy of R. Wright/ Maryland Fire and Rescue Institute/United States Coast Guard.*

Fire Main Piping

The piping directs water from the pumps to hydrant outlets at the fire stations. The piping must be large enough in diameter to distribute the maximum required discharge from two fire pumps operating simultaneously. The piping system consists of a large main pipe and smaller branch pipes leading to the hydrants. The main pipe is usually 4 to 6 inches (102 mm to 152 mm) in diameter. The branch pipes are generally 1½ to 2½ inches (38 mm to 64 mm) in diameter.

The water pressure in the system must be in the region of 50 psi (345 kPa) *{3 bar}* at the two hydrant outlets that are highest or farthest from the fire pump (whichever results in the greatest pressure drop) for cargo and miscellaneous vessels and 75 psi (517 kPa) *{5.17 bar}* for tank vessels. It would not be unusual to have water pressure greater than 100 psi (689.5 kPa) *{7 bar}* at some locations on large vessels. This requirement ensures that the piping is large enough to handle the pressures needed to provide effective fire streams at all locations.

Figure 8.26 Fire main isolation valve in passageway deckhead. *Courtesy of R. Wright/Maryland Fire and Rescue Institute/United States Coast Guard.*

Protect all sections of the fire main system on weather decks against freezing temperatures. Equip the system with isolation and drain valves so that water in the piping may be drained during cold weather (Figure 8.26).

Two basic fire main pipe layouts exist: *single-main* (also known as *direct system*) and *horizontal-loop* (also known as *ring-main*). Descriptions of these layouts are in the paragraphs that follow.

Single-main system. Single-main systems use one main pipe running fore and aft, usually at the main deck level. Vertical and horizontal branch lines extend the piping system through the vessel (Figure 8.27). On tankers, the main pipe usually runs the length of the vessel down its centerline. On some vessels, such as grain vessels of the North American Great Lakes or similar configurations, the main pipe is located along the port or starboard edge of the vessel's main deck. A disadvantage of the single-main system is its inability to provide water beyond the point of a break.

Horizontal-loop system. The horizontal-loop system consists of two parallel main pipes that are connected together at their farthest points fore and aft to form a complete loop (Figure 8.28). Branch piping extends the system to the fire stations. In the horizontal-loop system, a ruptured section of the main pipe can be isolated. The system can then deliver water to other parts of the system. Isolation valves are sometimes located on the main pipeline forward of each hydrant location to control the water flow if a break occurs.

International Shore Connection

At least one international shore connection to the fire main system is required on either side of the vessel. The connection must be in an accessible location and equipped with cutoff and check valves. A vessel

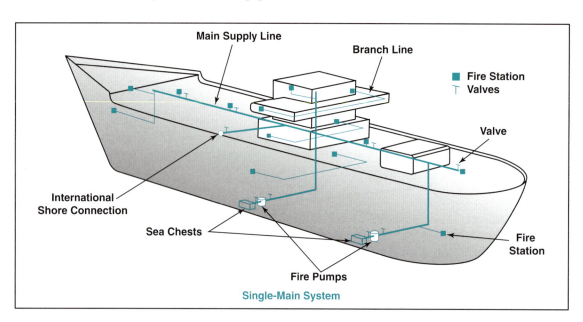

Figure 8.27 Single-main piping system.

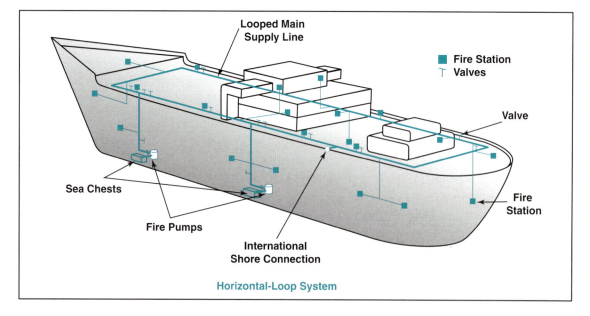

Figure 8.28 Horizontal-loop piping system.

on an international voyage must have at least one portable international shore connection available on either side of the vessel (Figure 8.29). International shore connections connect to matching fittings that are available at most ports and terminals throughout the world. These connections enable crew members to take advantage of the pumping capability of the shore installation or fire department/brigade at any port. The required international shore connections are permanently mounted on some vessels.

The international shore connection is also used when a vessel is in dry dock. Obviously the vessel's pumps are useless when the vessel is out of the water. The international shore connection can establish a water supply from ashore to pressurize the vessel's fire main system. It is important to supply sufficient pressure to maintain flow requirements and that pressures not exceed the designed pressure capabilities of the vessel's fire main system. Clearly post the maximum *do-not-exceed* pressure of the vessel's fire main system at the international shore connection. Contact port fire department/brigade personnel to schedule tests of the international shore connection and exercises in supplying water to a vessel's fire main system.

 ## Foam Suppression Systems

Some fixed suppression systems use foam as an extinguishing agent. Foam extinguishes mainly by smothering with some cooling action. Refer to Chapter 6, Water and Foam Fire Fighting Equipment, for other general foam information: foam characteristics, foam proportioning, and using foam as an extinguishing agent. A foam suppression system is used in locations where water in itself may not be an effective extinguishing agent. These locations include engine rooms and other large machinery spaces, flammable liquid handling or storage facilities, and aircraft hangars. Mechanical foam systems are commonly found in two varieties: *machinery space systems* and *deck foam systems*. Along with describing these two systems, procedures for maintaining, inspecting, and testing foam suppression systems are given in the sections that follow.

Machinery Space System

Mechanical foam systems are often installed for the protection of engine rooms, boiler flats, and other machinery spaces from the threat of fire. The primary job of these systems is to cover the involved area, whether it is below floor plates or in the boiler flat, with a smothering blanket of foam. In many cases, the use of low-velocity applicators or dry chemical units is necessary to extinguish fires of other combustibles above the floor plates or bilge area due to the three-dimensional nature of the fire. Refer to Chapter 12, Fire Fighting Strategies and Tactical Procedures, for additional information.

Wide variations exist in the design and operation of machinery space foam systems. An independent proportioning device supplies some systems, while others provide foam from a foam main that serves multiple systems. In either case, the proportioning device draws the foam concentrate from either a local source or a central foam supply room. The best information on these systems can be obtained from the prefire plans for the vessel (see Chapter 10, Prefire Planning).

In general, many of the systems are manually activated in response to a fire, although some have automatic or partially automatic activation devices that respond to fire conditions. Manually activated systems usually require the opening of valves, either directly or remotely, to allow the introduction of foam and water into the proportioning device and start the foam concentrate supply pump.

Foam application is generally achieved by special aspirating nozzles that aerate the foam and discharge it directly onto bulkheads or onto a special deflector,

Figure 8.29
International shore connection. *Courtesy of Captain John F. Lewis.*

which allows the foam to flow gently onto the surface of a burning liquid. Nozzles are spaced so that a foam blanket flows across the liquid and rapidly covers the entire area. The rate of discharge is adequate for the system to provide an effective blanket of foam within two to three minutes.

Without intervention, the system continues to apply foam until the supply of foam concentrate is exhausted. In the event that the system does not have an automatic shutoff feature tied into a foam supply-monitoring device, it is important to manually turn off the system when the foam supply is depleted. Continued operation applies pure water to the foam blanket, diluting it and destroying its effectiveness.

Additional hose connections (two hydrant outlets) are required outside any space protected by a fixed foam suppression system. These extra connections are required *in addition to* the hydrant outlets otherwise required of the fire main system. These connections give supplemental fire fighting capability to attack fire extending above the floor plates. However, excessive amounts of additional water can dilute the foam blanket below the floor plates. If required, apply water sparingly with a low-velocity applicator. Above all, avoid using straight steams that can quickly tear or break down the foam blanket as well as dilute it.

An operating diagram is provided in the foam supply room for systems that protect more than one space. The schematic diagram illustrates the arrangement of all piping and valves and indicates which valves are opened in response to a fire in each particular space. Complete and specific instructions, explaining all the steps required to operate the system, are also provided. Color coding of valves and corresponding colored markings on the diagram are also very helpful in determining which valves to open. Labeling valves as to their functions is helpful not only for operating the system but also for restoring it to service after a fire.

Deck Foam System

Tank vessels and many other types of vessels must be equipped with deck foam systems. Other vessels on which deck systems are beneficial and required by some authorities are aircraft carriers, vessels prone to spills of flammable liquids, and those capable of receiving helicopters on deck. The benefit of deck foam systems is that they provide blanketing capability for deck fires or spills from *outside* the hazard area.

Both foam fixed appliances (monitors) and hoses provide the discharge of foam, but the monitors are required to deliver at least half the application rate of the system. Monitors have a greater reach than hoses and require fewer crew members to put them into operation and staff them (Figure 8.30). The locations of hose and turret stations are important considerations for deck systems. The foam stations are situated so that the stream reach of each provides an overlap of protection.

A major requirement of deck foam systems is that the monitors deliver a foam solution at a minimum of 0.25 gpm per square foot (3 L/min/m²) of protected deck area. Capacity shall not be less than 330 gpm (1 250 L/min). The piping system for foam stations must have enough valves so that a rupture in the supply piping can be isolated and foam (either premixed or proportioned at the monitor) can still be delivered. For example, if the foam equipment is located aft, a fire could be attacked from aft, working forward.

Both the deck foam system and the individual foam stations require manual activation in response to a

Figure 8.30 A fixed foam monitor on a tanker. *Courtesy of Jon Swain, Texas A & M University System.*

fire. First, start the foam pump and any related water supply pumps. Once the equipment is operational, open the appropriate valves in the foam supply room. At this point, the system is producing foam and introducing it into the piping system. Put turret nozzles into operation by opening the monitor supply valves where each monitor is connected to the main. Air is entrained into the stream at each turret and nozzle to provide the finished product — mechanically produced low-expansion foam.

A deck foam system on a tanker is required to provide foam application at a rate of 1.6 gpm per 10 square feet (6.5 L/min/m^2) of the entire tank surface area for at least 5 minutes. For other spaces, the minimum rate of foam application is 1.6 gpm per 10 square foot (6.5 L/min/m^2) of total cargo surface area for 3 minutes. The minimum quantity of foam available in either case must provide 3 to 5 minutes of continuous full-flow operation. Some authorities require greater time, particularly on newer vessels. Total cargo tank deck area is determined by multiplying the length of the cargo tank spaces by the maximum beam of the vessel. A 6-inch (152 mm) blanket of foam applied for 5 minutes to tanks and for 3 minutes to other spaces is considered sufficient. The 3- to 5-minute figures are based primarily on the presumption that if a fire cannot be controlled by crew members in 3 to 5 minutes, it is unlikely that it can be controlled without resources from outside the vessel. See SOLAS Chapter II-2, Part D, Regulation 61, and Title 46 CFR 76.17 for specific requirements.

The use of water to supply a deck foam system must not deprive the fire main system of its required flow. Furthermore, foam mains are not allowed to be used for any other purpose. The purpose of a foam main system is to provide a simple system for a dedicated purpose. Normally the fire line and foam line crossovers remain in the closed position to prevent the loss of foam into the fire line in an emergency. To design a foam main that could be used for other purposes would make the system too complex to rapidly put into operation in an emergency. A more complex system could also be easily mismanaged so that foam would discharge into ballast tanks or other systems.

Maintenance, Inspection, and Testing

The best preventive maintenance for a foam suppression system is to thoroughly flush the proportioning equipment, mains, and application appliances after each use. Turn off the foam supply, and purge the system with clean freshwater until no foaming action occurs in the discharged streams.

Foam suppression systems are highly complex and require specially trained personnel to service and test them. Inspect all valves and alarms attached to the system semiannually. Inspect foam concentrates, foam equipment, and foam proportioning systems annually. Some foam concentrates have an extended shelf life of up to 20 years if they are stored under appropriate conditions. Refer to the foam manufacturer's specifications for specific shelf life and storage requirements. Nonetheless, perform qualitative tests on the concentrate on an annual basis to ensure that no contamination is present. Inspect the concentrate tank for signs of sludge or deterioration. Most of the standard inspection techniques that are performed for other fixed suppression systems apply to foam systems as well. Some of them are as follows:

- Inspect for obvious physical damage.
- Make sure that valves are in the appropriate open and/or closed positions.
- Check for obstructions in nozzles, turrets, or other openings.
- Agitate the foam and take a sample for testing.
- Check the level of the foam concentrate in the storage tank, and fill as needed.

 ## Inert Gas Systems on Tank Vessels

A type of system designed to prevent fires and explosions, rather than extinguish them, is the *inert gas system*. Inert gases (helium, neon, argon, etc.) have great stability and extremely low reaction rates. Every tank vessel built since 1975 of 100,000 or more dead weight tonnage must be equipped with an inert gas system to protect the cargo tanks. An inert gas system provides an inert gas mixture with an oxygen content not exceeding 5 percent by volume to the cargo tanks. The inert atmosphere excludes fresh air from the cargo tanks even when they are being mechanically washed. It is essential to maintain gas production during discharge so that the tanks are fully inerted before tank cleaning. Furthermore, the system operates in such a manner as to maintain an inert atmosphere in the protected tanks at all times — except during gas-freeing operations. These operations vent the tanks by using a combination of blowers and fans to remove both any residual gases from the cargo and inert gases to make tanks safe for crew members to enter. Tanks are tested to ensure that a safe environment is present before entry. Tanks are then inspected to ensure their structures are intact and clean before loading the next cargo. If tanks are not needed, they are left empty and gas free. The inert gas is introduced before loading cargo and again when unloading.

The standard components of an inert gas system include a gas generator, scrubbers, blowers, distribution lines, valves, pressure and vacuum relief valves, environmental instruments, alarms, and controls. Each of these components is discussed in the following sections.

Gas Generator

While some inert gas systems use the carbon monoxide from engine exhaust systems, the most common type of inert gas generator is the automatic oil-fired auxiliary burner. This device produces the inert gas at a rate equaling 125 percent of the total capacities of all cargo pumps that can operate simultaneously. This generator also delivers this rate of production while maintaining a gas pressure of 4 inches (102 mm) of water on filled cargo tanks during loading and unloading operations.

Gas Scrubbers

Scrubbers are installed to remove heat from the gas and reduce its sulfur and other solid particulate content. Scrubbers are not required if the inert gas is not heated or contaminated. At least two sources provide the water for the scrubbers, and the use of water must not interfere with the availability of water for any fire fighting purposes.

Blowers

Two independent blowers deliver the inert gas to the protected tanks. Together, the blowers deliver a combined flow equal to the production rate of the generator. The pressure delivered by the blowers cannot exceed the design pressure of the tanks.

Gas Distribution Line and Valves

Two nonreturn devices are installed on the inert gas system. One of these devices is a *water seal*. The water level for the seal is kept at an acceptable level at all times. Exercise care during freezing weather to prevent the water seal from freezing. The gas main has an automatic shutoff valve at the point where gas discharges from the generating and scrubbing equipment. The shutoff valve closes the line in response to a blower failure. The branch pipe serving each cargo tank is equipped with a stop valve that clearly indicates whether it is open or closed. Each cargo tank and the main line are fitted with pressure/vacuum relief valves. Some main delivery lines are fitted with a liquid pressure vacuum breaker. The proper amount of antifreeze and fluid level must be maintained for the breaker to operate correctly.

Environmental Instruments

Environmental instruments evaluate the quality and safety of the gas product by giving readings for oxygen concentration, gas pressures, and gas temperatures. Sensors for reading the environmental instruments are provided between the blowers and the delivery point. The operation of each instrument is independent of the others, and each gives continuous readings during the inerting process. Readouts for the instruments are provided at the cargo control station and on the engineering control panel or board. Portable instruments record measurements of the oxygen and hydrocarbon vapor concentrations when sampling the inerted atmosphere.

Alarms and Controls

Each inert gas system is monitored for oxygen content and gas temperature. Audible and visible warning alarms activate at the main propulsion control board when the oxygen content of the inert gas exceeds 8 percent by volume and gas pressure downstream of all nonreturn devices is less than 4 inches (102 mm) of water. Besides producing an alarm, the following conditions require that a corresponding control mechanism automatically turn off the blowers:

• Normal water supply at the water seal is lost.

• Gas temperature is higher than 150°F (65.6°C).

• Normal cooling water supply to any scrubber is lost.

◆ Galley Ventilator Washdown Systems

Several types of galley protection systems have been discussed in this chapter in the appropriate related section. One system that fits into a category by itself is the *galley ventilator washdown system*. It serves two purposes: (1) Prevents duct fires by keeping the ductwork clean (ductwork cleaning) and (2) Prevents fire from spreading to another location by way of the ductwork when a range top, grill, or deep fryer fire occurs (ductwork fire protection).

Ductwork Cleaning

During normal use of galley equipment (which can cause a buildup of grease, lint, and dust), grease-laden air passes through the vent hood and duct system either as a natural process of heat rising or by being drawn through the system by a ventilation fan. Within the duct system, the heated air zigzags around a series of baffles to escape. In the process of the air passing around these baffles, the grease collects on them. The grease then flows to a grease-collecting gutter (Figure 8.31). During cleanup after cooking, the washdown system sprays a solution of hot water and detergent from nozzles within the ductwork onto the gutter. This spray washes the grease out of the gutter and through a drain into a holding tank. The routine use of this system keeps the ductwork free from large accumulations of grease that would add fuel to a range fire or that could even ignite within the duct under high temperatures.

Ductwork Fire Protection

When a fire occurs on the range, a thermostat within the hood system releases a fire damper that blocks the flames from entering the duct. At the same time, the washdown system turns on a water spray and turns off the exhaust blowers. Together these actions prevent the fire from entering or spreading through the ductwork. In addition, the washdown system may have the capability of turning off the power or fuel source to the cooking equipment (Figure 8.32).

When the temperature in the duct falls below 250°F (121°C), the water spray feature reverts to its normal wash cycle, and then turns off. If the washdown system responds to a fire, resetting the fire damper and restoring power to the exhaust blowers easily restore the system to its normal condition once the fire is extinguished.

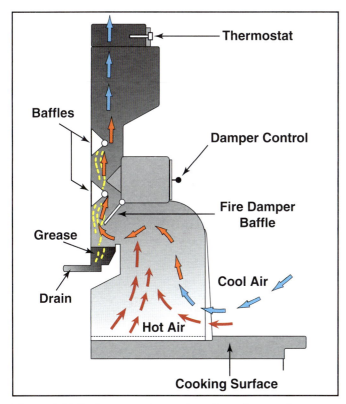

Figure 8.31 During normal operation of a galley ventilator washdown system, grease-laden air passes around a series of baffles where the grease collects and flows to a gutter.

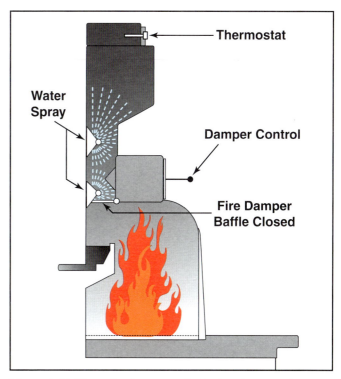

Figure 8.32 When the thermostat in the hood senses fire, the fire damper baffle closes and the spray nozzles activate, which keeps fire out of the ductwork.

Chapter 9

General Fire Fighting Procedures

Fires on vessels are among the most challenging fire emergencies to control. The variety of fuels and cargoes on board and the way in which products of combustion spread throughout a vessel are all problematic. In addition, a vessel's configuration complicates fire fighting operations and extinguishment. Steel decks and bulkheads usually surround a fire in a compartment, making the space difficult to ventilate. Materials burning in a lower cargo hold may be impossible to reach because everything stowed above the fire would have to be removed, which is impractical especially if the vessel is at sea. Fires located on weather decks may be easier to reach, but fire fighting operations are complicated by adverse wind, weather, and sea conditions.

Keep these considerations in mind while studying the material in this chapter. Specific areas and situations are discussed in Chapters 11 (Emergency Response Process) and 12 (Fire Fighting Strategies and Tactical Procedures). For now, the focus is on the general basic procedures used in fire fighting. If the basics are not mastered, the advanced tactical considerations will likely be of no use. This chapter gives an overview of a fire response from sounding the alarm and mobilizing the crew through fire attack and postfire activities (restoring resources, debriefing, etc.). The procedures described generally occur in a chronological order. Some exceptions exist, depending on the situation (for example, ventilation can take place at different times). Safety issues are always considered first.

Mobilizing the Crew

When a fire is detected (either by a person or detection system), the first actions are to sound the alarm and report the fire location in order to mobilize the crew. Crew members report to their emergency (muster) stations to receive instructions and assignments. A crew member who discovers a fire or the indication of fire reports the fire promptly to the bridge before taking any other action (see sidebar for an exception). Do not attempt to extinguish a fire, however small it may seem, until sounding the alarm by voice, telephone, pull box, etc. This point cannot be stressed enough. A delay in sounding the alarm could allow a small fire to become large. The longer a fire has to grow, the more swiftly it spreads. Of course, if two or more people discover a fire, only one needs to sound the alarm. The others stay and attempt to extinguish the fire with available equipment. Report all fires even if they extinguish themselves from lack of fuel or oxygen. The resulting investigation could uncover defects or conditions that, when corrected, would prevent future fires.

A Reporting Exception

A crew member who discovers a small fire in a metal wastebasket can cover it with a noncombustible lid, if readily available, before leaving to report the fire.

The extent of an alarm varies depending on the type of vessel. For instance, an immediate full-scale alarm may not be appropriate on a passenger vessel. To sound a general alarm before the extent of the fire is known and the ability to control it is determined could cause unnecessary and potentially deadly panic among passengers. In this situation, officers on the bridge determine when a general alarm is required.

Reporting the Fire Location

The crew member who sounds the alarm must give the exact location of the fire, including compartment and deck level. This information is important for several reasons: First, it confirms the location for the emergency response team. Second, it provides information about the type of fire to expect. The exact location may indicate the need to secure certain systems and what doors and hatches to close to isolate the fire. The more information that the crew member provides and the more accurate that information is, the better prepared the officers on the bridge can be to take appropriate actions. For example, when the *exact* location is provided along with *what* is burning, the officers will know which machinery stops, power shutoffs, and fuel shutoffs to use if appropriate. The nearest safe location for assembling emergency teams and evacuating passengers can also be determined if necessary.

Mustering

All crew members must report to their assigned emergency muster stations if it is safe to do so. Emergency situations can sometimes tempt individuals to freelance (act independently without authorization), particularly if they happen to pass the fire en route to their muster stations or if they were present when the fire was discovered. If it is safe to do so, only the individual who sounded the alarm remains at or near the fire location to brief the emergency response team(s). All other crew members report to their muster stations to receive assignments and instructions. All crew members not present and accounted for during a muster are assumed trapped, injured, or otherwise threatened by the fire. Crew members who decide to freelance and not report to their designated station may cause the initiation of an unnecessary search and rescue operation. Conversely, the mustering procedure provides valuable information to officers if vessel crew members or passengers *are* trapped or endangered. For this same reason, it is essential while in port to maintain a correct list of crew members aboard/ashore for accountability purposes.

◆ Locating the Fire

The location of a fire determines the methods used to extinguish it. Different procedures are needed in various places such as interior living spaces, interior ma-

chinery spaces, decks, and cargo holds. When an automatic detection system indicates the location of a fire, it is still necessary to confirm that location. Alarm panels on the bridge and in the control room are excellent resources. Often the information provided by these instruments can tell the fire response teams specifically where to search. If the detected fire is located in a cargo hold, it may be difficult — if not impossible — to determine the precise location of the fire within the hold. It may be possible only to identify the fire location by the hold number.

If the reported location is in a large area or zone, it may be necessary to further pinpoint the fire's location. The location is obvious if flames can be seen. However if only smoke is evident, the fire may be hidden behind a bulkhead or a compartment door. Observe certain precautions during attempts to find a fire's exact location. Examine a door or hatch before opening to check for fire (Figure 9.1). Discolored or blistered paint indicates fire behind the door or hatch. Smoke puffing from cracks at door seals or where wiring passes through the bulkhead is also an indication of fire. Feel the bulkhead or door with a gloved hand. If it feels hotter than normal, the fire is probably hiding behind it. If no heat is noticed with the gloved hand, evaluate the surface further with the back of a bare hand before taking further action. If fire is indicated, do not open the door or hatch to the area until crew members wearing full protective clothing and self-contained breathing apparatus (SCBA) and carrying a charged fire hose and appropriate nozzle are ready.

A fire burning in an enclosed space consumes the oxygen within that space. The fire seeks additional oxygen, and a newly opened door gives it a generous supply. When the door to the space is opened, air is pulled through the opening to feed the fire. As described in Chapter 1, Fire Science and Chemistry, a backdraft situation may result with explosive force as flames and superheated gases burst from the compartment. The fire may grow in size with explosive force. Anyone in its path could be severely burned. If the fire is not attacked with a fire stream, it can travel through the area uncontrolled. The longer the fire has been burning undetected, the more dangerous the situation. Therefore, cool the door with water before opening. Have everyone stand clear of the door or the path of its swing. Open

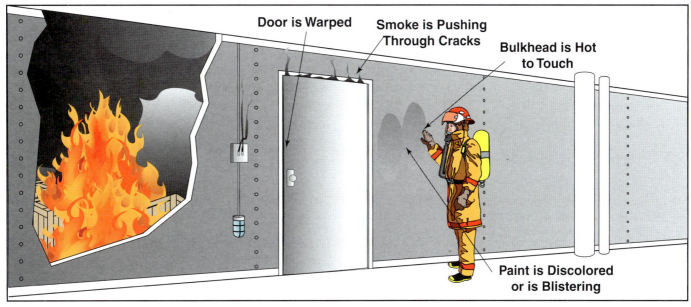

Figure 9.1 Signs of hidden fire: hot bulkhead, smoke pushing through cracks, warped door, and discolored or blistering paint.

the door from a position clear of the opening and opposite the hinges. It may be possible to extend a low-velocity fog applicator through a partially opened door to cool and knock down the fire in the immediate area before entering.

 ## Size-Up

Size-up is a mental process of evaluating all influencing factors before committing crew members and equipment to a course of action. It is an evaluation of the fire situation. An accurate size-up is crucial so that the required actions may be initiated. Failure to perform an accurate size-up can lead to an inadequate or inappropriate fire attack. It is important to relay accurate information to the bridge because other factors such as whether the vessel is in port or at sea, the proximity of other vessels, the nearness to land, and weather conditions could impact a response. The first crew members to arrive at the fire scene might extinguish a small fire. They would probably perform a partial size-up and begin the attack immediately. Larger fires would require a coordinated attack using more crew members and equipment. See Chapter 11, Emergency Response Process, for information on size-up and incident management systems.

When making a size-up, the on-scene leader or officer in charge of the fire team quickly and accurately answers and considers the following series of questions and factors:

- **What are the facts?**
  - Initial report
  - Visual factors
  - Reconnaissance
  - Search and rescue

- **What are the probabilities?**
  - Life hazards
  - Explosions
  - Damage
  - Extension of the fire

- **What resources are available?**
  - *Personnel resources:* crew members, outside assistance
  - *Extinguishing resources:* foam and/or water agents, portable/semiportable fire extinguishers, fixed fire suppression systems
  - *Safety resources:* Personal protective equipment, SCBA

- **What is the best strategy?**
  - Offensive versus defensive
  - Proper extinguishing agent and method for the type of fire or hazard
  - Indirect attack versus direct attack or fixed fire suppression systems

- **What tactics are necessary to achieve the strategy?**
  - Protection of responding crew members
  - Accessing the fire

— Containing the fire

— Delegation of tactical aspects

• **How are strategies implemented?**

— Fire extinguisher teams

— Fire hose teams

— Fixed fire suppression systems

 ## Safety During Emergency Response

Emergency response procedures require teamwork, which is a crucial factor for safety. The first and foremost concern is the protection of life. Fire fighting is an inherently dangerous task, but steps can be taken to protect those engaging the fire such as wearing personal protective equipment (including SCBA), using the buddy system, using protective fire stream patterns, having a means of escape, and ensuring personnel accountability. If the situation is so hazardous that crew members are jeopardized even with these safety measures, choose a more defensive alternative that ensures the safety of all aboard.

Personal Protective Equipment

The appropriate personal protective equipment for fire fighting was described in Chapter 4, Safety Equipment. Do not permit crew members to engage in fire fighting operations until they are wearing full protective equipment, including SCBA (Figure 9.2). An adequate number of properly equipped crew members must be available to attack a fire. It is not necessary to wet down crew members before they attack a fire. Research shows that personal protective equipment performs better and causes less fatigue for the wearer when it is dry.

Temperatures can exceed 1,000°F (538°C), and large concentrations of smoke and dangerous fire gases result even when a fire burns for only a short time. Crew members who are not sufficiently protected against these hazards cannot force an attack against a fire. They may have to retreat or face burns, exhaustion, or asphyxiation. Fire team members who become casualties place a burden on an already limited fire fighting force.

Buddy System

To ensure safety, the buddy system requires that at least two individuals perform functions together as a team in a fire area. One person should not engage in fire fighting alone. If something goes wrong, chances are that at least one member of a team will be capable of getting both to safety. Perhaps more importantly, two minds are better than one, and together they may avoid a dangerous situation that a single individual might overlook.

The buddy system is particularly important for fire hose teams. Not only is it necessary to have at least two persons deploy and operate even the smallest fire hose, but the additional security of a second person or partner is essential. Three crew members in full protective equipment are the minimum recommended to safely operate a 1½-inch (38 mm) fire hose. For each operating fire hose team, an equally staffed backup fire hose team is required. The minimum

Figure 9.2 This marine firefighter is properly equipped and ready to advance a fire hose from the fire station. *Courtesy of Maritime Institute of Technology and Graduate Studies.*

number of crew members required for a 2½-inch (64 mm) fire hose is three to four, which also requires an equally staffed and equipped backup fire hose team. Provide one *additional* crew member on fire hose teams for *each* turn the hose makes and for *each* change in level, up or down. The number of crew members available influences these guidelines. A best-practice staffing recommendation is at least a primary attack fire hose team, a secondary attack fire hose team, and a backup fire hose team.

Protective Fire Stream Patterns

Produce protective fog fire streams with either adjustable nozzles or marine all-purpose nozzles when facing intensive heat (see Chapter 6, Water and Foam Fire Fighting Equipment). The fog pattern helps hold back the fire and heat and provides water-curtain protection for crew members. Use the fog stream carefully to prevent unintended production of large volumes of steam that can also burn fire hose team members.

A protective fire stream pattern can also save a person's life when faced with a flash fire situation. Recall in Chapter 6, Water and Foam Fire Fighting Equipment, the discussion of the operation of the adjustable and marine all-purpose nozzles and the importance of knowing how to adjust the pattern to protect the fire hose team from a sudden flash fire. Practice and drill with the nozzle frequently enough to instinctively know how to obtain the protective pattern. If the wrong choice is made, the straight stream produced can do little to provide protection to crew members.

Means of Escape

Perhaps the most important safety consideration for the attack hose teams is to always ensure a means of escape. Never enter a hostile environment without first knowing a reliable escape route. A fire situation is a dynamic event, and many things can occur that change the situation unexpectedly. The damage caused by a fire can interfere with lighting, hydraulic systems, communications systems, and even the water supply. It is the responsibility of every crew member involved to be aware of these potential problems and alert other team members to hazards.

In Chapter 3, Vessel Types, Construction, and Arrangement, it was learned that automatically closing watertight doors could isolate crew members, cut off an escape route, and damage or literally sever a fire hose (Figure 9.3). Know how to manually open watertight doors and how to chock them open when entering a space through them. Dealing with watertight doors is only one of many considerations to make when planning a means of escape. An escape trunk (enclosed vertical shaft with a ladder) also gives an escape path from the low areas of a vessel (Figures 9.4 a and b).

Another consideration is to provide a means of escape from the vessel. Make survival crafts ready for evacuation concurrent with the fire fighting response. Swing out lifeboats and lower to the embarkation deck level. Usually a designated team such as the backup response team or first aid team prepares the lifeboats so that they are ready if needed.

Accountability

An extremely important aspect of fire fighting operations is the concern for crew member accountability. The concept of mustering is an old but valuable means of accounting for personnel. Historically, once crew members were accounted for, it was not a concern to continue monitoring their whereabouts and well-being. Thus, personnel accountability systems correct this gap in safety during an emergency response.

Many types of accountability systems exist. Chapter 4, Safety Equipment, discusses SCBA accountability systems. Chapter 15, In-Port Fire Fighting and Interface with Shore-Based Firefighters, gives some simple systems for knowing who is aboard and who is

Figure 9.3 A fire hose cutout panel can prevent damage to a fire hose. *Courtesy of R. Wright/Maryland Fire and Rescue Institute/United States Coast Guard.*

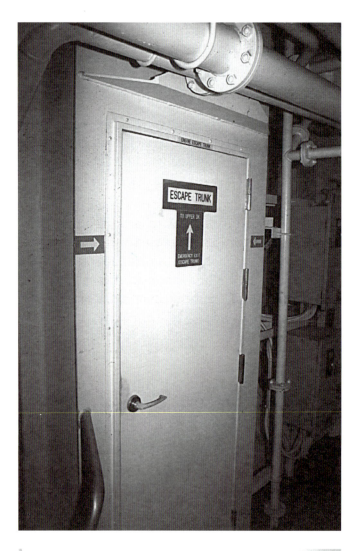

ashore when in port. SCBA systems can be as simple as having an accountability officer record the names and times of those entering and leaving the fire area in a logbook. A tag system where crew members leave a tag from their protective equipment with the accountability officer is another example. Some land-based fire departments even use bar codes that scan the pertinent information from members of a fire team going into and out of a dangerous area. The accountability method must be reliable and used consistently regardless of the type chosen.

◆ Search and Rescue

The rescue of trapped crew members/passengers is an extremely important aspect of every fire fighting operation. Once crew members have gathered at their muster stations and are accounted for (mustered), a search for missing persons can begin. Passengers are also assigned a muster station and lifeboat, and the officer in charge has a list of passengers assigned to that station. If individuals are in the fire area, an initial quick search (primary) may find them before they sustain injuries or die. Rescue may be delayed because of adverse circumstances. For example, suppose someone is trapped in a compartment that is located beyond the fire. If some crew members can get past the fire while others attack it, the rescue may take place immediately. If the fire cannot be controlled easily or quickly, it may be best to attack and control the fire before attempting the rescue. Perform a thorough secondary search after the fire is extinguished if persons are still missing. Secondary searches are more comprehensive and take longer, but they ensure certainty of any victim's whereabouts. Closely examine areas that were filled with smoke and heat.

Sometime the decision on when to attempt a rescue is a difficult one. If the rescue attempt is delayed, an attack could push the fire into an area where one or more individuals are trapped. A holding action may be feasible while an alternative route is used to make the rescue. The decision involves the twofold problem of protecting lives and protecting the vessel. A delay in controlling the fire, due to imprudent rescue attempts, could result in an uncontrollable fire, forced abandonment, and loss of the vessel (and possibly loss of many more lives). The nature of the situation could force a decision to attack the fire in a manner that might be detrimental to trapped individuals but

Figures 9.4 a and b (a) Escape trunk. *Courtesy of Captain John F. Lewis.* (b) Escape scuttle. *Courtesy of R. Wright/ Maryland Fire and Rescue Institute/United States Coast Guard.*

would save the vessel and many other crew members. An intense engine room fire onboard the Royal Australian Navy's supply ship *HMAS Westralia* on May 5, 1998, describes the possible scenario (see Appendix A, Case Histories). After an aggressive but unsuccessful attempt to contain the fire using fire hoses, the decision was made to close the space and activate the carbon dioxide suppression system, although at least one crew member was thought to be in the space. Afterwards, four crew members were found dead. However, medical reports indicated they died from smoke inhalation well before the carbon dioxide system was activated.

Various search procedures are available, but safety concerns are important issues. The safety of the team performing the search must always come first, but once that safety is ensured, every possible effort is made to locate trapped or isolated victims. The search team may take a charged fire hose for protection. Although cumbersome, the fire hose not only gives protection from fire but also may be used as a guideline to return to the original place. The search team can use a guideline (steel-wired rope) as a way of returning to safety when searching in smoke or darkness where there is little or no danger of encountering fire.

Ensure that crew members are familiar with finding their way around the vessel. They can even practice in darkness in their own quarters or down the passageway to the nearest exit. Crew members can practice and drill on search and rescue procedures in unfamiliar spaces. Practice crawling while searching because visibility may be poor in fire conditions. The air is cooler and smoke is less dense near the deck. Establish communication techniques and signals within a team, with other rescue teams, and with fire officers. See Chapter 14, Fire and Emergency Training and Drills, for more information.

Some particular search patterns are appropriate in accommodation spaces, but in machinery spaces, no one pattern is totally possible or successful due to the complex structures. Thoroughly cover each space before proceeding to the next. Establish coordination with other search teams so that efforts are not wasted by searching the same area twice. For example, one system on some cruise vessels is to place a rolled towel from a cabin at the door after it has been searched. Use the senses of sight, touch, and sound when searching. Tapping a bulkhead or pipe may attract a victim's attention. Always check escape areas. In accommodation spaces, always look in closets/lockers, bathrooms, under beds, behind furniture and drapes, and under blankets.

It makes no difference in which direction crew members perform a search of a space. Going clockwise or counterclockwise, they are just as likely to locate a victim. The key thing is to *always* stay in contact with a bulkhead or other structure. When entering a space in a clockwise fashion, keep the left hand in contact with the bulkhead. Use that hand to sweep up and down the barrier for possible openings or indications of danger. Likewise, for a counterclockwise search, keep the right hand on the bulkhead. Continue to sweep up and down, feeling for openings, and identifying objects along the way. If crew members never change hands, they will always come back to the point of entry. Losing contact with the bulkhead, overlooking an opening, or changing hands could cause disorientation and wandering.

A large number of rescue methods or techniques are available to bring a victim to safety. Assess the situation and the physical layout to determine which technique and equipment to use. Some victims may have to be carried up or down stairs or ladders or raised or lowered from a platform. The following are some basic techniques that can be used to remove a victim from danger:

- *Two-arm drag* — One person can rescue a victim who may be either conscious or unconscious as long as the victim is not seriously injured. Kneel, lift the victim from behind, and prop the victim with a leg against the victim's back. Reach under the victim's armpits and grasp the victim's forearms. Stand and drag to safety (Figure 9.5).

- *Clothes drag* — One person can rescue a victim who may be either conscious or unconscious. The victim can be moved on horizontal surfaces without bending the victim's body. Drag the victim to safety by holding onto the clothes (shirt/blouse/coat collar) (Figure 9.6).

- *Two-person extremities carry* — More than one person may be available to rescue a victim who may be either conscious or unconscious. One rescuer holds the victim in the same way as the two-arm drag. The other rescuer faces the victim, stands

Figure 9.5 Rescue method: two-arm drag.

Figure 9.6 Rescue method: clothes drag.

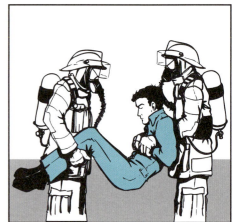

Figures 9.7 a–c Rescue method: two-person extremities carry. (a) Rescuers position themselves in front of and behind the victim. The rescuer behind holds the victim in the two-arm drag position. (b) The rescuer in front lifts the victim's legs. (c) Rescuers stand and carry victim to safety.

between the victim's legs, and lifts one leg with each hand. Both rescuers then carry the victim, feet first, to safety (Figures 9.7 a–c). The rescuer at the victim's head guides the second rescuer as they move to safety. It is best that rescuers face each other during the evacuation so that the victim is not in danger of being injured by the SCBA cylinder unit. If rescuers are not wearing SCBA, the second rescuer may face away from the victim when carrying the victim's legs.

• *Bowline drag*—One person can use this technique to rescue a victim who may be either conscious or unconscious. The victim can be quickly moved on horizontal surfaces without bending the victim's body. Place a rope with a bowline knot around the chest of the victim. Tie the hands of the victim together at the front. Drag the victim to safety by the rope (Figure 9.8).

 Fire Attack

Make the actual fire attack as soon as possible to gain immediate control and to prevent or minimize the extension of fire to *exposures* (areas of the vessel that are adjacent to the fire area on all six sides of the involvement: four bulkheads, space above, and space below). A fire is usually controlled when extinguishing agent is applied to the seat of the fire from initial

Figure 9.8 Rescue method: bowline drag.

fire hoses (and backup fire hoses if they were required), the agent penetrates to the seat of the fire, and the fire is effectively cooled. However, if a fire has already advanced to the point of threatening to spread to other spaces, the first effort must be to control that threat by *establishing boundaries* (protecting exposures by cooling) to prevent a fire from extending beyond its original space. Three attack methods (direct, indirect, and combination) are available for making a fire attack, depending on the circumstances.

Fire teams need practice on handling fire hoses to perform and communicate efficiently in emergency situations. Basic fire hose commands aid the communication process among team members. Fire teams also train to manage water effectively and to have an understanding of the effects of loose fire-fighting water on vessel stability.

Establish Boundaries

Establishing boundaries by protecting exposures prevents a fire from extending beyond the space in which it originated. If accomplished effectively, a fire is usually controlled and extinguished without spreading to other portions of the vessel. To protect exposures, surround a fire on all six sides. Position crew members with fire hoses or portable/semiportable fire extinguishers to cover the flanks and the spaces above and below the fire. The officer in charge must also consider fire travel through venting, ducts, pipe and cable chases, etc. Dispatch crew members to examine and protect all openings through which fire might enter other spaces. Halt the spread of fire by all means. Establishing boundaries is obviously very taxing on crew member resources. However, to attack a fire without first considering the need for establishing boundaries is to invite disaster. Protecting exposures from fire spread requires knowledge of the dangers of heat conduction, radiation, convention, and direct flame contact (Figure 9.9). Therefore, establishing boundaries for an involved compartment can include any of the following actions:

- Cool bulkheads.
- Close doors, hatches, portholes, or similar openings.
- Shut down ventilation or exhaust systems from the involved space.
- Remove combustibles from adjacent spaces.

Attack Methods

A fire attack can be direct, indirect, or combination, depending on the fire situation. The stage of fire development is usually the key indicator of the method chosen. Direct, indirect, and combination attacks differ widely in how they achieve extinguishment, but all are efficient when properly performed.

Direct Attack

In a *direct attack,* a fire team advances to the immediate fire area and applies an extinguishing agent directly to the seat of the fire. The initial attack on a fire may be made with one or more portable/semiportable fire extinguishers or two or more well-coordinated fire hose teams. It may not be difficult to get to the immediate fire area if the fire is small and has not developed into a more advanced stage or fully developed fire (Figure 9.10). A direct attack on a fully developed fire is sometimes coupled with ventilation procedures (see Ventilation section). As a fire increases in intensity, the heat, gases, and smoke produced increase the difficulty of locating and reaching the seat of the fire. In these instances, an indirect attack (described in the next section) may be more effective.

Indirect Attack

An *indirect attack* is employed when the fire development has created a superheated atmosphere in which working in close proximity to the seat of the fire is impossible. An indirect attack with water directs a fire stream to the deckhead above the fire to generate steam that cools the space. An indirect attack by activating a fixed fire suppression system from outside the fire area is chosen for spaces that are not ordinarily occupied by crew members. The success of an indirect attack depends on complete confinement of the fire. Closing doors and hatches and shutting down ventilation systems cut off all possible means of fire travel. The indirect attack is then made from a low point of entry or other perimeter location.

The indirect attack with water causes heat to convert a fog fire stream to steam, which cools the heated atmosphere to a temperature below which combustion cannot continue. One technique for this type of indirect attack is to make a small opening into the enclosed fire space, insert a nozzle, and inject a fog spray to the superheated deckhead. The water partially cools the atmosphere as it is heated to the boiling point and then absorbs a tremendous additional amount of heat when it turns to steam, cooling the space dramatically (although not cool enough yet for a person to enter). Finely divided particles of water produced by a fog stream pattern expose a much greater surface area of water to the heat than other stream patterns. The more finely divided the water particles, the greater the amount of heat absorbed.

Another key factor in an indirect attack with water is that it is applied to the superheated deckhead. Less steam is produced if water is applied to a cooler area. When water is applied to the hot-

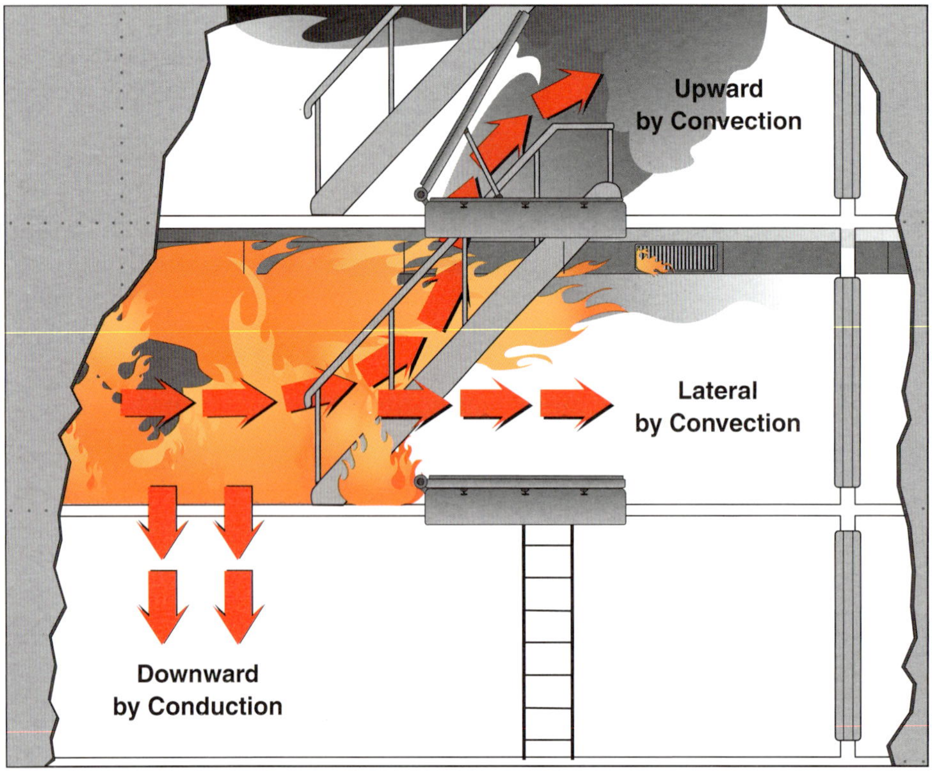

Figure 9.9 Fire can spread upward, laterally, and downward on board a vessel.

(Figure 9.9 labels: Upward by Convection; Lateral by Convection; Downward by Conduction)

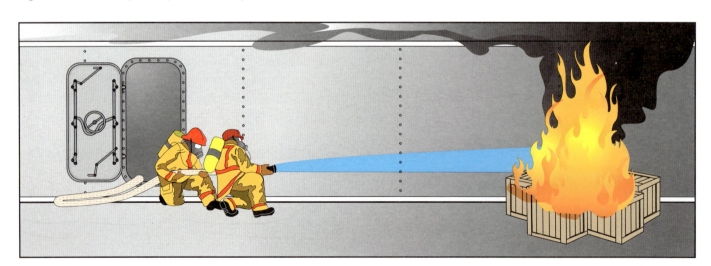

Figure 9.10 Direct attack: Apply extinguishing agent directly to the seat of the fire.

ter areas above the fire, the water flashes to steam immediately and provides an instant cooling of the fire atmosphere. Several factors are essential for a successful indirect attack with water. The more successfully each of the following variables are achieved, the more steam is available to cool the fire area and the more quickly the fire is extinguished:

- Enclose the fire area so that the steam does not escape through a ventilation opening.

- Use fine particles of water for maximum surface area coverage and greatest cooling effect.

- Apply water to the superheated deckhead.

    CAUTION: Wear full personal protective equipment (including SCBA) to prevent steam burns.

When the decision is made to activate a fixed fire suppression system, the fire space is sealed so the agent and fire cannot escape nor can air enter to feed the fire. Close all doors and hatches, stop fans, and close vents. Remove combustibles located next to the space and cool hot bulkheads and decks with water. The time needed for complete extinguishment after system activation may be hours or days. See Chapter 8, Fixed Fire Suppression Systems for more information.

Combination Attack

The combination attack with water uses the steam-generating technique of deckhead-level attack along with a direct attack on materials burning near the deck level. Move the nozzle in a *T, Z,* or *O* pattern, starting with a solid, straight, or penetrating fog stream

directed into the heated gases at the deckhead. Then lower the nozzle to attack the combustibles burning near the deck (Figure 9.11).

Fire Hose Commands

Fire fighting requires teamwork, and teamwork is best achieved through frequent, regular practice. Drill regularly on working together as a team to handle fire hose. Deck fires, in particular, require that fire hose teams work together efficiently and flawlessly to ensure fire containment and to prevent injuries. A wrong move by only one member of a hose handling team could result in a loose fire hose nozzle, sudden exposure to a fire that had been pushed back, or ineffective application of an extinguishing agent.

Execute hose commands in a smooth, fluid motion. Communication is critically important. Be aware of what every member of the hose team is about to do. Never leave a position or perform an action without advising the others. Become skilled at responding to the following basic hose commands and executing them simultaneously as a team unless an individual team member is directed to execute an action alone:

- *Forward*—Advance the fire hose forward, one step at a time.

- *Left* — Shift the fire hose to the left, one step at a time.

- *Right* — Shift the fire hose to the right, one step at a time.

- *Prepare to back out* — *Last Member:* Move to a rear position to keep slack out of the hose as the hose

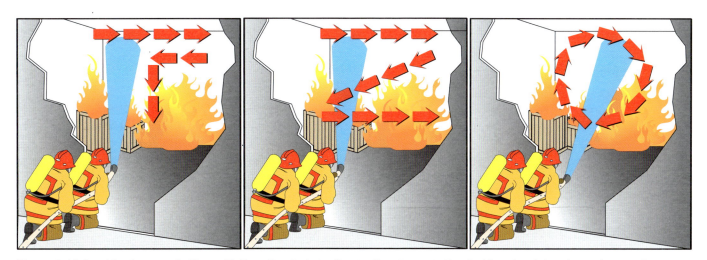

Figure 9.11 Combination attack: Use a *T, Z,* or *O* pattern to direct a fire stream to the deckhead and then lower the nozzle to attack the seat of the fire.

team retreats. Communicate intended actions to the other members of the hose team before executing them.

- *Back out* — Retreat, one step at a time. *Rear Member:* Take up the slack hose.

- *Hold position* — Remain in place (usually while a specific task is achieved such as closing a valve).

- *Down on one knee* — Move together into a position that lowers the level of the hose and fire stream.

- *Straight stream* — *Nozzle Operator:* Switch discharge pattern to a straight stream.

- *Full fog* — *Nozzle Operator:* Adjust discharge pattern to full fog, the widest pattern available.

- *Nozzle on* — *Nozzle Operator:* Open the nozzle.

- *Nozzle off* — *Nozzle Operator:* Close the nozzle.

Water Management/Dewatering

Unconfined, loose fire-fighting water can impair the stability of a vessel. Make every effort to manage water properly and limit its accumulation in large compartments and cargo holds. Begin these efforts with the use of water patterns that allow maximum cooling with minimal quantities of water. Give preference to fog sprays over solid or straight streams. Use only as much water as absolutely necessary. Proper management of fire fighting water is a salvage/loss control action also; it prevents unnecessary water damage to property. See Overhaul Operations section.

Dewatering (removing fire-fighting water) procedures begin as soon as water is used for extinguishment. The basic principle is to move water overboard or to the lowest possible point in the vessel where it can be pumped out. The lack of portable dewatering equipment on some merchant vessels may create some problems, but fixed pumps, ejectors, or eductors can be used. See Chapter 13, Shipboard Damage Control, for more information on vessel stability and dewatering.

 ## Ventilation

Ventilation is the action taken to release combustion products trapped within a vessel and vent them to the atmosphere outside the vessel and replace with fresh air. It can be done before the attack on a fire, during the attack, or immediately after the attack, depending on circumstances. Ventilation is performed for a number of reasons: If a fire is vented promptly and properly, smoke, heat, and gases are diverted away from potential victims, fire teams, and uninvolved combustibles. Besides protecting life, controlling ventilation also prevents fire spread.

Most fire fatalities do not result from burns, but rather from asphyxiation by combustion gases or lack of oxygen. Before smoke or heat becomes apparent, deadly carbon monoxide (CO) and other noxious gases can seep into compartments. These gases easily overcome people who are asleep. Even conscious individuals who are trapped by fire can easily succumb to the effects of CO and other products of combustion. Just as importantly, fire fighting teams have a superheated, noxious atmosphere to face. Use of ventilation techniques can make the fire atmosphere much more tolerable for those charged with the responsibility of entering the fire environment and containing the fire.

Several ways of ventilating a vessel are available. If possible, position the vessel to take advantage of the wind to carry smoke and heat away. *Vertical ventilation* methods (opening at the highest point) use existing vents or holes created to allow products of combustion to escape through the top of the space. *Horizontal ventilation* methods (opening at a point parallel to the horizon) use doors or holes created to channel these products from one side of the space. A combination of vertical and horizontal ventilation is also possible. *Assisted* (also called *forced* or *mechanical*) *ventilation* methods use fans (positive/negative pressure) or fire streams (hydraulic) to force smoke and gases from spaces. These various methods are discussed in the sections that follow.

Vertical Ventilation

Vent the smoke and hot gases generated by a fire vertically to the outside if possible. The configuration of most vessels can make this a difficult task. As a fire intensifies, the combustion gases become superheated. If they ignite, they spread a fire very quickly. In the ideal situation, the gases are released at a point directly above the fire just as an extinguishing agent is applied to the fire (Figure 9.12). Some engine rooms have mechanical access hatches directly overhead that aid ventilation greatly.

Figure 9.12 Vertical ventilation: Smoke and gases are vented to the atmosphere at a point above the fire as extinguishing agent is applied to the fire.

Horizontal Ventilation

Achieve horizontal ventilation by opening windward and leeward doors to create airflow through the spaces where combustion products are collecting. Fresh air flowing in through a windward opening moves the combustion products out through the leeward openings (Figure 9.13). Open the leeward openings first. Open portholes also, although small portholes are not very effective in removing smoke and heat, unless assisted by some other method or means. See Assisted Ventilation section.

Combination Vertical and Horizontal Ventilation

When a fire is below deck, it is difficult to move smoke and heat out of the vessel. In some instances, a com-

bination of vertical and horizontal ventilation may work. A horizontal flow of air may sometimes be created over a hatch on the deck above the fire. This airflow can produce a Venturi effect that pulls smoke and heat upward from the lower deck (Figure 9.14). A properly placed ventilation fan helps move the air more rapidly. Close the openings to uninvolved areas to keep out contaminated air. These doors remain closed until ventilation is complete.

Assisted Ventilation

Ventilation is assisted with fans/blowers or fire streams. Properly positioned portable fans/blowers can move smoke-contaminated air out of compartments, along passageways, and up through deck openings. Place the fans/blowers to either push (positive

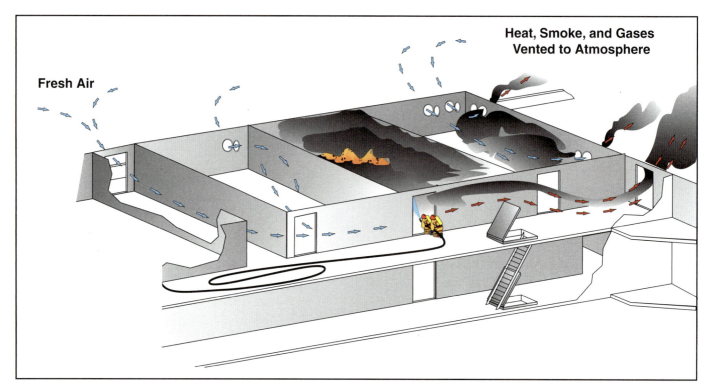

Figure 9.13 Horizontal ventilation: Fresh air entering through windward doorways and portholes pushes heat and smoke out leeward doorways and portholes.

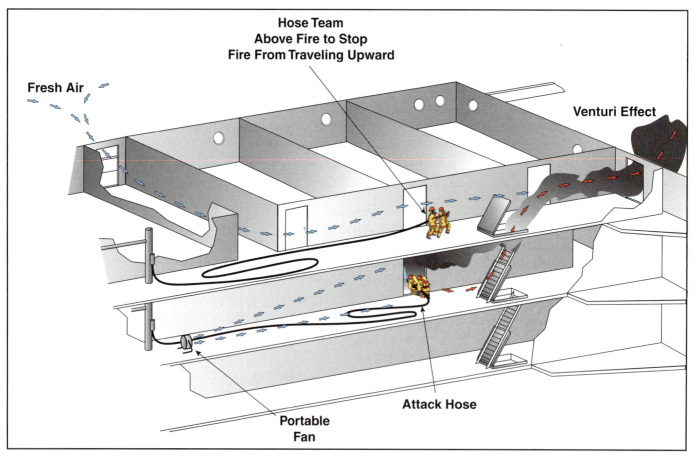

Figure 9.14 Combination ventilation: Airflow is created above the fire. It pulls combustion products up from the involved deck and out the doorway.

pressure) or pull (negative pressure) air in order to establish airflow from the contaminated area to the outside. Positive- and negative-pressure applications can complement each other with a "push/pull" technique. In some instances, the vessel's mechanical air intake system can be used in conjunction with portable fans/blowers. Even though not designed for fire fighting ventilation purposes, some air conditioning systems have recirculation features. Positive-pressure ventilation is far more effective at moving air within and out of a space and is a safer operation than negative-pressure ventilation. Positive-pressure ventilation can be set up outside the fire space during ongoing interior attack operations. After a fire is extinguished, smoke can also be removed by directing a fire stream out an opening, which is the basic technique of hydraulic ventilation.

Positive-Pressure Ventilation

Positive-pressure ventilation involves introducing fresh air into a confined space at a rate faster than it is exiting, thus creating a slight positive pressure within the space (Figure 9.15). Positive pressure counteracts the pressure being generated by the combustion process. Three basic applications for positive-pressure ventilation are possible in the marine fireground environment: (1) removal of smoke after extinguishment, (2) prevention of smoke and heat spread into the boundary areas around the fire, and (3) removal of heat and smoke during a fire to allow a fire team to make a quicker, safer, and more efficient attack. Using positive-pressure ventilation during a fire attack (referred to as a *positive-pressure attack*) allows fire entry attack teams to access the fire area faster and more safely and use less water. However, several problems

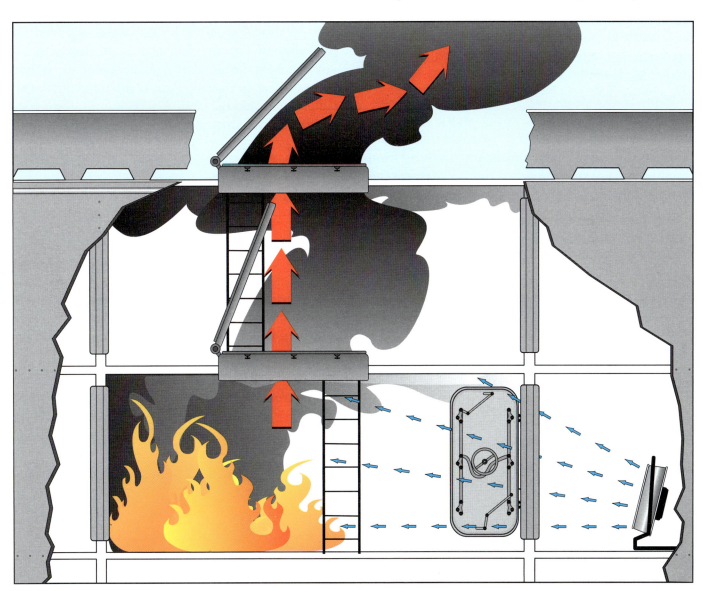

Figure 9.15 Positive-pressure ventilation: A portable fan introduces fresh air into a space faster than it is exiting.

such as forward moving smoke explosions, extension of the fire to uninvolved boundary areas, and backdraft situations must be considered before using positive-pressure ventilation as part of a fire attack plan. When positive-pressure ventilation is performed correctly, the likelihood of fire being pushed into unaffected areas is minimal.

Before considering positive-pressure ventilation for any fire fighting purpose, several basic issues must first be addressed. Size-up must include determining the ventilation process required to mitigate the incident. A thorough knowledge of the vessel's construction and arrangement (layout) is also essential. Knowing where ventilation ducts and wire and pipe chases go and what openings are available for the ingress and egress of pressurized/forced air are critical factors. Select a ventilation egress point as near to the fire location as practical. Single, series, parallel, and/or hybrid configurations of positive-pressure ventilation fans/blowers can be established at predetermined ingress points.

Using positive-pressure ventilation within designated fire boundaries assists fire attack teams because pressurizing the boundary areas to the seat of a fire removes radiant heat and steam produced from cooling with fire streams. As a result, these teams can maintain their positions longer and more safely.

When using positive-pressure ventilation during postextinguishment activities, select an exit as close to the fire compartment as practical, and seal all other locations of ventilation egress. Start the blowers/fans, and then open the entrance/ingress point to start the movement of air. Positive-pressure ventilation removes smoke and by-products of combustion faster than conventional methods, thereby improving visibility and atmospheric conditions. Overhaul can proceed in a safer and cooler environment.

Hydraulic Ventilation

To remove smoke from a compartment after a fire is extinguished, employ the basic technique of hydraulic ventilation by directing a fire stream fog pattern out any available opening. Smoke and combustion gases are pulled from the area as the fog pattern creates Venturi airflow dynamics. For best results, use the largest fog pattern (80 percent coverage of the egress space) that will fill the opening yet allows little or no water flow into the compartment. Pay attention to the quantity and location of water flow so as not to add unnecessary amounts of water to the interior of the vessel.

 ## Crew Member Rehabilitation

During a shipboard fire emergency response, crew members may encounter extreme heat, dehydration, and fatigue. Identify the need for rehabilitation (rehab) and measures that are required to prevent injuries to crew members under these circumstances. Most crew members can perform strenuous fire fighting duties for an average of 15 minutes; therefore, use a general guideline of 15 minutes of response activity and 30 minutes of rest. If this guideline is not used effectively, the response capabilities of the few crew members aboard are limited.

During a prolonged response, emergency response crew members generally require water and food replenishment. Locate rehab facilities in an area that is covered from the elements and has adequate cooling or heating as necessary. Supply the rehab area with water for drinking, food if necessary, and an area where an individual may remove all equipment and rest. A person not allowed to properly rehabilitate may have diminished capabilities on the next entry as well as risk medical problems such as heat exhaustion, heatstroke, and dehydration — all of which can be medically dangerous to crew members if no immediate advanced medical help is available.

Medically evaluate each person in rehab to identify crew members who could possibly be injured in the next entry. Obtain information on vital signs (blood pressure, pulse, body temperature, etc.) and overall general appearance at a minimum. Do not serve tea, soft drinks, and caffeine-based products. They are considered diuretics that cause the body to excrete or dispose of water, which can cause severe dehydration.

 ## Overhaul Operations

Overhaul is the process of searching for and extinguishing any hidden or remaining fire once the main body of the fire is extinguished. It is actually a combination of several procedures: (1) examining for hidden fire and extinguishing any found, (2) preserving evidence for fire cause determination, and (3) conducting salvage/loss control operations. Carelessness during overhaul can allow remaining hot spots to reignite with renewed vigor, destroy evidence, and damage property needlessly.

Overhaul includes salvage/loss control (the property conservation component of handling a fire emergency): an ongoing process considered throughout an emergency. Salvage actions can actually begin early in an incident with confining a fire to prevent damage to other areas and continue with a careful fire attack that does not cause unnecessary damage. Along with preventing damage, recognizing the origin of a fire and source of ignition is critical to fire cause determination. Exercise care to preserve evidence.

Overhaul can be a dangerous procedure that can lead to injuries. Often injuries are attributed to a letdown after a fire is controlled or an attitude of diminished caution that can lead to carelessness and a lack of regard for personal safety. Be aware of the structural condition of the affected area when performing overhaul. Always wear personal protective equipment and SCBA because fire gases can accumulate in hidden spaces. Set up portable lights if needed.

Examination and Extinguishment

The objectives of the examination are to find and extinguish hidden fire and hot embers and to determine whether the fire has extended to other parts of the vessel. Conduct this important aspect of fire fighting as diligently as the initial attack on the fire. Make use of three senses — hearing, sight, and touch. Maintain use of full protective equipment and SCBA during the overhaul process. More carbon monoxide is produced during this stage of a fire than when it is free burning. Many have died needlessly when they removed their protective breathing equipment too soon, thinking the danger had passed.

Trace the length of all duct systems. Look into them to investigate the extent the fire traveled. Thoroughly inspect all deckhead spaces, decks, and bulkheads in the same manner. Especially examine where wiring or piping penetrates through bulkheads or decks. Fire can travel through very small openings. Carefully check smoke-blackened seams and joints. Expose areas that are charred, blistered, or discolored by heat until a clean area is found. If fire is discovered, wet the area until it is completely extinguished.

Pull apart and examine any materials that might have been involved with fire, including mattresses, bales, crates, and boxes. Turn over all debris, but do not throw smoldering materials overboard (they may blow back on lower decks). Do not throw plastics overboard because these products pollute the marine environment. Remove materials that might reignite, especially bedding, baled cotton, and bolts of fabric, from the fire area. Place these materials on a weather deck with a charged fire hose staffed and ready to extinguish any new fire.

Fire Cause Determination

After a fire is extinguished, the cause is determined in order to avoid a similar fire, identify hazardous conditions or practices, or perhaps find evidence of an arsonist. Even if not trained in fire cause determination, a vessel's officers can often ascertain the cause of a fire because of their knowledge of the vessel. Look for the path of fire spread. If an accelerant (such as gasoline) was used, the resulting burn pattern is quite distinctive, which would indicate arson. Fire spread through conduction is common on board. Paint blistered by conduction looks very different from paint damaged by flame impingement.

Careless smoking, cooking, heating appliances, electric wiring, or arson usually cause fires in the accommodation areas on board. Engine room fires are common and frequently involve fuel from a fractured line spraying over a hot manifold. Economizer or stack fires, scavenge fires, and turbocharger fires are not unusual, and most engines and engine rooms have systems to combat these hazards. The vessel's engineers are the most knowledgeable persons to identify the cause of engine room fires.

Take photographs before removing debris if possible. Make notes identifying what is pictured and where it was found. Take care to safeguard evidence. If feasible, close off the affected area until a shore-based investigation team can examine the scene, especially if arson is suspected or a fatality has occurred.

When interviewing vessel personnel, include a current crew list. Establish the identity of interviewees by reference to seafarers' documents — identity cards, passports, discharge books (records of sea service), certificates of competency, and other documents. Prepare a list of questions to ask so all interviews are consistent, for example *"Was the fire door open when you arrived?"*

Take statements from all involved persons, and record all actions in the casualty book, a log of occurrences that resulted in injury or death to vessel personnel or damage to the vessel. The casualty book is a source of valuable information if a formal investigation is conducted, which is almost certain for insurance purposes, death or injury inquiries, or suspected arson.

Salvage/Loss Control

Salvage/loss control is the practice of minimizing damage through operating procedures associated with fire fighting and recovery efforts during and after an incident. Proper management of fire-fighting water is a salvage/loss control action: It prevents unnecessary water damage to property and excess water accumulating where it could affect vessel stability. After a fire is extinguished, continue salvage by taking whatever actions are necessary to prevent any further damage. Salvage is also an aspect of damage control (see Chapter 13, Shipboard Damage Control) and the cleanup operation after a fire, including saving water- and smoke-damaged goods. Basic procedures include removing damaged materials, other debris, and excess water. Correct any unsafe conditions. For example, remove hanging lagging, place boards with exposed nails/screws in containers, and secure hanging wires to make the fire area as safe as feasible.

◆ Postfire Activities

Before a fire can be declared completely extinguished, the master of the vessel must ensure that certain essential steps were taken. A thorough examination of the immediate fire area must show the following procedures were conducted:

- Paths of extension were examined and opened where necessary.

- Smoke and combustion gases were removed by ventilation techniques.

- A complete overhaul of all burned material was completed.

- An oxygen indicator verified the fire area was safe for crew members to enter without breathing apparatus.

Conduct a postfire muster to account for all crew members and passengers. Although any missing crew members were identified at the initial muster and accountability systems tracked them during the emergency response, it is still important to account for everyone before allowing anyone to go off duty (stand down). Take statements and reports before releasing crew members. In port, account for any visitors from ashore, and obtain their statements before allowing them to leave the vessel. Even if the fire was small and no shore agencies were involved, enter the details in both the vessel's logbook and the casualty book. Other postfire activities include posting fire watches, restoring resources, maintaining seaworthiness, and conducting debriefing sessions.

Assign Fire Watches

When extinguishment is complete, assign at least one *fire* or *reflash watch* (person trained to detect fire) to check for reignition and to sound the alarm if it occurs. More than one reflash watch may be needed if the fire was extensive. Assign another reflash watch to patrol the exposures and the paths of extension. Reflash watches are particularly important after a fire when a vessel's fire fighting systems may be inoperable or have not yet been restored. A reflash watch is kept on duty for a minimum of 30 minutes.

Restore Resources

Do not let down the fire prevention and readiness guard just because a fire has occurred. A second fire could occur just as easily after one is extinguished as it could at any other time. Before leaving the fire scene, restore all equipment to its original readiness state by performing the following actions:

- Replace fire hose used in fire fighting with dry hose. Clean and dry the used hose, and roll it for storage or place it back at a fire station. This is especially important with unlined hose (Figure 9.16).

- Clean nozzles and install them on the dry hoses.

- Recharge or replace portable fire extinguishers.

- Clean breathing apparatus equipment, clean and sterilize facepieces, and refill or replace air cylinders. Stow the entire SCBA unit so that it is ready for the next emergency. Ensure additional (spare) breathing air cylinders are available.

- Inspect and restore fixed fire suppression systems (if used) (Figure 9.17). Replace sprinklers as needed if a sprinkler system was activated.

- Reset activated fire detection systems. Replace any disposable parts as needed.

Figure 9.16 Place clean and dry fire hose at the fire station after a fire. *Courtesy of R. Wright/Maryland Fire and Rescue Institute/United States Coast Guard.*

Maintain Seaworthiness

Make sure the vessel is seaworthy (watertight), and monitor the integrity of affected spaces. Initiate a thorough inspection to determine whether the fire damaged the vessel's structural members. The high temperatures associated with fire can cause decks, bulkheads, and other structural members of a vessel to warp or become structurally unsound. High temperatures can weaken steel plating and support members considerably. This weakening may not be apparent unless there is visible deformation. Carefully inspect all cases where structural weakness is suspected. Support weakened members by shoring or with strongbacks. Immediately undertake any other repairs necessary to secure the well-being of the vessel. Conduct any necessary dewatering operations.

Conduct Debriefing

Hold a debriefing session after a fire is completely extinguished and fire protection equipment has been restored to service. Hold this session before releasing crew members to return to their normal duties. Crew members have extinguished a fire on a vessel at sea and have every reason to be proud of their accom-

Figure 9.17 After a fire, inspect and restore sprinkler systems as necessary. *Courtesy of R. Wright/Maryland Fire and Rescue Institute/United States Coast Guard.*

plishment. Discuss anything pertaining to the fire. Encourage suggestions and recommendations, and record them as the official record of the postfire debriefing. Include resulting worthwhile ideas in the prefire plan (see Chapter 10, Prefire Planning). Consider the answers to the following questions as items for discussion while the details are still fresh in everyone's minds:

- What was the situation?

- What was done well?

- What could have been done differently?

- Could the fire have been prevented? If so, how?

- Could other equipment or resources have prevented the fire or enabled a better response. If so, which ones?

Chapter 10

Prefire Planning

A prefire plan is exactly what its name indicates — a plan for fighting fire that is worked out before a fire occurs and includes possible actions and procedures. A prefire plan contains a collection of plans for different spaces, different types of fires, and different circumstances. A prefire plan includes a set of drawings and written descriptions of compartments or zones, showing detailed locations of fire hazards, equipment, and devices. A prefire plan provides the crew with a plan of action before a fire occurs. The value of prefire planning is hard to dispute; yet many vessels do not have such a plan.

Confusion sometimes arises with use of the word *plan* because a vessel has two types of plans: (1) drawings and (2) written descriptions or instructions. The fire control plan, general arrangement plan, and deadweight and capacity plan are all drawings, and they contain information that may be of value in planning an emergency response. A prefire plan is a written description of a predetermined set of actions that may be used in responding to a fire in a given space. It may seem unrealistic to consult a plan during a response, but a well-prepared plan ensures that required actions are carried out in the proper sequence and that no essential action is forgotten during the stress and confusion of an emergency. For example, if it is required to isolate the power to a space, then it is an advantage for a crew member to learn exactly where the breaker is located from the prefire plan instead of having to ask an electrician or engineer. A well-prepared plan also provides alternative strategies and tactics, allowing emergency responders to rapidly decide which option to use in attacking a fire.

The most comprehensive and well-designed prefire plans are of little value if they are not used for training,

and they are even less valuable if they are not used when a fire actually occurs. An effective prefire plan must be current, tried, proven, and understood. Gather the information needed, using this chapter as a guide. Synthesize this information into a plan of action. Keep the prefire plan current, practice it, and make sure it works. Use the prefire plan to train and drill the crew. Regular practice leads to understanding. If the prefire plan meets these criteria, the crew will perform well in an emergency and achieve the desired outcomes.

This chapter describes the major elements of a vessel's prefire plan and use of the prefire survey form, the fire control plan, general arrangement drawings, and other documents that may be of value in designing a response to fire on board. Gathering information is the first step toward building a prefire planning manual. Sample prefire survey forms are shown, and an example of how a station bill is used in organizing an emergency response is given. Prefire plans must be continually updated through evaluation and review processes.

 ## Gathering Information

Technical information about the vessel and its systems must be gathered before a fire occurs. Consider the purpose and design of the vessel and its hazardous conditions. Base all preplanning on worst-case conditions. Consider the situation when a vessel is in port. List the resources (personnel, technical expertise, and equipment) that are available, how emergency services are contacted, and how long it takes for these resources to arrive on board.

A written prefire survey form is used to collect and organize prefire planning information and is a building block in preparing the prefire plan. The survey form contains emergency response information for each space on the vessel. Vessel drawings (fire control plan, general arrangement, etc.) and other information (such as that found in the vessel's stability and loading book and system technical manuals) support information summarized on the survey forms. System technical manuals and operating instructions contain specific information such as how to operate a fixed fire suppression system. Collect all this information for use in planning responses to potential fires.

When all the information on hazards and vessel fire suppression systems has been gathered for each space on the vessel, review the station bill and muster list to ensure that crew members assemble in safe, appropriate locations and are assigned to duties and tasks that make maximum use of the vessel's fire fighting equipment and systems. Once documented, this information becomes the vessel's prefire plan.

Fire Control Plan

A *fire control plan* is a set of drawings for each deck of the vessel; it contains information on vessel arrangement and fire suppression systems and locations of fire fighting and lifesaving equipment, but it does not say how to extinguish a particular fire (Figure 10.1). The *International Convention for Safety of Life at Sea (SOLAS)* requires an up-to-date fire control plan on board and also requires that it be available in more than one location (both port and starboard sides of the accommodations area). Additional copies are usually available on the bridge, in the master's office, in the chief mate's office, in the engineering control room, and at the fire control station. Although the primary purpose of the fire control plan is to assist the vessel's crew in fire suppression efforts, it is vital that this plan be available to responding land-based firefighters. Land-based firefighters are unlikely to have a preplan for the vessel and will depend on the fire control plan to form strategy and tactics if a vessel's crew is disabled or in need of rescue.

SOLAS requires that fire control plans be in the official language of the flag state. However, they must also be translated into either English or French. The plans, either in a rolled set or a booklet, are located in a prominently marked, red weatherproof enclosure outside the deckhouse adjacent to the gangway (Figure 10.2). Signs may indicate the plan's location.

The fire control plan begins with a set of deck arrangement drawings. The plan must show all the required information; it cannot refer to any other documents. Fire control plans are required to contain the following listed information:

- *Fixed fire suppression systems*
  - Locations of the fire main, hydrant fire stations, manifolds, and valves; locations and capacities of fire pumps
  - Location of spaces protected by fixed fire suppression systems, locations of flooding agent cylinders, and locations of release controls for these agents

Figure 10.1 A fire control plan is a set of drawings that contains information for each deck of the vessel. *Courtesy of R. Wright/Maryland Fire and Rescue Institute/United States Coast Guard.*

Figure 10.2 The fire control plan can be located in a prominently marked, red weatherproof tube with a sign indicating its location. *Courtesy of R. Wright/Maryland Fire and Rescue Institute/United States Coast Guard.*

— Locations of deck monitors and the arcs or areas of coverage

— Locations of foam proportioning equipment, controls, and hose outlets; amount and type of foam used and the stowage locations of foam containers

— Types of sprinkler systems installed, areas of coverage, and locations of control valves

— Location of the international shore connection

— Locations of portable fire fighting tools/equipment such as hose nozzles, spanners, axes, self-contained breathing apparatus (SCBA), and protective clothing

- *Portable/semiportable fire suppression equipment*
    — Locations of portable/semiportable fire extinguishers; classes and types of fire extinguishers, consistent with requirements of the flag state authority

- *Ship construction features*
    — Locations of fire-resistive bulkheads and decks
    — Locations of fire-retardant bulkheads and decks
    — Locations of watertight bulkheads
    — Locations of watertight doors, including location of all local and remote controls for watertight doors

- *Fire detection systems*
    — Locations of smoke and heat detectors
    — List of zones and areas served by each detector
    — Locations of alarm and control panels
    — Locations of alarm pull stations

- *Ventilation system*
    — Locations of ventilation fans and areas served; locations of fan controls and whether or not the fans are reversing
    — Locations of dampers and areas served; locations of damper controls

- *Means of access and egress*
    — Locations of normal paths of travel
    — Locations of companionways between decks and doors between horizontal areas
    — Locations of escape hatches and escape trunks

Other Plans/Drawings

A *general arrangement plan* (drawing) gives the general vessel layout (Figure 10.3). Because it almost always has an elevation view of the vessel, it is quite useful in emergency response. Fire control plans rarely have an elevation view (although some do). Other drawings that may aid in prefire planning include the deadweight and capacity plan, the piping diagram, electrical drawings, and stability information charts (see Chapter 13, Shipboard Damage Control) (Figure 10.4).

Figure 10.3 The vessel arrangement plan gives the general vessel layout. *Courtesy of R. Wright/Maryland Fire and Rescue Institute/United States Coast Guard.*

Figure 10.4 A color-coded piping chart aids prefire planning. *Courtesy of R. Wright/Maryland Fire and Rescue Institute/ United States Coast Guard.*

Station Bill and Muster List

A *station bill* assigns emergency duties by rank (Figure 10.5). It provides an effective outline for assigning response stations and emergency duties to vessel personnel. It also gives a quick visual reference for crew members so they can find where to go in the event of an emergency. Each vessel has its own station bill. For example, a station bill may show the second mate having the following emergency duties depending on the situation:

- *Abandon ship* — Second in command of No. 2 lifeboat

- *Man overboard* — In charge of the rescue boat

- *Fire on board* — In charge of a fire hose team and third in command of the fire fighting team

- *Collision, grounding, oil spill, etc.* — Other specified duties

Adding the names of vessel crew members to the station bill gives a *muster list* that states where each person is to go during a specific emergency and what each person's responsibilities are, subject to orders from command (the master) (Figure 10.6). Some may think a station bill and muster list are sufficient components for prefire planning, but unless information is gathered for specific situations and the crew has practiced responding to these scenarios, then all that the station bill and muster list achieve are gathering personnel and resources. Further information and direction are needed to effectively respond to a fire. The example given on pages 267–268 examines the use of a station bill for organizing an emergency response for a fire at sea.

 Prefire Survey Form

The document that connects the vessel's fire fighting resources with the personnel listed on the station bill is the *prefire survey form*. The form contains information pertinent to a fire response in any given space. Not every space has the same requirements. Questions such as whether to ventilate, where to start an attack, and how to shut off power may be answered by using the information on the survey form. The form may indicate the effect of free surface from fire fighting water collecting in the space or the effect of fire extending beyond the space. For example, the form may indicate that one of these effects may make the port lifeboat inaccessible. Thus decisions can be made quickly. The information for each space is given on a single page. Combining all the forms gives a comprehensive, ready-reference overview of the vessel. Consider the following major areas when preparing a survey form:

- *Construction considerations* — Classifications of divisions, construction modifications, space size, etc.

- *Arrangement considerations* — Locations of fire stations, fire equipment and systems, exits, etc.

- *Paths of extension* — Identifications of exposures, fire spread possibilities, hazards, etc.

Table 10.1 on page 269 lists items that can be included on a prefire survey form along with an example of the information needed for a third mate's cabin. Use this information to prepare prefire survey forms for every space on the vessel. Use the sample blank survey form in Figure 10.7 (page 270) as a guide. A completed form is given in Figure 10.8 (page 271).

Construction Considerations

The key requisite for successful prefire planning is an in-depth, complete review and understanding of the vessel's construction and fixed fire fighting systems. All officers and crew members should become familiar with the construction of the vessel. Different areas of the ship require different types of construction (for example, Classes A, B, or C bulkheads). Some types may contain a fire for some time. See Chapter 3, Vessel Types, Construction, and Arrangement. Knowing the type of construction in the area of a fire benefits the officers in charge because they can then estimate how long it may take a fire to penetrate to the next compartment. For example, transverse bulkheads subdivide a vessel into zones: Above the main deck, access through these bulkheads is through fire doors; below the main deck, access is through watertight doors that are fire-resistant. Review and inspect the major subdivisions or zones of the vessel. Consider the following factors when preparing prefire survey forms:

- Modifications not shown on the fire control plan that might alter integrity or restrict access (Take steps to update the fire control plan in these cases.)

- Points in the zone that could allow fire extension into other zones, for example, pipe chases, wireways, and ventilation ducts

Station Bill

General Instructions

1. Each person shall familiarize themselves with their assigned location in the event of an emergency immediately upon boarding the vessel.
2. All crew members shall be thoroughly familiar with the duties they are assigned to perform in the event of an emergency.
3. Each person shall participate in emergency drills and shall be properly dressed including a properly donned life preserver or exposure suit.
4. In all vessels carrying passengers, the Stewards Department shall be responsible for warning passengers, seeing that passengers are

properly dressed and have correctly donned their life preservers or exposure suits, assembling and directing passengers to their appointed stations, keeping order in passageways and stairways, controlling passenger movements, and ensuring a supply of blankets is taken to the lifeboats.

5. The proper chain of command is indicated by the sequential numbers assigned to each department. Should a key person become disabled, the next senior member of that department shall take the disabled person's place.

6. The Chief Mate shall be responsible for the maintenance and readiness of all lifesaving and fire fighting appliances and equipment above the main deck. The First Assistant Engineer shall be responsible for the maintenance and readiness of all lifesaving and fire fighting appliances and equipment on the main deck and below.

_____

Master's Signature

Fire and Emergency

Instructions

1. Any person discovering a fire shall notify the bridge by sounding the nearest available alarm and then take all initial actions as appropriate.
2. Upon hearing the fire and emergency signal, all airports, watertight doors, fire doors, scuppers, and designated discharges shall be closed and all fans, blowers, and ventilating systems shall be stopped. All safety equipment will be prepared for immediate service. OMED Numbers 9 and 10 shall check to ensure this item is completed after they report to their station.
3. Upon seeing a person overboard, immediately throw a life preserver (with a light attached if at night) and notify the bridge by reporting "MAN OVERBOARD PORT (STARBOAR) SIDE." In all cases keep the person in sight.
4. Any extra persons shall report to the Hospital Treatment Room.

Signals

Fire and Emergency Signal (————)
   The fire and emergency signal shall be a continuous blast of the whistle for a period not less than 10 seconds followed by a continuous ringing of the general alarm for not less than 10 seconds.

Man Overboard Signal (- - -)
   The man overboard signal shall be the letter O sounded several (at least 4) times on the ship's whistle followed by the same signal on the general alarm.

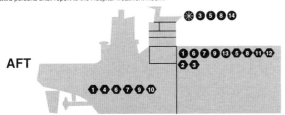

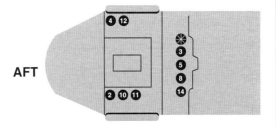

Abandon Ship

Instructions

1. All persons indicated in the diagram on the left should use Lifeboat #2. All persons indicated in the diagram on the right should use Lifeboat #1.
2. Any extra persons should muster at Lifeboat #1.

Signals

Abandon Ship Signal (. . . . . . ————)
   The abandon ship signal shall be at least 7 short blasts followed by one long blast on the ship's whistle followed by the same signal sounded on the general alarm.

Boat Handling Signals
   All boat handling signals shall be sounded on the ship's whistle and shall mean the following:
   (-) One short blast means to lower the lifeboats.
   (- -) Two short blasts means to stop lowering the lifeboats.

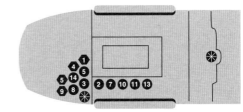

Personnel and Duties

Deck Department Functions

Symbol	Position	Fire & Emergency	Abandon Ship
✳	Master	In Command	In Command Lifeboat No. 1
❶	Chief Officer	In Charge Emergency Squad	In Command Lifeboat No. 2
❷	Second Mate	In Charge STBD Boat Deck	In Charge Lowering No. 1
❸	Third Mate	Assist Master	Provide New Equipment
❹	Third Mate	In Charge Port Boat Deck	In Charge Lowering No. 2
❺	Bosun	Emergency Squad	Operate Winch No. 1

Engineering Department Functions

Symbol	Position	Fire & Emergency	Abandon Ship
❶	Chief Engineer	In Charge Engine Room	Boat Engineer No. 1
❷	First Engineer	In Charge Emergency Squad	Boat Engineer No. 2
❸	Second Engineer	Emergency Squad	Assist Boat Engineer No. 2
❹	Third Engineer	Engine Room Assist	Assist Boat Engineer No. 1
❺	Third Engineer	Emergency Squad	Assist as Directed

Figure 10.5 The station bill assigns emergency duties by rank. The sample shown is only a portion of a station bill.

Emergency Muster List by Location

Station No.	Rating	Name	Emergency Station	Boat No.	Abandon Ship Station
Bridge					
A	Master		Bridge In Command		
1	Chief Officer		On Scene In Charge		
3	Third Mate		On Bridge — Communications		
6	AB#1		Relieve the Wheel		
Engine Control Room					
4	Chief Engineer		Engine Spaces — In Charge		
5	Third Engineer		Engine Control Room		
Fire Team No. 1					
2	Second Mate		Fire Team #1 Leader		
7	AB #2		Fire Team — Nozzle		
8	AB #3		Fire Team — Fire Hose		

Muster List

Station No.	Rating	Name	Emergency Station	Boat No.	Abandon Ship Station
A	Master		Bridge In Command		
1	Chief Officer		On Scene In Charge		
2	Second Mate		Fire Team #1 Leader		
3	Third Mate		On Bridge — Communications		
4	Chief Engineer		Engine Spaces — In Charge		
5	Third Engineer		Engine Control Room		
6	AB #1		Relieve the Wheel		
7	AB #2		Fire Team — Nozzle		
8	AB #3		Fire Team — Fire Hose		

Figure 10.6 The muster list tells each person where to go during a specific emergency and what each person's responsibilities are. The sample shown is only a portion of a muster list.

Emergency Response Organization Using a Station Bill

Even with reduced numbers on board, sufficient personnel are available to form a fire (emergency response) team, backup (support) team, first aid team, and command team. A smaller crew can still have a command group, an emergency group, and a support group. Vessels with fewer crew members may have sufficient numbers for only two teams or groups: emergency response and support. However large or small the numbers, the concept is the same, even for vessels with only six or seven people on board. In that case, the ship's company is organized to form one team that provides a comprehensive response to an emergency.

Imagine a representative vessel with a master, three deck officers, chief engineer, three engineering officers, bosun, No. 1 oiler, six deck ratings, three engine ratings, and four crew members in the catering department, giving a total of twenty-three persons. The vessel's station bill shows the following:

Personnel	
Command Team	Master Chief Engineer Deck Officer of the Watch (OOW) Deck Rating (helm) Engine Room Officer Rating in Engine Room
Fire Team	Chief Officer Second Engineer Bosun (Petty Officer) Ratings (3 deck and 2 engine room)
First Aid Team	Catering Department (4 crew members)
Backup (Support) Team	Deck Officer Third Engineer No. 1 Oiler (Petty Officer) Deck Ratings (2)
Duties	
Initiation of Response	When the fire alarm sounds, the OOW performs the following duties: ♦ Call the master.

	♦ Start the fire pump. ♦ Announce the location of the fire while sounding the general alarm. ♦ Add a second steering motor. All these actions are planned in advance for all foreseeable situations (both at sea and in port) and are written down as operational guidelines.
Organization	
Command Team	For simplicity, the Command team functions as follows: ♦ ***Master, OOW, and Rating*** — Staff the bridge. (The OOW maintains a written record of events during the response.) ♦ ***Chief Engineer*** — Rove to coordinate activities. (The chief engineer serves as another set of eyes and ears for the master and is well briefed to *assume command of the emergency response* should the master become incapacitated.) ♦ ***Engineer and Rating*** — Maintain engineering services. ♦ ***Radio Officer (if available)*** — Staff the communications function. (Distress calls that give the latest position may be sent by pressing one button on many vessels.)
Fire Team	Either the chief officer or second engineer serves as team leader or backup, depending on the fire's location. For example: ♦ The second engineer is the leader when the fire is in the engine room. ♦ The chief officer is the leader when the fire is on deck. If we assume a minimum staffing of two persons per fire hose with each wearing personal protective equipment (complete protective clothing and SCBA), then enough people are available for two fire hoses with one backup hose team (assuming enough

	equipment is available). When responding to an accommodation fire, for example: ♦ The chief officer is in charge of the fire attack. ♦ The second engineer is in charge of ventilation and support. The staffing is scanty but manageable. For an interior attack, more people are used when more personal protective equipment and fire hoses are available.
First Aid Team	The first aid team has two immediate duties during the emergency: ♦ Prepare to receive casualties (two people are sufficient for this task). ♦ Ensure additional provisions are in the lifeboats. (During training, ensure that what is taken to the lifeboats is what would be required in a real emergency response.)
Backup (Support) Team	While all other activity is taking place, the backup (support) team performs the following duties: ♦ Prepare the lifeboats concurrently with the fire fighting response: Swing out both boats and lower to the embarkation deck. (It is simple to leave the toggle painters permanently rigged and connected to a wire pennant to prevent chafing, thus reducing the need for the team to carry the painter out in an emergency.) ♦ Stand by to assist the fire team. ♦ May control ventilation on some vessels or for some situations.

The result of the efforts in this example is that the chain of command is intact with replacements ready when necessary: The fire team is working, the backup team is standing by, and the first aid team is standing by. Lifeboats have provisions and are ready to be lowered if required. It has all taken only a few minutes, and preparations are complete. If the fire team is successful, everyone can return to his or her normal duties. If crew members have to abandon ship, they may be able to return later, which has happened in many successfully fought vessel fires.

Table 10.1
Prefire Survey Form Information

Item	Example
Name of Space	Third Mate's Cabin
Location	Boat Deck, port side; window onto Boat Deck
Nearest Fire Equipment	20-pound (9 kg) dry chemical fire extinguisher and 50-foot (15 m) fire hose with adjustable nozzle connected to hydrant at Fire Station 8 in athwartships passageway, port side aft
Accessibility	Nearest additional hoses are at Fire Station 9 (starboard side), Fire Stations 4 and 5 (one deck up), or Fire Stations 10 and 11 (one deck down). (See fire control plan for details.)
Fuel Load	Class A combustibles
Fuel Shutoffs	N/A
Ventilation Shutoffs	**Accommodation:** Switches at Fire Control Station, on Bridge, and in Engine Control Room (shuts down all ventilation for accommodation block)
Electrical Shutoffs	Circuit Breaker No. 5, electrical panel inside closet on port side aft of Poop Deck, one deck down (isolates all of port side Boat Deck and Captain's Deck)
Vents	Fiddley Deck, port side aft
Exposures	Battery Room above on Captain's Deck, partly overlapped on forward end

Adjacent Spaces:
    Forward
    Aft
    Port
    Starboard
    Above
    Below
	
	Second Mate's Cabin
	Hallway
	Open deck
	Hallway
	Battery Room and Emergency Generator Room
	Bosun's Cabin

Size and Capacity	• 25 x 15 feet (5 m by 8 m) in two sections plus bathroom: 3,750 ft^3 (106 m^3) • Estimated 40 gpm (151 L/min) fire hose and nozzle from Fire Station 8 • Free surface effect is negligible for this compartment (See Chapter 13, Shipboard Damage Control, for calculations.)

Divisions:
    Deck and Deckhead
    Bulkheads
    Doors
	
	Class A
	Class A on port side; Class B15 forward, aft, and starboard
	Class B15

Drains	Bathroom scupper only, 1½ inches (38 mm)
Fixed Fire Suppression System	N/A
Smoke Detector	One in deckhead of cabin and also one in hallway
Heat Detector	One in Engine Change Room across the hall

Prefire Survey Form

Name of Space _____ Location _____

Adjacent Spaces: Forward _____ Aft _____ Port _____

    Starboard _____ Above _____ Below _____

Nearest Fire Equipment _____

_____

Accessibility _____

_____

Size and Capacity _____ Fuel Load _____

Diagram:

Shutoffs: Fuel _____

      Ventilation _____

_____

      Electrical _____

_____

Vents _____ Drains _____

Exposures _____

Divisions: Deck and Deckhead _____ Doors _____

    Bulkheads _____

Fixed Fire Suppression System _____

Fire Detection Systems: Smoke Detector _____

    Heat Detector _____ Others _____

Other Information _____

Figure 10.7 Use this blank prefire survey form as a guide. Prepare a survey form for every space on the vessel.

Prefire Survey Form

Name of Space _Third Mate's Cabin_   **Location** _Boat Deck, port side; window onto Boat Deck_

Adjacent Spaces: Forward _Second Mate's Cabin_   **Aft** _Hallway_   **Port** _Open Deck_

     Starboard _Hallway_   **Above** _Battery Room and Emergency Generator Room_   **Below** _Bosun's Cabin_

Nearest Fire Equipment _20-pound dry chemical fire extinguisher and 50-foot fire hose with adjustable_
nozzle connected to hydrant at Fire Station 8 in athwartships passageway, port side aft

Accessibility _Nearest additional hoses are at Fire Station 9 (starboard side), Fire Stations 4 and 5 (one deck_
up), or Fire Stations 10 and 11 (one deck down). (See fire control plan for details.)

Size and Capacity _25 x 15 feet in two sections plus bathroom: 3,750 ft³._
Estimated 40 gpm fire hose and nozzle from Station 8.   **Fuel Load** _Class A combustibles_

Diagram:

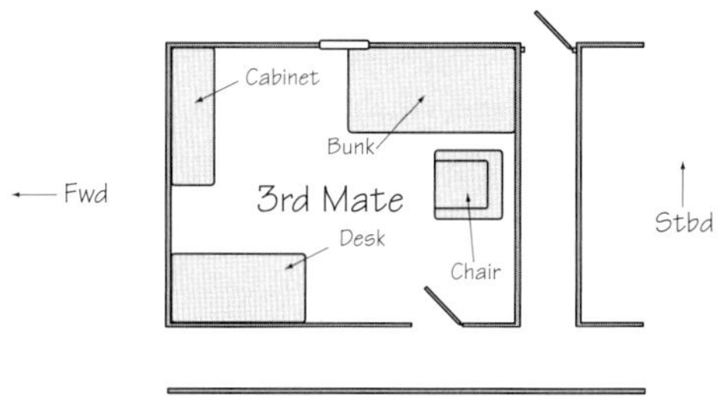

Shutoffs: Fuel _N/A_

     Ventilation _Accommodation: Switches at Fire Control Station, on Bridge, and in Engine Control Room_
(shuts down all ventilation for accommodation block)

     Electrical _Circuit Breaker No. 5, electrical panel inside closet on port side aft of Poop Deck, one deck_
down (isolates all of port side Boat Deck and Captain's Deck)

Vents _Fiddley Deck, port side aft_   **Drains** _Bathroom scupper only 1½ inches_

Exposures _Battery Room above on Captain's Deck partly overlapped on forward end_

Divisions: Deck and Deckhead _Class A_   **Doors** _Class B15_

   Bulkheads _Class A on port side; Class B15 forward, aft, and starboard_

Fixed Fire Suppression System _N/A_

Fire Detection Systems: Smoke Detector _One in deckhead of cabin and one in hallway_

   Heat Detector _One in Engine Change Room across the hall_   **Others** _N/A_

Other Information _Free surface effect is negligible for this compartment._

Figure 10.8 A prefire survey form completed for a third mate's cabin. See Chapter 13, Shipboard Damage Control, for free surface effect calculations.

- Areas that might be weakened by fire

- Areas that might retain fire fighting water and the fire surface effect if those areas do retain water

- Location of primary and secondary accesses and whether or not these accesses are normally locked

Arrangement Considerations

The prefire plan illustrates the vessel's interior and exterior arrangement. Consider the following factors when preparing prefire survey forms:

- Locations of exits, doors (types), and fire stations

- Locations and types of specialized equipment and systems (foam eductors, fixed fire suppression systems, detection systems and their controls, etc.)

- Use of spaces: cabin, freezer, jail, radio room, elevator, etc.

- Bulkhead ratings and vertical separations

Paths of Extension

Identify potential paths of extension on the prefire survey forms. Include ventilation ducts, electrical cable runs, and piping runs that pass through the compartment. Wind and weather conditions can influence fire behavior and the emergency response, but extension by conduction is usual in shipboard fires, and possibilities must be considered. Identify possible avenues of fire spread, and develop methods or procedures to prevent it. For example, a fire in a cargo hold or tank could possibly extend into a pump room and from the pump room to the engine room. However, if the pump room were fitted with a carbon dioxide fire suppression system that is activated to form a protective barrier or buffer, then fire would not extend in that direction. Another method of preventing fire spread in this example would be to have fire hose teams cool the bulkhead of the involved tank or hold.

Complete Prefire Plan

When all the information, plans (drawings), forms, and booklets are gathered together, they form the prefire plan. Assemble the plan in a three-ring, loose-leaf binder to allow easy updates when modifications are necessary. Keep copies on the bridge, in the engine room, in the deck office, and in the emergency headquarters or fire control station. Also keep copies where the fire control plan is usually located (port and starboard of the accommodation). When in port, keep a copy at the top of the gangway with a current crew list and a copy of the dangerous goods manifest. Keep these documents in a container, box, or tube marked *For Use of Fire Department/Brigade.*

Evaluation and Review of Prefire Plans

The prefire plan is continually reviewed and tested during training (see Chapter 14, Fire and Emergency Training and Drills). Base a drill on a simulated fire in a particular space, for example, the third mate's cabin. Respond to that space as though a fire was there. Observe and review the effectiveness of the response. Review the prefire survey form for the space, and revise it when necessary.

Emergency preplans are most valuable and effective when they are current, tried, proven, and understood. Test the total prefire plan at least once a year. Request feedback from fire team leaders and crew members regarding the value of the prefire plan. Note any changes to the prefire plan (discovered during practice) on all copies.

The prefire plan is a dynamic document that must be used and continually reviewed to maintain its effectiveness. It is the link between personnel and equipment and between training and response. The task of preparing a complete prefire plan may seem daunting, but it is accomplished one step at a time.

Chapter 11

Emergency Response Process

Any response to an emergency must be a team effort in order to ensure success. When all parties involved understand the organizational structure and their roles within that structure, confusion is held to a minimum. Confusion must be avoided if the vessel master and officers (Command) are going to have any chance of directing the appropriate resources to the defined goals (strategies) and objectives (tactics). Additionally, and most importantly, the use of an existing defined fire team organization results in a safer operation for all involved.

This chapter discusses the priorities and considerations in assigning emergency response positions and functions and factors to consider in responding. These factors include size-up, communications, crew member/crew officer responsibilities, stowage and handling of hazardous materials, emergency response priorities, and an overall view of actions taken to successfully extinguish fire on board.

Organizational and Emergency Response Structures

The organizational structure for fire response takes into account the existing organizational structure for day-to-day operation of the vessel (Figure 11.1). Like an industrial plant, a vessel is organized into various departments, each under control of a department head (chief mate or chief officer, chief engineer, and chief steward). The chief mate or chief officer is second to the master in the chain of command and is usually in charge of safety, lifesaving and fire fighting

equipment, and the training of the crew. The station bill lists duties for major emergencies: fire, man overboard, and abandon ship (boat stations). However, more specific information is needed for fire emergencies than can be provided on the station bill and muster list (the list of names and ranks or ratings on board). See Chapter 10, Prefire Planning, for more information.

It is up to the master to determine how the existing organizational structure can be modified for emergency response, taking into account the nature of the vessel, the size of the crew, and the abilities of officers and crew to serve in particular roles. One emergency organizational structure example is shown in Figure 11.2. There must be flexibility for the organization to respond effectively to changing circumstances. The location of a fire determines which team serves as the primary attack team and which serves as the support team. For example, if the fire were in the engine room, the chief engineer would lead; if the fire were on deck or in the accommodations area, the chief officer or first mate would direct operations. In order to provide a unified response, the master (backed by the chief engineer) has overall command of the response actions, while the chief officer or second engineer (depending on where the fire is located) leads the suppression effort.

Test, evaluate, and review emergency organizational structures to establish the best structure for the vessel. While emergency response cannot be totally prescriptive, pre-incident planning, allied with relevant training and drills, gives the best approach to quick and effective response with sufficient flexibility to react to changing circumstances. In preparing and

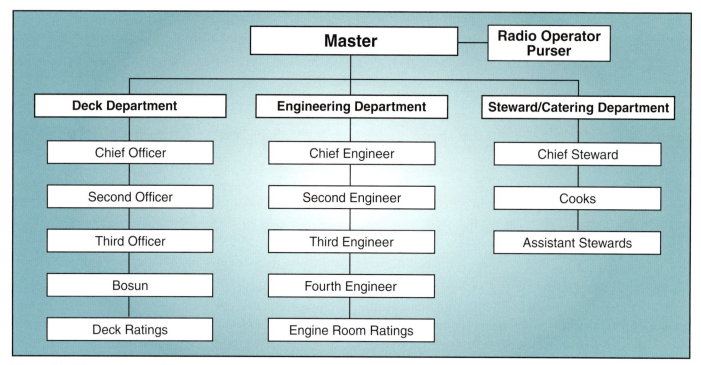

Figure 11.1 The basic vessel organizational structure.

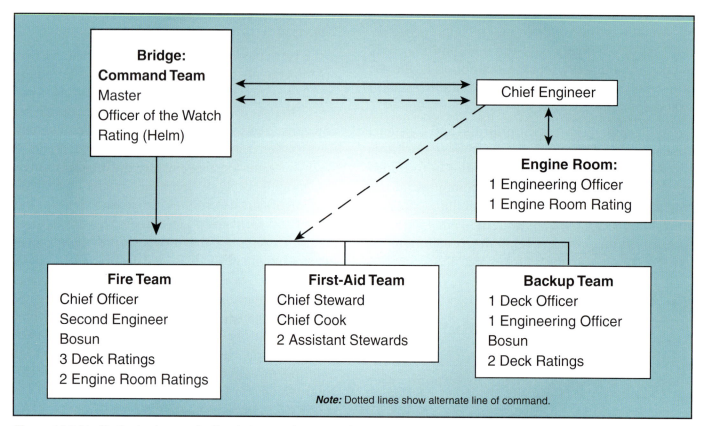

Note: Dotted lines show alternate line of command.

Figure 11.2 Modify the basic organizational structure for a vessel emergency.

training the emergency teams, it is possible that some members will be involved in or possible casualties of the incident itself. All team members should be cross-trained so vacant positions can be competently filled.

Crew Member Responsibilities

Working within the organizational structure, each member of the crew is assigned specific duties and responsibilities. Crew members should perform their assigned tasks and work routines with a conscious effort to prevent fire and other emergencies through good housekeeping and safe work habits. Statistically, the largest single cause of fires on board a vessel is human error, for example, welding without a proper fire watch, improper use of equipment, or smoking in bed. Participate actively and with a positive attitude in the emergency training exercises conducted on board. Train to the next level when possible. Understand and participate in the prefire planning of the vessel to allow the fullest contribution of information and details so that everyone can perform to the level of their competency during an emergency.

Emergency response responsibilities may vary, depending on how the crew is organized. It is impor-tant that individuals are assigned positions in the emergency organization that suit their physical abilities and talents, rather than just arbitrary as-signment of duties based on rank. The following boxes list the duties assigned to various officers and crew members specific to emergency response and do not include normal operational duties or re-sponsibilities. However, the safety officer position is ongoing and not just an emergency scene posi-tion. Generally a senior officer (the chief mate or chief engineer) fills this position, with responsibili-ties often delegated to one of the junior deck offic-ers. Regardless of the rank (which varies on vessels and in companies), the position of safety officer is always filled.

 Master

♦ *Ensure (through one or more of the junior offic-ers) that all fire fighting and emergency equip-ment is in proper working order.*

♦ *Supervise and periodically review the prefire planning for the vessel.*

♦ *Appoint a safety officer, and supervise activities to ensure that all personnel are properly trained.*

♦ *Utilize all resources available through prefire planning, training, meetings, inspections, drills, and reference sources.*

♦ *Ensure proper communications, both on board and with other agencies.*

♦ *Control the operation and use of all shipboard fixed fire fighting systems.*

♦ *Coordinate the efforts of shipboard fire teams with an overall emergency plan. Generally, this coordination (of strategy) will originate from the bridge with the chief mate or an engineering officer supervising tactical operations at the scene or in proximity to the fire.*

♦ *Decide if it is necessary to abandon ship. When the crew is ordered to abandon ship, the master ensures that proper procedures are imple-mented. Although the crew may be summoned to boat stations by ringing alarm bells or sound-ing the ship's whistle, the master verbally gives the final order to abandon ship.*

♦ *Coordinate all activities with outside organiza-tions when appropriate (for example, fully par-ticipate in the incident management Unified Command with the local fire department/brigade when in port). See Chapter 15, In-Port Fire Fighting and Interface with Shore-Based Firefighters.*

Chief Officer or First Engineer

♦ *Report to scene or general vicinity of the fire, and take overall command of fire suppression activi-ties (supervise tactical operations) if the fire is in his/her own area, or act as second in command if it is not.*

♦ *Oversee the implementation of the chosen fire suppression strategy.*

♦ *Monitor the actions and status of tactical units. This monitoring necessitates the establishment of effective communications with all tactical units.*

♦ *Ensure the safety of all emergency response team members.*

♦ *Keep the vessel's master informed of the state of the fire and status of fire fighting and rescue efforts.*

Engineering and Deck Officers

♦ Perform as assigned, and supervise the tactical activities of the crew members under their command.

♦ Ensure the safety of the crew members under their command as they carry out tactical operations.

♦ Maintain communications with the overall scene leader (for example, chief mate, chief officer, or first engineer) and keep him/her informed as to fire and team status.

 Chief Steward/Catering Officers

♦ Ensure that the first-aid party is organized and that a hospital or an alternative location is prepared for victims.

♦ Ensure that sufficient supplies (food, blankets, etc.) are brought from stores and stowed in lifeboats.

♦ Provide backup teams.

 Safety Officer

♦ Report directly to the master on all matters concerning emergency equipment and emergency training of the crew.

♦ Inspect and report any deficiencies on all shipboard emergency equipment.

♦ Prepare and conduct a useful, dynamic emergency training program for the vessel.

♦ Review, evaluate, and assist in assigning personnel to emergency teams.

♦ Ensure the correct, proper posting and updating of station bills.

♦ Perform in the emergency organization as assigned.

Incident Management Systems

A vessel's crew responds to emergencies and operates using the vessel's Command structure. A formal man-agement structure such as the Incident Command System (ICS) or Incident Management System (IMS) was originally developed to coordinate multiple agencies that may become involved in a response to an incident. The IMS has proven to be an effective means of managing emergency response by single as well as multiple agencies. It has been adopted and recommended by the International Maritime Organization (IMO) for use during emergencies on board vessels. Although the prefire plan is a form of incident management, it is not referred to as such and is not a substitute for IMS. Incident management systems are discussed further in Chapter 15, In-Port Fire Fighting and Interface with Shore-Based Firefighters.

Size-Up

Size-up is a mental process of evaluating all influencing factors before committing personnel and equipment to a course of action. Just as passage planning is a process of organizing resources and personnel to get the vessel from one place to another, size-up is a method of identifying the factors involved in an emergency. Any fire presents a complicated and rapidly changing situation. For example, a small fire in a berthing compartment, if discovered early, can be extinguished with a handheld fire extinguisher. The same fire may rapidly develop into a dense, smoke-filled atmosphere requiring self-contained breathing apparatus (SCBA), personal protective equipment, and rescue procedures.

The early operational phase of any fire response is often accompanied by considerable confusion, excitement, and in some situations, panic. An officer in charge must quickly survey and analyze the situation. The officer uses the prefire plan information, evaluates the information obtained during the size-up, applies basic principles, decides what action should be taken, formulates a plan of operation, and exercises command. Success or failure in an emergency situation depends, to a major degree, upon the ability of an officer to perform these essential functions in a practical, rapid, and skillful manner. This requires education, training, discipline, and experience. Success in a fire situation demands capable, resourceful, and aggressive leadership, not only from the officer in charge but also from the entire chain of command.

The person in charge of the response to a fire aboard ship must ensure that adequate resources (both crew members and equipment) are allocated. In allocating those resources, place the response efforts at least one step ahead of the fire. It is better to make an initial response to the report of a fire with all available resources than to send a couple of people to investigate and allow the fire to grow. Experienced firefighters always prefer to return unnecessary personnel and apparatus rather than respond with minimum resources and risk delay in controlling a rapidly growing fire.

The information available for size-up can come from any of the following sources:

- Initial report
- Automatic detection equipment
- Visual factors
- Reconnaissance
- Prefire plan

Initial Report

The initial report may be an alarm bell, or it could be a detailed report on the fire's location and fire conditions. Instruct crew members to provide as much pertinent information to the bridge as possible. Emphasize the importance of accurate information. Cases of misreading a compartment number and sending a fire team to the wrong place where no fire was found have occurred. Meanwhile the fire had time to grow. This information should include but is not limited to the following items:

- Name of the person making the notification
- Location of the fire
- Type of fire if it can be determined
- Some indication as to the extent of the fire
- Identification of any persons trapped or threatened by the fire and in need of immediate assistance
- List of actions already taken in an effort to confine or extinguish the fire

Automatic Detection Equipment

Smoke, heat, and flame detectors are of great value in detecting fires at an early stage so that they may be quickly extinguished (Figure 11.3). Investigate all alarms. Even if unfounded or caused by error, do not ignore an alarm with the attitude *"It's a false alarm."*

There are no "false alarms," merely responses with "no action to be taken." Automatic detection equipment reports the following information:

- *Location of alarm (zone)* — Location indicators may tell where the fire is located, although the area of greatest heat may not be the area of fire origin. Search rooms and spaces above, below, and on all sides of the alarm location for signs of fire.

- *Type of alarms sounding (smoke or temperature)* — Alarm type may give an indication of the intensity of the fire. A smoke detector may detect a fire in its beginning stages. Activation of a heat detector may indicate a more rapidly developing fire condition.

- *Number of zones reporting* — Reports from more than one zone may indicate a large fire or that fire doors are open, allowing smoke and heat to spread.

Visual Factors

The officer in charge of the fire scene should make an attempt to visually assess the situation. A few very simple visual clues could provide a very comprehensive picture of current fire conditions. Useful visual indicators include but are not limited to the following examples:

- *Smoke* — May indicate fire location. Note smoke conditions such as intensity, color, etc. If smoke is issuing from vents, use the prefire plan to find the spaces that those vents serve.

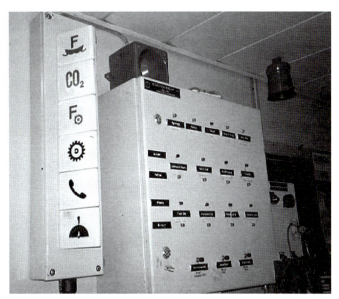

Figure 11.3 An engine room annunciator panel with audio and visual alarms. *Courtesy of R. Wright/Maryland Fire and Rescue Institute/United States Coast Guard.*

- *Visible flame* — May indicate the fire origin (if small) or extension of the fire (particularly extension through conduction).

- *Visibly hot bulkhead* — Indicates high heat (blistered or charred paint, distorted metal, etc.). A quick and effective method of checking heat conditions from a safe distance is to apply water spray to bulkheads and note any steam generation. If an insulated Class A or Class B bulkhead is visibly hot, a significant fire condition is likely on the other side of the bulkhead. Check the fire control plan and the prefire plan to identify the class of a hot bulkhead.

Reconnaissance

If the sources given in the previous sections do not provide an adequate picture of the situation, some form of reconnaissance (exploratory survey) may be needed. A team (at least two people wearing full protective equipment including SCBA) capable of properly assessing the situation may have to be sent to the scene of the fire to evaluate conditions. The purpose is to ascertain the location and size of the fire, whether any persons are at risk, what vents or hatches are open or closed, and if any immediate exposures are evident. Because team members are equipped with full protective equipment, they can make an immediate rescue should they encounter a victim.

Prefire Plan

As described in Chapter 10, Prefire Planning, the prefire plan provides a wealth of information useful during the initial stages of an emergency response. It gives guidance on confinement, the isolation of electricity and fuel, the manipulation of ventilation, and so on. It can also identify hazards in the fire area and give indications as to the fuels involved or in danger of involvement.

◆ Emergency Response Priorities

Life safety is the first and main emergency response priority. It is essential to prepare lifeboats and life rafts concurrently with the fire response so that they are ready if needed for retreat or to abandon ship. Many crews have left their vessel during a fire and returned when conditions improved.

Both dewatering and stability priorities are addressed during a fire response. As mentioned in Chapter 10, Prefire Planning, the potential of free surface effect may be evaluated in advance. Dewatering activities may be planned and initiated concurrently with fire attack. Also see Chapter 13, Shipboard Damage Control, for more information on stability and dewatering.

Control of ventilation is critical in an emergency response. Ventilation to unstaffed spaces involved in fire would normally be shut down to slow fire growth. In some cases, it should be maintained. For example, if persons are trapped in cabins, they may be safe as long as air is being delivered to the spaces (Figure 11.4).

Additional emergency response considerations could be the following:

- Whether at sea or in port with land-based fire fighting assistance available

- Time for assistance to arrive (for example, other vessels, port fire department/brigade, or initiation of the Automated Mutual-Assistance Vessel Rescue [AMVER]) System

- Exposures, such as hazardous materials, either on board or adjacent to the vessel if in port

The master, through the officers and crew, must decide what actions to take to resolve an emergency incident. No two emergency responses are the same, but they all contain most of the basic elements for successful resolution. The precise actions taken will

Figure 11.4 Ventilation fan controls. *Courtesy of R. Wright/ Maryland Fire and Rescue Institute/United States Coast Guard.*

vary along with the circumstances. Actions that were appropriate in initial response may be irrelevant at a later stage. A successful Command process ensures that all relevant actions are taken at the proper time. See Chapter 12, Fire Fighting Strategies and Tactical Procedures, for more information.

In addition to the classic priorities of life safety, exposure protection, fire confinement (including type of attack and ventilation), fire extinguishment, and fire overhaul, the following aspects of the Command process are also implemented:

- Continually review, reevaluate, and revise size-up.
- Cautiously make the decision to abandon the vessel.
- Report and record emergency actions.
- Conduct crowd-control management when passengers are carried.

Review, Reevaluate, Revise

Size-up begins when the alarm sounds. Some might say it starts at the prefire planning stage. It does not end until the emergency has come to an end. A fire is a dynamic event. Conditions are constantly changing. The situation needs to be constantly reassessed. Suppression efforts need to be constantly evaluated and occasionally may have to be revised.

Decision to Abandon Ship

The decision to abandon ship is one that must not be made lightly. Abandoning ship in a fire situation means that the fire is beyond the capabilities of the crew. If the fire cannot be contained (confined or isolated), and the resources are not adequate to yield a positive outcome at that time, it is time to abandon ship. As stated earlier, it may be possible to return if conditions improve, and every effort should be made to save the vessel. If the decision is made to abandon the vessel, allow adequate time for the safe evacuation of everyone on board. Sufficient extra stores (food and blankets) must be placed in the lifeboats for all aboard.

Recording and Reporting

Just as with routine activities, the actions taken during an emergency and the outcomes achieved must be recorded. It is essential to record the actions. Station a scribe at the Command location (usually the bridge), and alert this person to the activities performed so that they can be properly recorded. The master confirms that the scribe has recorded the actions as they occur.

Crowd Control (Management)

On commercial vessels, emergency response usually requires everyone aboard to participate. When passengers or other people are carried, they must be instructed where to go, when, and what they may or may not take with them in an emergency. Control of a large number of people, many of whom may be untrained or even unfit for evacuation, requires calm, confident leaders who have a clear knowledge of objectives.

Passengers may be required to stay in one area during a fire response, or they may have to be evacuated. Evacuation involves donning life jackets and moving to lifeboats or evacuation chutes. Passengers must be dissuaded from trying to gather their luggage if abandoning ship is required. See Chapter 12, Fire Fighting Strategies and Tactical Procedures, for more information on crowd control.

 ## Stowage and Handling of Hazardous Materials

Consider and prepare to command a fire incident involving hazardous materials and dangerous goods. This text alone is not adequate to fully address the very real possibility of handling a hazardous materials incident. Some hazardous materials react with water; some may react with other extinguishing agents. Different people and organizations have defined hazardous and noxious substances in many different ways. Some authorities define hazardous materials as *"any substance or material in any form or quantity that poses an unreasonable risk to safety and health and property when transported in commerce."*

The inclination may be to think only of cargo when considering hazardous materials, but they may also include common items stored on the vessel for everyday use. For example, drums of toluene in a paint locker could pose a serious threat, as could boiler water chemicals. The vessel's fuel for propulsion and auxiliary power may become a hazardous material if it is not contained properly.

Hazardous materials are stowed in designated spaces in accordance with the *International Maritime*

Figure 11.5 Material safety data sheets describe hazardous materials on board. Post them where crew members can readily see them. *Courtesy of R. Wright/Maryland Fire and Rescue Institute/United States Coast Guard.*

Dangerous Goods (IMDG) Code. Containers should have identification placards on them. The dangerous goods cargo manifest is another source of information. Material safety data sheets (MSDSs) describing hazardous materials on board are posted to give crew members more detailed information (Figure 11.5). Copies should be immediately available for use in an emergency response. Recommended fire fighting tactics and methods of spill containment can be planned in advance.

 Communications

For the emergency organization on a vessel to function effectively, it must have an efficient means of communication. During an emergency, communication is important from the beginning and throughout the emergency because it provides the forum through which the whole organization's command and control structure may function.

Good communications equipment and proper reporting procedures can greatly assist the team leader in decision-making. In times of emergency, the initial action is crucial to the success or failure of the emergency response. Crew members should be properly trained in effective communication and understand what information is important. Lengthy discussions and verbosity should be discouraged. It is important to always speak clearly, concisely, and slowly, especially during times of emergencies.

Crew members should be well acquainted with the types of communications systems available on board the vessel and familiar with proper operation through practice. Special procedures such as codes, protocols, and echo communications are available and should be used when appropriate to enhance effective communication. The crew should be aware of the problems that may be encountered with various means of communication and take the necessary actions to eliminate those problems when possible through familiarity, practice, and maintenance. The following sections list the requirements for good communications, means of communication on board, and how best to use these means in emergencies.

Oral Communication Process

It is essential that crew members understand that communicating is of little value if the receiver does not receive the message or does not understand it. *Communication* is a process by which information is exchanged among individuals through a common system of symbols, signs, or speech. Communication (in this context) refers to the oral exchange of information among people, the equipment used to communicate, and the language of communications. The fact that we have to discuss the topic of communication suggests that it is a problem area and that we cannot take for granted that the ideas we "send out" are the very same ideas that are "taken in" and understood. This miscommunication is a common problem, which is evident from the anonymous statement: *"I know you believe you understand what you think I said, but I am not sure you realize that what you heard is not what I meant."* (Figures 11.6 a–c)

Perhaps the most vital of the elements of communication is the language, that is, the words used and the meaning taken from the words used. We have trouble communicating orally for a number of reasons; some of these reasons are as follows:

- Not hearing fully because of the following:
  — Our own hearing level
  — Noise interference because of sender's location, receiver's location, or transmission medium
- Not listening properly because of the following:
  — Distraction
  — Boredom

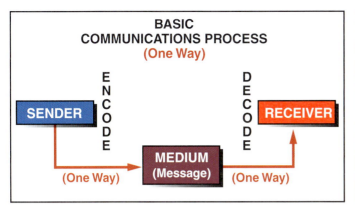

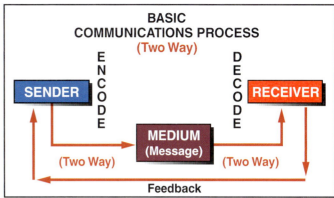

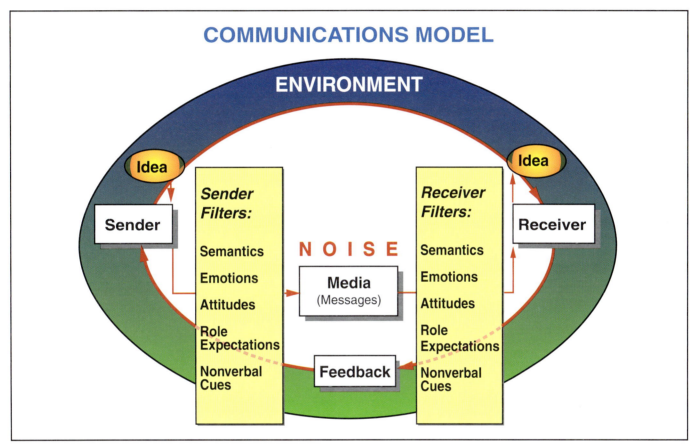

Figures 11.6 a–c (a) *(top left)* A basic one-way communication model consists of three components: *sender, medium,* and *receiver. Encoding* is when the sender transforms behaviors into words (or other media) that have meaning. *Decoding* is when the receiver extracts meaning from the sender's words (or other media). (b) *(top right)* Effective communication is an active two-way process that creates an understanding between sender and receiver. (c) *(bottom)* The complete communications model considers filters to understanding that affect the messages of both parties. *All diagrams courtesy of Maryland Fire and Rescue Institute.*

- Not understanding the words that are used because of the following:

  — Lack of common language

  — Unfamiliar technical language

  — Poor speaking ability

  — Words spoken hurriedly because of stress, excitement, etc.

  — Words garbled when spoken while wearing SCBA

If these and other communications problems (such as unfamiliar and/or faulty communications equipment) are combined with the chaos sometimes found at the scene of an emergency, there is indeed a communication problem. Under normal circumstances, the most frustrating element at the scene of an incident is waiting for information. The time that the emergency response team is exposed to unknown danger potentials while doing a survey of the incident scene is of vital importance. It is imperative that the emergency response team spends as little time as possible in the danger area and gets as much information as possible to the bridge. This information forms the basis of most response actions. Proper communications minimize loss of property and danger to crew members on the emergency response team. Make sure every crew member is proficient in communication skills. Log entries should be kept of response actions taken and the times they occurred. This log entry not only gives better on-scene control but also is essential for postincident review and is required for legal purposes (Figure 11.7).

To ensure the best possible communications between emergency response teams and Command, the following points should be decided before entry to the site of an incident:

- Identify the location of the fire or incident and the requirements for maneuvering the vessel.

- Ensure that all members of the emergency response team know in advance exactly what emergency hand, audio, or radio call signs are in use.

- Place backup team members where they can keep visual contact with emergency response team members so that hand signals will be visible if radio communications fail.

- Learn and use proper terminology for various parts of the vessel (avoid slang terms).

- Use the International Phonetic Alphabet when using radio communications and the International Code of Signals, which is especially important when identification is critical. See Table 11.1. Contact the International Phonetic Association for information about the International Phonetic Alphabet (see Appendix B, References and Supplemental Readings).

Radio transmissions are sometimes garbled due to location, protective clothing, face masks, or dead spots. In the case of dead spots, the only alternative is to move to a more optimal location for transmission of data, which may mean moving a short distance to get clear of the problem area. Radios used in a flammable vapor atmosphere must be of the intrinsically safe type. The following points may be taken for granted but are vital to free-flowing radio communications:

Table 11.1
Alphabet and Number Codes

Letter = Word	Letter = Word	Number = Word
A = ALPHA	N = NOVEMBER	0 (ZERO) = NADA
B = BRAVO	O = OSCAR	1 (ONE) = UNA
C = CHARLIE	P = PAPA	2 (TWO) = BISO
D = DELTA	Q = QUEBEC	3 (THREE) = TERTE
E = ECHO	R = ROMEO	4 (FOUR) = QUARTO
F = FOXTROT	S = SIERRA	5 (FIVE) = PANTA
G = GOLF	T = TANGO	6 (SIX) = SOXI
H = HOTEL	U = UNIFORM	7 (SEVEN) = SEPTA
I = INDIA	V = VICTOR	8 (EIGHT) = OCTO
J = JULIETTE	W = WHISKY	9 (NINE) = NONE
K = KILO	X = X-RAY	10 (TEN) = DECA
L = LIMA	Y = YANKEE	
M = MIKE	Z = ZULU	

Source: *International Code of Signals,* The Stationery Office, London, England, 1991.

Figure 11.7 A deck log is important for postincident review of actions taken and for legal requirements. *Courtesy of R. Wright/Maryland Fire and Rescue Institute/United States Coast Guard.*

- Keep the channel clear for messages related to the incident.

- Keep messages brief and accurate.

- Speak clearly, slowly, and distinctly.

- Acknowledge understanding by repeating the message.

- Minimize use of very high frequency (VHF) Channel 16 and other calling frequencies. Use Channel 16 to make contact, and then move to another channel.

A person may have trouble being understood when communicating by radio while wearing an SCBA. A solution to this problem is to place the radio microphone pickup directly on the outside surface of the hard plastic facepiece view plate. Speaking slightly louder than normal is heard much more clearly than shouting. The following are several options or variations on the described technique. Practice each of these, and determine what works best for particular SCBAs and radios:

- Use the facepiece method (described in the previous paragraph).

- Hold microphone against the speaking diaphragm (if provided).

- Hold the microphone against the throat.

- Hold the microphone against the exhalation valve.

- Use an internal microphone (available from some manufacturers).

- Use a throat microphone.

Communications Systems

The most common types of systems found on board a vessel are briefly described in the following list. See Table 11.2 for possible communication difficulties with these systems.

- *Internal communications* — Fixed system that is routed to all areas of the ship for talk-back (two-way) or open communications via a transceiver to each compartment.

- *Public address systems* — Fixed, direct (one-way) communication mechanism that broadcasts to all areas of the ship via a speaker system.

- *Telephones* — Fixed system that allows the user to speak person-to-person at a specific station (Figure 11.8). Most vessels have a telephone system and a backup system (sometimes two). The backup system may be fitted to a few stations only such as the bridge, engine control room, steering gear room, and fire control station (Figure 11.9).

- *Hand/visual signals* — An alternative when voice communication is difficult because of machinery noise, distance, and language barriers, etc. They must be understandable to crew members.

- *Face-to-face/messengers/runners* — Facilitate communication when most practical, no other means is available, or the information must be kept

Figure 11.8 A telephone system is often found in an engine control room. *Courtesy of R. Wright/Maryland Fire and Rescue Institute/United States Coast Guard.*

Figure 11.9 A backup telephone system can be battery-powered or sound-powered. The sound-powered phones pictured are on a cruise vessel bridge to use in case the vessel has a power loss. *Courtesy of Robert E. "Smokey" Rumens.*

Table 11.2
Possible Communication System Difficulties

Electronic Communications	Telephones	Face-To-Face Methods	Voice Pipes	Lifeline Commands	Ship's Public Address Systems	Vessel Alarms	Hand/Visual Signals
• Vessel construction features (decks and bulkheads can cause dead spots) • Low batteries • Noisy areas • Weather conditions • Poor maintenance • Interference	• Noisy areas • Damage to wiring • Poor maintenance • Electrical service failures • Human factors: stress, excitement, etc.	• Language barriers • Information overload/memory • Noise • Wearing self-contained breathing apparatus (SCBA) • Elapsed time from receiving the information until conveying it • Environmental conditions • Human factors: stress, excitement, etc.	• Noisy areas • Obstructions • Poor maintenance • Not available in all areas	• Lack of knowledge of commands • Lack of training	• Lack of power • Poor maintenance • Limited to one-way communication only • External noises	• Loss of system • Noise level • Poor maintenance • Lack of training • Limited access to controls	• Misinterpretation • Distance • Environment (smoke, weather, etc.)

confidential. It is recommended that all face-to-face communications be written as well so that they can be communicated properly. Verbal messages should be brief and accurate; technical information and numerals should be written.

- *Vessel's alarms* — Equipment for bells, whistles, other sounds, and lights to notify crew members. In vessel traffic or close quarters situations, it may be necessary to sound the signals on both the bells and the internal systems to prevent confusion with other vessels.

- *Electronic communications* — Common equipment on vessels includes fixed radios and portable radios; may have built-in automated distress communications capability such as the Global Marine Distress Safety System (GMDSS) (Figure 11.10).

- *Voice pipes* — Allow direct communication between compartments via a pipe array; seldom used but may be a reliable alternative.

- *Lifeline (tethering) commands* — Preset series of signals that may be incorporated into the use of lifelines. Standard signals are explained using the OATH acronym as follows:

  — *Okay:* One short pull

  — *Advance:* Two short pulls

  — *Take Up Slack:* Three short pulls

  — *Help:* Four short pulls

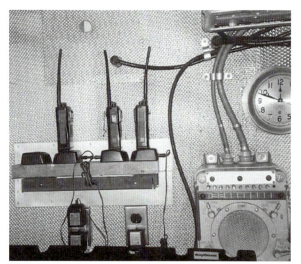

Figure 11.10 Portable radios are commonly found on vessels. *Courtesy of R. Wright/Maryland Fire and Rescue Institute/United States Coast Guard.*

Fire Fighting Strategies and Tactical Procedures

There is no standard formula for fighting fires on board vessels. However, with adequate prefire planning, fire training, and careful size-up and by using the guidelines in this chapter for the specific circumstances, a set of instructions can be developed that will be successful in most fire situations. Even with all the different types of ships at sea, common procedures for shipboard fire fighting exist. Prefire planning is a vital step in controlling fire. Realistic fire training is very important to the seafarer along with knowing and understanding the arrangement of the vessel and the vessel's systems.

A *strategy* is an overall plan for incident attack and control. Strategies are identified by three methods: (1) prefire planning, (2) particular emergency situation, or (3) officers in charge. The specific tasks/duties that are completed in order to meet the overall strategy are known as *tactics*. The guidelines in this chapter assist in handling various types of fire emergencies and are discussed in general. The key is to develop a plan by sizing up the situation, communicating the information, and completing the plan of action. The information in Chapter 13, Shipboard Damage Control, goes hand in hand with the tactical procedures for fighting fire on board.

The decision to abandon ship is a serious one, but may become the only option in a catastrophic fire situation. Abandoning ship may become the best option in the interest of crew safety, and the possibility exists that the crew may be able to reboard the vessel at a later time. Lifesaving appliances are a primary concern when crew members can no longer maintain control of a fire situation. Make all lifesaving appli-

ances ready at the outset of a fire for the possibility of abandoning ship.

This chapter deals with strategy and tactics for onboard fire fighting and gives tactical guidelines for different vessel types (bulk, container, tanker, etc.) and various spaces (galley, engine room, accommodations, etc.) and types of fires. Procedures vary for small, medium, or large fires. Incident strategy options, general tactical procedures, and tactical priorities are also discussed.

◆ Incident Strategy Options

There are three strategy options: offensive, defensive, or abandon ship. From the outset, and continuously throughout the incident, the person in charge determines the appropriate strategy. The strategy chosen depends on several variables, including the size of a fire, resources available, and the fire's location.

An *offensive* strategy is one in which an aggressive attack is made on the fire with expectations of success in extinguishing the fire, while limiting the danger to fire fighting crew members. The goal is to not only stop the fire's progress, but to extinguish it and limit the damage to that which has already occurred when the attack is begun. An offensive strategy requires adequate resources and a tenable environment. An offensive strategy does not necessarily mean an interior, face-to-face fire attack. It could employ the use of fire hoses, or it could be accomplished by using fixed fire fighting systems or foam monitors.

A *defensive* strategy is one in which the goal is to contain the fire and protect exposures. A defensive strategy is chosen when resources are inadequate and/or exposing personnel to extreme danger would be necessary to achieve extinguishment. If conditions and lack of resources are extreme, this could mean moving the vessel away from exposures (when in port), dropping anchor (when at sea), and abandoning ship. Other defensive strategies involve a holding action on the fire to prevent spread to other parts of the vessel or to external exposures (whether in port or at sea).

The decision to *abandon ship* is one that is not made lightly. In some situations, ships have been abandoned only to be reboarded by crew members when conditions on board improved. If a fire cannot be contained (isolated) and the resources are not adequate to yield a positive outcome, it is time to abandon ship. Make every effort to save the vessel. Stay with it as long as possible. However, allow adequate time for the safe evacuation of everyone aboard.

 ## Tactical Priorities

Based on the overall strategy chosen, tactical priorities are established based on current conditions of the incident (extent of fire, threat of spread, trapped or missing passengers or crew, etc.). The *RECEO* model is a commonly accepted list of priorities. Many such aids to memory exist, but all models put rescue (life safety) first. The *RECEO* model as it applies to marine situations is listed, and then each item is briefly discussed in the sections that follow. See Chapter 15, In-Port Fire Fighting and Interface with Shore-Based Firefighters, to see how the *RECEO* model enables

the mariner to provide support to fire department/brigade operations.

Rescue (Life Safety)

Rescue is the *life safety* component of the emergency. Above all else, rescue takes priority. Fortunately, in some scenarios, rescue may not be required; all crew and passengers may be accounted for. If rescue is a requirement, crews advancing into the fire area may achieve it. In fact, a fire team may have to take exposure-protection or fire-extinguishment actions in order to execute a rescue. But in all cases, command officers make sure that the objective of these actions (rescue) is not lost.

Exposures

Exposure protection is the first priority in the incident stabilization/fire suppression component of an emergency. Command officers make sure exposure protection has been conducted before ordering attack on an advanced fire. Exposures include the following:

- *Onboard* — Cargo (particularly hazardous materials or dangerous goods), adjacent spaces (possibly six sides to consider), ductwork, cable runs, and piping runs (Figures 12.1 a and b)

Figures 12.1 a and b Onboard exposures (a) *(left)* Cargo. *Courtesy of R. Wright/Maryland Fire and Rescue Institute/United States Coast Guard.* (b) *(below)* Hazardous area. *Courtesy of Firefighter Luke Carpenter, Seattle (WA) Fire Department.*

 ### RECEO Model

- **R**escue (life safety) — *Rescue those who are endangered.*

- **E**xposures — *Protect vessel areas, other vessels, or structures that are threatened.*

- **C**onfine — *Contain the fire, and prevent its spread (may include ventilation).*

- **E**xtinguish — *Control and extinguish the fire.*

- **O**verhaul — *Check for and extinguish hot spots (prevent reignition/reflash); investigate the fire cause; conduct salvage/loss control operations.*

- **In port** — Terminal structures, wharf or pier structures, other vessels, and vehicles (trucks, railcars, automobiles, and cargo handling equipment/apparatus) (Figures 12.2 a–c)

Determine the combustibility of exposures and the effect heat from the fire and water used for fire fighting will have on the exposures. Protect exposures in one of two ways: Move the exposures if possible, or protect them in place. The most effective way of protecting exposures is to move them. If they are immovable, apply fire streams directly to the exterior of the exposure to provide cooling, which prevents the spread of fire to the exposure.

Confine

The second priority in the incident stabilization/fire suppression component of an emergency is fire confinement. Fire confinement can be viewed as protection of interior exposures. Identify the fire location during size-up. Once the fire is located and isolated, crews can then advance to extinguish it. Property is not necessarily conserved when attacking a fire before it is confined. In fact, an attack on an unconfined fire can spread smoke and flames into uninvolved areas. Whatever is burning is already damaged, so save what is unburned by confining the fire. Leave the attack on the fire for the extinguishment step.

Even though ventilation is listed as part of the confine priority, it may occur at any point. Ventilation may assist in achieving any of the other tactical priorities. Ventilation can be a key component of incident stabilization; it is often required to facilitate the previous actions. Ventilation may be necessary to provide

Figures 12.2 a–c In port exposures (a) Terminal facilities. *Courtesy of R. Wright/ Maryland Fire and Rescue Institute/United States Coast Guard.* (b) Pier. *Courtesy of R. Wright/Maryland Fire and Rescue Institute/United States Coast Guard.* (c) Cargo handling equipment. *Courtesy of Robert E. "Smokey" Rumens.*

a safe atmosphere for both those in need of rescue and those performing rescue functions. Ventilation can help protect exposures by allowing a fire to spread vertically rather than horizontally, which also helps to confine the fire. Ventilation can be absolutely critical in extinguishing a fire. It provides a cleaner, cooler atmosphere in which crew members may operate. It can greatly enhance visibility in locating the seat of the fire and assist crew members in finding their way about.

Ventilation can mean one of three actions: (1) the provision of fresh air into a fire area, (2) a means for the hot fire gases and smoke to escape, or (3) the sealing off of fresh air from a fire area such as when using a carbon dioxide (CO_2) system or a similar situation. Command officers evaluate the need for ventilation and determine when it is used. Command officers also make sure crew members do not ventilate a fire area when fresh air is contraindicated. In some cases, ventilation may not be possible until the overhaul stage of operations.

Extinguish

Only after a fire has been confined, should attempts be made to extinguish it. This is true, of course, only in the advanced fire situation. With an incipient-stage fire, all of the preceding objectives are met by an immediate attack on the fire once the alarm has sounded. Extinguishment may be achieved in a number of ways, depending on the specific fire situations as addressed in this chapter. Some ways include smothering, the use of fixed fire suppression systems, turning off the fuel supply, indirect and direct attacks by fire hose teams, and others (Figure 12.3).

Plan for additional crew members to relieve those combating a fire because an emergency taxes the human body greatly. A crew member who has been working vigorously, feeling no sense of fatigue, can suddenly be overcome by exhaustion and collapse.

Overhaul

Overhaul is another important priority in the incident stabilization/fire suppression component. Overhaul is the process of making sure a fire is extinguished and preventing reignition/reflash. When a fire is properly overhauled, it will not reignite. Overhaul includes fire cause determination and salvage/loss control operations. Provide rested crew members to perform

overhaul because it is very physically demanding work. Rotate crew members to allow for rest periods.

Overhaul includes *salvage* (the property conservation component of handling a fire emergency): an ongoing process considered throughout the emergency. Salvage actions can actually begin early in an incident with confining the fire to prevent damage to other areas and continues with a careful fire attack that does not cause unnecessary damage. Proper management of fire fighting water is a salvage action; it prevents unnecessary water damage to property. After a fire is extinguished, salvage continues by taking whatever actions are necessary to prevent any further damage. Salvage is also an aspect of damage control (see Chapter 13, Shipboard Damage Control) and the cleanup operation after a fire, including restoring systems and salvaging water- and smoke-damaged goods.

◆ General Tactical Procedures

The following sections describe general fire fighting tactics that apply to all fire situations and vessel types, specifically sounding the alarm, emergency actions, and reentry considerations. Tactics for specific spaces and vessel types are addressed in separate sections later in this chapter. Depending on the size of fire, area involved, and situation encountered, crew members may take all or most of the actions described. Crew members with any doubt whatsoever regarding emergency procedures on board their vessel should not hesitate to contact the officer in charge for information.

Figure 12.3 Activating a carbon dioxide suppression system is one way to extinguish an onboard fire. *Courtesy of R. Wright/Maryland Fire and Rescue Institute/United States Coast Guard.*

Sound the Alarm

Immediately sound the fire alarm, and inform the bridge of a fire, stating calmly and clearly the location and extent of the fire, the material burning, the present exposures, what action has been or is being taken, and any other pertinent information. When discovering smoke coming from a space, sound the alarm, and shut the doors and ventilation systems to the area. When the general alarm sounds, proceed to assigned stations without delay. Take necessary, reasonable, or prudent initial actions to protect life and the vessel.

Emergency Actions

After an alarm sounds, crew members report to their emergency stations and stand by for briefing on the situation and their assignments. The following actions (initial, attack, control, and postfire) are taken when responding to a fire. These actions are listed in order of priority; however, some reordering may be needed based on the real-time assessment made during fire fighting operations. Some actions are performed (possibly simultaneously) in accordance with tactical priorities established after sizing up the situation.

 Initial Actions

♦ *Start fire pumps, including emergency (backup) pumps, generators, and additional steering motors. The engineers usually perform these actions when the general alarm sounds. On some vessels these actions may be started from the bridge.*

♦ *Account for crew members/passengers.*

♦ *Control and/or secure ventilation (fans/blowers and dampers).*

♦ *Confine/isolate the fire by closing doors, hatches, and other openings to the fire area. Use the vessel's zoning capabilities.*

♦ *Isolate power to the involved area/space.*

♦ *Operate fuel shutoffs (deck stops) if required (Figure 12.4).*

♦ *Establish staging area for necessary equipment and crew members in a safe location near the fire area.*

♦ *Assemble additional crew members in proper protective equipment including self-contained breathing apparatus (SCBA); make ready to relieve crew members currently fighting the fire.*

♦ *Protect exposures. Set fire boundaries on all six sides (top, bottom, and four lateral sides).*

♦ *Consider use of fixed systems to establish boundaries. For example, release carbon dioxide into a pump room to prevent extension from a hold fire.*

Figure 12.4 An initial emergency action is to turn off fuel supplies. *Courtesy of Maritime Institute of Technology and Graduate Studies.*

 Attack Actions

♦ *Attack the fire. See direct, indirect, and combination attack procedures in Chapter 9, General Fire Fighting Procedures.*

♦ *Continue to size up the fire and to gather and report information.*

♦ *Ventilate fire area to the advantage of fire fighting teams and to the disadvantage of the fire.*

- *Practice good water management to avoid stability problems.*

- *Alert other vessels in the area of the need for assistance if necessary; proceed to the nearest port if possible.*

- *Provide appropriate medical attention for those in need.*

 Control Actions

- *Report the situation, and update the bridge when the fire is controlled.*

- *Search for fire extension to other areas (hidden fires).*

- *Perform overhaul/salvage/loss control.*

 Postfire Actions

- *Perform fire cause determination. Take photos and preserve the scene in as near to original condition as possible in case of serious injuries, fatalities, arson, or possible future legal action.*

- *Remove smoke and gases. Reestablish the vessel's normal ventilation system.*

- *Maintain fire/reignition (reflash) watch.*

- *Replace or service all used fire equipment.*

Reentry Considerations

If a fire is clearly extinguished, the means of access can be secured fully open, but entry to the space without wearing SCBA must be postponed until oxygen levels within the space are back to normal (at least 21 percent). Follow proper confined-space entry procedures. Monitor the space for oxygen content by using remote or handheld portable instruments. Ensure that those doing the testing are wearing SCBA, have a clear field of vision, and can be observed from outside the space by a backup team. Resist the urge to hurriedly enter to check the damage before oxygen levels are confirmed.

 ## Tactical Guidelines by Type of Space

Work within the framework established from size-up, determine which strategy is appropriate (offensive, defensive, or abandon ship), and determine tactical priorities. The following sections outline actions to fight small (also called *incipient*), medium, or large fires in specific spaces (accommodations, galley, and engine room/machinery/stack). Do not be misled by the terms *small, medium,* and *large* to describe the extent of a fire because even the smallest fire can lead to a disastrous situation. In general, the terms are described as follows:

- *Small* fires are controlled with portable or semiportable fire extinguishers.

- *Medium* fires require a fire hose team attack and/or fixed fire suppression system attack.

- *Large* fires require multiple control measures (fixed fire suppression systems, multiple hose team attack, vessel maneuvering, outside resources, etc.) because they are beyond the ability of a single fire hose team to control.

Accommodation-Space Fires

Accommodation fires include those in berthing spaces, office areas, storage areas (rope, stores, etc.), laundries, or interior vessel areas with predominantly Class A fuels (ordinary combustibles). These areas also could include Class C fuels (energized electrical equipment) such as office equipment, computers, laundry equipment, etc. Depending on the size of the fire, area involved, and situation encountered, crew members may take all or most of the actions listed in the following sections after sounding the alarm and notifying the bridge.

Small Fires

Swift and intelligent actions when a fire is first discovered may avert disaster. Many serious fires resulting in loss of life were found when still small enough to be easily extinguished yet were ignored until they grew so large they could not be controlled. Sound the alarm, notify the bridge, and attempt control. Even closing a door may retard fire spread long enough for help to arrive. Take the following actions for small accommodation-space fires:

- Control with portable extinguishers or automatic sprinkler system.
  - Use water, foam, or dry chemical for Class A fires.
  - Use carbon dioxide or halogenated agents for electrical fires.
- Perform applicable postfire activities.

Medium Fires

Although the difference between small and medium fires may seem subjective, it depends as much upon its location as its size. A small fire on an open deck may qualify as a medium fire if it is in a cabin because the atmosphere inside may be hazardous. A fire that cannot be safely approached because of either size or location is more than small. Full personal protective equipment (PPE) is required. Take the following actions for medium accommodation-space fires:

- Use proper entry procedures (full protective gear and proper fire hose techniques).
- Provide a minimum of two fire hose teams.
  - Attack (primary)
  - Backup (secondary)
- Conduct a primary search (rapid but thorough search performed either before or during fire suppression operations).
- Locate the seat of the fire.
- Conduct a coordinated attack.
- Avoid the situation where fire hose teams/streams come from opposite directions and face each other.
- Attack from upwind position if attacking from outside/exterior. Position the vessel to take advantage of the wind. See sidebar on the next page.
- Use cross ventilation in front of advancing fire hose teams, where possible, to push/discharge steam and products of combustion away from/out of the space. Accomplish by coordinated opening of leeward and windward sides.
- Extinguish/control the fire with direct, indirect, or combination attack methods, depending on specific conditions.
- Conduct a secondary search (a thorough search conducted after fire suppression operations are complete).

- Perform applicable postfire activities.

Large Fires

A large fire either is beyond the ability of fire hose teams to control or it overtaxes available personnel. Locations would include large-volume spaces such as passenger vessel assembly areas, casinos, entertainment areas, or dining areas. Multiple spaces may be involved. Small and sometimes medium fires are usually extinguished rapidly; large fires require time and persistence before success is gained. Take the following actions for large accommodation-space fires:

- Position the vessel to take advantage of the wind and to put the fire area downwind of uninvolved areas of the vessel if possible. See sidebar on page 292.
- Use a multiple fire hose team attack if personnel and resources are available (large crew member complement or personnel from other vessels).
- Extinguish/control the fire compartment by compartment.
- Use a coordinated attack. Avoid the situation where fire hose teams/streams come from opposite directions and face each other.
- Have properly equipped backup/relief teams ready to relieve exhausted attack teams (Figure 12.5).
- Maintain a continuous fire suppression effort until the fire is controlled (while considering vessel stability and the need for dewatering). A stop in the fire attack may allow a fire to regain its original intensity or spread further.
- Set fire boundaries; protect exposures.

Figure 12.5 Medium and large fires require backup teams to continue a fire attack. *Courtesy of Firefighter Luke Carpenter, Seattle (WA) Fire Department.*

- Fight fire from outside the involved area through doors, hatches, failed portholes, and other openings.

- Make holes where needed and practical to apply suppression agents.

- Consider the need to abandon ship if the fire is uncontrollable, and wait until reboarding becomes possible.

 Maneuvering the Vessel

Starting a second steering motor is an initial response action to a fire emergency. It is used to enable the vessel to maneuver during the emergency response. It may be possible to swing the vessel so that the fire is on the lee side, thus reducing smoke and directing heat, smoke, and flames away from the vessel. The effectiveness of this strategy depends on the vessel's location and room to maneuver, wind speed, vessel speed, and location of the fire. For example, a fire in the forecastle might be attacked while the vessel steams astern into the wind.

Galley Fires

Galley fires may include those in cooking equipment such as range top burners, grills, deep fryers, broilers, ovens, etc. Fires may also develop or spread into vent hood and duct systems. Other potential galley fire situations include those involving liquefied petroleum gas (LPG) (fuel for stoves) and alcoholic beverages on passenger vessels. Such fires are generally treated as Class B fires, depending on the energy source for the cooking equipment and the material burning. For fires involving ovens or storage that are Class A fire situations, follow the appropriate actions described in the Accommodation-Space Fires section. A Class C fire is possible if the cooking equipment is electrical. Depending on the fire's size, area involved, and situation encountered, crew members may take all or most of the actions listed in the following sections after sounding the alarm and notifying the bridge.

Small Fires

Control small fires with portable extinguishers or by automatic or manual fixed suppression systems (Fig-

ure 12.6). Take the following actions for small galley fires:

- Isolate power and/or heat source at cooking equipment (on/off switch).

- Isolate power at breaker box or appropriate electrical panel.

- Secure fuel supply (for example, LPG) if applicable.

- Smother fire with container lids if possible.

- Use fire blankets to smother if a fire is small enough.

- Check the vent hood and duct system for fire extension.

- Let cooking equipment cool.

- Repair cooking equipment, or evaluate the safety of reusing cooking equipment.

- Evaluate potential contamination of food stocks, and dispose of them if necessary.

Medium Fires

When portable extinguishers or fixed fire suppression systems fail to control a fire or the fire spreads, provide a minimum of two fire hose teams: attack team

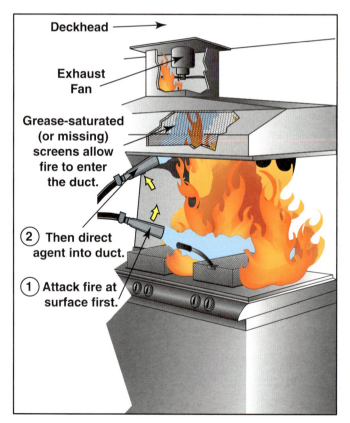

Figure 12.6 Fighting a galley fire: First attack the fire at the surface with an extinguishing agent. Second, direct the agent into the duct.

(primary) and backup team (secondary). Take the following actions for medium galley fires:

- Protect exposures/prevent fire spread.

- Shut down electrical power/fuel/heating element.

- Secure/close dampers in exhaust ducts. Shut down exhaust fans.

- Conduct postfire activities, including the servicing of equipment and considering contaminated food stocks.

Consider the following methods for deep fryer fires; other methods are available.

- *Deep fryer fire control method No. 1: applicator or nozzle*

  — Using applicator or nozzle, cool the sides of cooking equipment without getting water into burning fat/cooking oil.

  — Extinguish fire with dry chemical portable extinguishers.

  — Continue to cool cooking equipment.

  — Conduct postfire activities, including the servicing of equipment and considering contaminated food stocks.

- *Deep fryer fire control method No. 2: adjustable nozzle*

  — Approach deep fryer from a distance behind a wide fog/spray stream.

  — Contain and control the fire with a narrow spray cone pattern.

  — Minimize water spray in hot fat.

  — Surround immediate fire area with spray cone.

  — Discharge dry chemical from portable extinguishers in a narrow fog/spray pattern into spray cone until fire is extinguished.

  — Continue cooling to prevent reignition.

  — Conduct postfire activities, including the servicing of equipment and considering contaminated food stocks.

Large Fires

A large fire may involve galley vent hoods and duct systems, and fire may spread into other or adjoining spaces/areas. See Accommodation-Space Fires section for information on fires in adjoining spaces. Extinguish galley deep fryer fires as described in the previous Medium Fires section. Take the following actions for large galley fires:

- Secure/close dampers in exhaust ducts, and shut down exhaust fans.

- Set fire boundaries on all six sides.

- Manually discharge hood and duct fixed fire suppression system if available. If system is automatic, ensure it discharges.

- Open and inspect the entire hood and duct system, including areas where ductwork passes through and terminates, and check for fire extension and spread.

- Open the involved hood and duct system if fire was not controlled with a fixed suppression system (may have to cut into or penetrate bulkheads) and extinguish with water spray, dry chemical, or a combination of the two agents.

- Conduct appropriate postfire activities.

Engine Room/Machinery Space Fires

These spaces include all machinery spaces, boiler rooms, engine rooms, etc. The types of fires encountered are most often Class B (flammable and combustible liquids) and Class C (electrical). Class B fires may be either spill fires where liquids are standing or pooled (such as in the bilge) or spraying, three-dimensional fires where pressurized liquids are flowing due to ruptured piping, fire hoses, or fittings.

Machinery spaces probably have the greatest number and concentration of electrical equipment and circuitry. Although electrical fires are discussed in this section, they can be encountered anywhere on a vessel and can be handled in the same manner. Class C fires may involve high or low voltages and alternating or direct currents (AC or DC). Know electrical safety considerations (see Chapter 2, Onboard Fire Causes and Prevention, and Chapter 3, Vessel Types, Construction, and Arrangement).

Engine rooms and other machinery spaces can present some of the most dangerous fire fighting situations (Figure 12.7). Because these spaces are typically located low in the vessel (similar to a basement), fire teams may have to descend through intense heat and smoke with zero visibility to fight a fire. Depending on the specific circumstances, the person in charge evaluates the situation to make the crucial decision of whether to enter the space. Use extreme caution when entering the space. If it is too hazardous, the choice may be to close (button up) the space and attack the fire by controlling ventilation and using fixed fire suppression systems if available. Depending on the situation encountered, crew members may take all or most of the actions listed in the following sections after sounding the alarm and notifying the bridge.

Small Fires

Depending on the fuels involved, take the following actions for small engine room/machinery space fires:

- *Spraying, three-dimensional/pressurized liquid Class B fires:*
  - Address safety considerations before making the attack.
  - Secure the fuel supply.
  - Isolate electrical power.
  - Contain fuel runoff.
  - Extinguish the fire with portable or semiportable fire fighting equipment using CO_2, dry chemical, or halogenated agents.

- *Bilge liquid spill (standing/pooled) Class B fires:*
  - Secure the fuel supply.
  - Control the fire/fuel spread through the bilge.
  - Extinguish the fire with portable or semiportable fire fighting equipment using CO_2, dry chemical, halogenated, or foam agents (Figure 12.8).

- *Electrical Class C fires:*
  - Isolate power if possible.
  - Extinguish the fire with portable or semiportable fire fighting equipment or local application fixed suppression system (not an entire machinery space system) using CO_2, dry chemical, or halogenated agents.

Medium Fires

Generally, a medium fire is too large for fire extinguishers to control. A fire becomes a medium fire if extinguishers are not successful. As a result, fire teams need to make an attack with fire hoses. A minimum of two fire hose teams (primary attack team and secondary backup team) is recommended. Based on the size-up, the person in charge may consider using a fixed suppression system for this size fire. For Class B fires, foam is the recommended extinguishing agent. Another choice is to use semiportable dry-chemical extinguishers with water cooling from fire hose teams. Take the following actions for medium engine room/machinery space fires:

- Secure the fuel, electrical power, and engines or other machinery in the space.

- Attack the fire with properly protected and equipped fire hose teams. Teams enter together *through the same point of entry.*

- Attack the fire from the lowest point possible. Consider alternative routes to approach the fire (escape

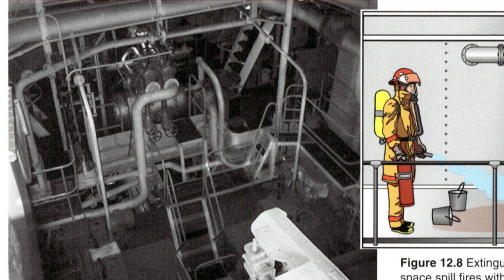

Figure 12.7 Fires in engine rooms present dangerous situations. *Courtesy of Firefighter Luke Carpenter, Seattle (WA) Fire Department.*

Figure 12.8 Extinguish small engine room/machinery space spill fires with portable or semiportable fire extinguishers.

trunks, shaft alley, etc.). Teams may be forced to attack from above, but attempt this attack only if it is the only tactic left before discharging a fixed suppression system.

- Control ventilation to the advantage of fire teams.

Large Fires

Large fires in engine rooms or other machinery spaces are generally beyond the ability of fire teams to extinguish or overtax available personnel and equipment resources. A large fire requires a defensive fire fighting effort. Most burning Class B materials flow to the bottom of an engine room to the bilge. One technique is to apply enough foam so that it cascades down to the bilge, extinguishing most of the fire and cooling as it flows. An alternative strategy is to apply an adequate amount of foam or other agent to control the fire from outside the space. Plan these types of attack in advance as a contingency in the prefire plan. These types of extinguishment techniques could deplete extinguishing agent supplies. Take the following actions for large engine room/machinery space fires:

- Prepare to activate the fixed fire suppression system for the affected space.

- Activate fuel shutoffs.

- Evacuate the space.

- Shut down ventilation. Close all vents, doors, and hatches.

- Provide boundary cooling. Inert adjacent spaces (such as pump rooms or cofferdams) if possible.

- Follow specific procedures approved by the master and chief engineer to activate the fixed fire suppression system.

- Check for proper fixed fire suppression system activation and application.
  - Did the agent reach the space?
  - Did the agent reach other protected spaces instead of the fire area?

- Monitor interior conditions and temperatures.

- Allow for a proper waiting time period (soaking time), which may be hours or *days.*

- Do not open the space until there is no question that the fire is extinguished and cooling has occurred. Determine extinguishment and cooling by checking boundary temperatures, visible signs of

bulkhead temperatures, thermometer and pyrometer readings, closed-circuit television, and thermal imaging devices.

- Perform overhaul after the fire is extinguished and adequate cooling has taken place. Enter the space in full personal protective equipment and SCBA.

- Open the space and ventilate.

- If the fixed fire suppression system did not succeed in controlling the fire, reevaluate the situation and develop a new strategy. Continue defensive actions.

- Close the space, set fire boundaries, cool boundary bulkheads and decks, move combustibles away from fire boundaries, and proceed to nearest port if the vessel has power and steering.

- Prepare the lifeboats, and abandon ship if necessary.

 Dead Ship Situation

A large fire in an engine room requires turning off or shutting down engines and other vital equipment, which creates a dead ship situation. Do not underestimate the seriousness of a dead ship. In order to extinguish this size fire, the space may have to be closed for a great length of time — maybe 24 hours or longer. Because a dead ship has no means of propulsion or maneuverability, the safety of the crew and passengers and the vessel itself and other vessels nearby is jeopardized. If a large fire requires a dead ship situation, the vessel's officers and crew must realize that the vessel may have to remain dead for an extended period, and they must prepare to deal with the associated hazards.

Stack Fires

Stack fires may include the economizer (type of boiler in the stack), funnel, uptake, casing, and fiddley (vertical space from engine room to the stack). These fires are generally caused by poor maintenance and the buildup of propulsion system combustion deposits (carbon) in the exhaust system. Fires are less frequent and less intense when these areas receive frequent and thorough washing and blowing of the economizer tubes to remove soot and carbon. When these fires do happen, they can generate great amounts of heat and cause significant damage to the economizer. Therefore, extinguish this fire as quickly as possible to reduce damage. A stack fire may also extend to clothes spread for drying and other combustible storage in the fiddley area. Handle these fires with procedures similar to those listed for accommodation spaces. Two extinguishment methods are discussed in the following sections.

Traditional Extinguishment Method

The traditional extinguishing method for a stack fire in its early stage of development includes the following steps:

Extinguishing a Stack Fire (Early Stage)

Step 1: *Turn off the main engine(s) to reduce exhaust flow and temperature. In low-temperature fires, let water circulate to remove heat.*

Step 2: *Initiate the water or steam wash system (maintenance system that cleans the stack and economizer tubes) or initiate the CO_2 or halogenated agent fixed suppression systems. Do not blow the economizer tubes; it could intensify the fire.*

When a stack fire exceeds temperatures of 1,300 to 1,500°F (704°C to 816°C), the possibility of a hydrogen fire exists. Hydrogen fires occur when water subjected to extremely intense heat disassociates into hydrogen and oxygen and becomes fuel for the fire rather than an extinguishing medium. If the fire indicates high internal temperatures; that is, stack temperatures of 1,000°F (538°C) or greater or the glowing of the stack casing, initiate the following steps:

Extinguishing a High-Temperature Stack Fire

Step 1: *Turn off the main engine to reduce exhaust flow and temperature.*

Step 2: *Deploy fire hoses to cool the outer casing and prevent fire extension.*

Step 3: *Drain water from the economizer tubes.*

Step 4: *Seal the stack and inject CO_2 to extinguish the fire.*

CAUTION: To prevent water in the system from feeding the fire in hydrogen fires, stop the force feed pumps to the waste heat boiler. Isolate and drain the economizer and generating tubes at the headers.

If this method does not extinguish the fire, employ the *fuel removal method,* which in reality allows the fire to consume the fuel. Cool all boundaries until well after the fire has burned out. Obviously, this method causes the greatest amount of damage, and the possibility of fire extension or spread is very serious.

Nontraditional Extinguishment Method

Shoreside fire departments have had great success in chimney and stack fire extinguishment by applying positive-pressure ventilation (PPV) to a chimney while injecting an A-B-C-type fire extinguishing agent (monoammonium phosphate is preferred) at the base of the chimney. The agent interrupts the chain reaction and extinguishes the fire. In addition, the agent coats all surfaces and reduces the possibility of reignition. Because there is no need to seal the stack, shut down the engine, or drain the economizer tubes, this procedure can be initiated very quickly, allowing minimal damage to the economizer. Investigate and thoroughly inspect all spaces and areas adjacent to the fire area for fire spread or hidden fires. Cleanup is performed by the water or steam wash system that cleans the stack and economizer, followed by inspection and repair. Employ this extinguishment method in a stack fire by performing the following steps:

Extinguishing a Stack Fire With a Positive-Pressure Technique

Step 1: *Slow main engine to maintain exhaust flow, which reduces stack temperatures somewhat.*

Tactical Guidelines by Type of Vessel

Determining tactical guidelines by type of vessel is the same as determining them by type of space: Work within the framework established from size-up, determine which strategy is appropriate (offensive, defensive, or abandon ship), and determine tactical priorities. Depending on the circumstances, the cargo carried or the vessel type or both may be factors in the tactics used. The same definitions given earlier apply for the terms *small, medium,* and *large* when describing fire extent. The following sections outline actions to fight fires involving specific vessel types such as bulk and break bulk, container, ro/ro or car carrier/ferry, tanker, gas carrier, passenger, and miscellaneous/small craft.

Bulk and Break Bulk Vessel Fires

On bulk vessels, unique types of cargo fires are encountered such as fires in bulk cargo in holds, cargo ignited by heat buildup from spontaneous heating, and fires following dust explosions. Moisture in cargo may cause it to expand or swell. Fire detection may be delayed due to the size and configuration of the vessel, a deep-seated fire location, and the fact that cargo spaces are not staffed except during loading or unloading operations.

Break bulk vessel fires present additional challenges because of the diverse and unique construction of these vessels. Multiple cargo types exist, and consideration must be given to potentially incompatible cargoes that may come together to create a fire or to make one worse during an emergency. Consider the access problems created by tween decks and partial loads. Depending on the fire's size, area involved, and situation encountered, crew members may take all or most of the actions listed in the following sections after sounding the alarm and notifying the bridge.

Small Fires

Small cargo fires on bulk vessels are not likely at sea. In order to be considered a small fire, it would have to be discovered in the early (incipient) stage when the hatches are open, generally during cargo handling. A small fire would only involve the surface of the bulk material. If a small fire were discovered after the hatches are closed, crew members would have to open the hatches to access it. Introducing air would accelerate a fire. Open hatches may also warp from the heat and then be impossible to close.

It may be possible to attack break bulk fires in their early stages with fire hose teams and/or portable extinguishers. Approach from the lowest point possible, and exercise extreme caution. A properly protected and equipped backup team is required, and a means of escape is always maintained.

Medium and Large Fires

In medium and large fires, the type of cargo involved determines options, strategies, and tactics, which include proper ventilation. Fires in bulk and break bulk vessels are handled in the same way. Take the following actions for medium and large fires in bulk and break bulk vessels:

- Provide boundary cooling.

- Determine the best extinguishing procedure for the cargo type.

- Fill (press-up) hopper tanks, wing tanks, and double bottoms with seawater to displace vapors and improve stability (subject to stress and stability considerations).

- Provide cooling streams over the hatch covers.

- Flood or soak the involved material with the appropriate extinguishing agent. Consider the following potential vessel stress and stability issues with this technique:
    — Steam explosions may result if water or foam is introduced.
    — Material may be water reactive.
    — Wet cargo may expand or swell.

- Close the hold, set fire boundaries, provide boundary cooling where possible, and proceed to the nearest appropriate port for handling the type of cargo involved. Remove the involved cargo in port.

- Monitor temperatures in adjacent compartments.
- Prepare to activate the fixed fire suppression system if available.
  — Close the hold.
  — Set fire boundaries.
  — Provide boundary cooling where possible.
- Activate fixed fire suppression system (authorized by person in charge).
  —Check for proper discharge.
  —Allow adequate wait/soaking time.
  —Monitor interior conditions and temperatures.

Container Vessel Fires

Container vessels usually have deep holds in which intermodal containers (capable of transport by more than one carrier) are loaded into cell guides. The holds may be filled with containers extending up several levels above the main deck. Containers come in various types and sizes but generally are constructed of steel. They are very strong and present a difficult access situation. Containers may hold multiple cargoes and hazardous materials/dangerous goods.

The possibility exists that a cargo (such as an acid) may leak from one container and react with other containers, the vessel, and other cargoes. Very often crew members are unable to immediately determine the exposures. Check the dangerous cargo manifest (DCM) and dangerous cargo locator chart to see what hazards are in the area. Consult the International Maritime Organization (IMO) *International Maritime Dangerous Goods (IMDG) Code* for recommendations on how to deal with a specific dangerous goods situation.

Fires occurring on container vessels can include all types of cargo and all classes of fire. Class C fires may also occur in generator sets and circuits providing power to refrigerated units (reefers) (Figure 12.9). Generator sets are often diesel-fueled, which presents a further Class B fire potential. Fires may be deep-seated in ordinary combustibles, and delayed discovery of a fire is a possibility. Depending on the fire's size, area involved, and situation encountered, crew members may take all or most of the actions listed in the following sections after sounding the alarm and informing the bridge.

Small Fires

The chance of discovering a fire in a container small enough to put out with a portable extinguisher is unlikely. A small fire outside the container would more likely be discovered, and the chance of extinguishment with portable equipment would be greater. Take the following actions for small fires in container vessels:

- Isolate power to the affected area.
- Extinguish the fire with portable fire extinguishers. Dry chemical is the recommended agent.

Figure 12.9 Refrigerated container units (reefers) on container vessels use extensive electrical systems. *Courtesy of Captain John F. Lewis.*

- Secure fuel and power. Diesel generator fires on reefer containers are similar to vehicle fires.

Medium and Large Fires

For tactical purposes, medium fires do not exist in container vessels. Treat all smoking containers as having a large fire potential. Any fire involvement inside a container or multiple containers or smoke showing from a container is treated as a large fire. Take the following actions for medium and large fires in container vessels:

- Identify cargo in the affected and surrounding area, below hatch covers, and in holds/cells.

- Isolate the hold/cell. Secure all openings.

- Use fixed fire suppression system.

- Set fire boundaries. Inert adjacent spaces (pump rooms or cofferdams) if possible.

- Proceed to the nearest appropriate port that handles containers.

For hatchless container vessels or where containers on deck are above the hatch covers, the best tactic may be to protect exposures and take a defensive posture. Take the following actions:

- Make access if possible. Usually it is not possible to open container doors due to the way they are latched and the way the containers are lashed.

- Maneuver the vessel to the advantage of the fire teams and the disadvantage of the fire.

- Cool the container if possible. A minimum of two 2½-inch (64 mm) fire hoses is recommended. Cool boundaries on all six sides. Consider using unmanned, remote fire streams.

CAUTION: The use of piercing nozzles to fight container fires can be very dangerous because what is on the other side of the container wall may not be known. The same problem exists with cutting holes with a fire axe to insert a nozzle. Attempt these tactics only with extreme caution.

Ro/Ro, Car Carrier, and Ferry Fires

Fires in these types of vessels present several new problems. The most obvious and most common problem is dealing with vehicle fires. Vehicle fires have a number of unique hazards, including air bags, driveshafts, pneumatic bumpers, high-pressure components, tires, magnesium parts, and fuel tanks. On ferries, vehicles carry unknown cargoes that may be hazardous (Figure 12.10). Fires in the interior (passenger compartment) of vehicles are generally Class A. However, fires involving the fuel system, engine compartment, and possibly the cargo space are usually Class B. A fuel fire may result in a flowing flammable liquid (running spill) fire that can spread over decks to other vehicles and spill down to scuppers. Tires may also be involved. They are susceptible to exploding and are difficult to extinguish.

Other problems associated with vehicle-carrying vessels are large, open decks and multiple *layers* of vehicles in a multideck situation. Due to the nature of large, open decks, vessel stability is a major consideration. The vessel's hydraulic systems for ramps and elevated decks may present a pressurized, three-dimension fire situation similar to those found in engine rooms.

Be alert for vehicle fires with LPG, compressed natural gas (CNG), or liquefied natural gas (LNG) fuel systems (recreational vehicles are particularly suspect). The most serious fire problem in these types of vessels is the unknown — unknown cargo in individual trunk spaces and mixed cargoes in transport vehicles.

Depending on size, area involved, and situation encountered, crew members may take all or most of the actions listed in the following sections after sounding the alarm and notifying the bridge. Be aware of

Figure 12.10 Vehicle fires present unique hazards (pneumatic bumpers, high-pressure components, unknown cargoes, etc.) on ro/ros and ferries. *Courtesy of Firefighter Luke Carpenter, Seattle (WA) Fire Department.*

overall strategic options before considering tactical options. These options may include the following:

- Access the vehicles to extinguish a fire with a fire team in proper PPE including SCBA, using fire hoses and/or portable fire extinguishers.

- Extinguish a fire using the vessel's fixed fire suppression system (deluge sprinklers, CO_2, etc.), and then move in to overhaul the fire (PPE with SCBA is required).

Gaining access *to* a vehicle may be a challenge. When vehicles are tightly packed into a deck, crew members may have to crawl between or around other vehicles to reach a fire. Gaining access *into* a vehicle may also present a problem. The involved space has to be opened to control and extinguish a fire. Do not assume that all doors are locked or that hoods and trunk lids can only open by prying. *Try before you pry,* but when forced entry is necessary, use the following techniques:

- Shatter safety glass windows with a sharp, pickhead axe or more easily with a spring-loaded center punch tool.

- Use the usual mechanism to open hoods. If a remote release cable has burned, mechanisms can sometimes be operated by reaching through or behind a grill to grasp the remaining end of the cable.

- Open trunks by breaking or removing a taillight assembly to gain access to the mechanism.

Small Fires

Take the following actions for small fires in ro/ro, car carrier, and ferry vessels:

- Use portable extinguisher(s) to control a fire in individual vehicles that are partially involved. Use dry chemical (preferred), CO_2, or foam agents. Extinguish engine compartment fires by discharging a portable fire extinguisher under and behind the radiator.

- Preestablish an escape path.

- Provide backup fire hose teams (minimum of two: attack or primary and backup or secondary).

- Keep significant ventilation systems for the deck in operation to maintain ventilation. Turn them off only when using a fixed fire suppression system.

- Be alert for a potential running fire.

- Disconnect the batteries on burned vehicles. Fire damage can cause a vehicle to suddenly start. First cut or remove the *ground cable*; then cut or disconnect the *positive cable*.

- Use absorbents to control fuel, lubricating oil, and transmission fluid spills.

- Stay clear of vehicle bumpers, shock absorbers, and air bags because these are possibly pressurized components. Approach the vehicle from the side.

- Approach from upwind on open deck.

Medium Fires

Medium-sized fires are those that cannot be controlled with extinguishers such as a fully involved vehicle, a large vehicle/truck, a fuel spill, and multiple vehicles. Hydraulic fires involving vessel components (doors, decks, and ramps) are also considered medium fires. These fires require fire hose teams. Provide two hose teams (additional if possible) as a minimum: attack team (primary) and backup team (secondary). Class B foam is the recommended extinguishing agent. Do not mix water streams with foam streams. Take the following actions for medium fires in ro/ro, car carrier, and ferry vessels:

- Access vehicle(s) using the easiest route possible. The fire hose may catch on tires of other vehicles. Consider advancing dry fire hose, and then pull some extra slack to allow for advancing before the fire hose is charged.

- Keep the ventilation system operating, depending on fire conditions. Monitor the situation.

- Approach vehicle(s) with a protective stream in operation.

- Apply short bursts of agent (foam or water) and reassess.

- Consider the use of fixed fire suppression systems.

These types of vessels present the possibility of pressurized, three-dimensional, spraying fires involving the hydraulic systems required for opening doors, raising and lowering decks and ramps, etc. It is usually not possible to shut off or bleed off the hydraulic pressure in these systems. If possible, turn off or isolate valves supplying a broken component. However, the fuel will continue to flow after shutdown until the pressure is depleted.

These types of fires present a serious slip hazard from spilled hydraulic oil. Maintaining position while staffing a pressurized fire hose can be a challenge. Other hazards include the possibility of flying shrapnel produced from ruptures or system failure. System failure can also cause decks to drop or tailgate doors to open.

Use a combination water and dry chemical fire attack to envelop the hydraulic fluid pressure-fed fire with a narrow fog/spray pattern, and discharge dry chemical into the pattern. This method is similar to deep fryer technique No. 2 given earlier. Use the following instructions for hydraulic fires on ro/ro, car carrier, and ferry vessels:

- Protect exposures.
- Use dry chemical or halogenated agent extinguishers.
- Use fixed fire suppression systems if available.
- Recover spilled hydraulic fluid with absorbents.

Large Fires

Large vehicle fires are those that are beyond the capabilities of fire hose teams to extinguish and usually include multiple vehicles. Vessel stability considerations are very important. Take the following actions for large fires in ro/ro, car carrier, and ferry vessels:

- Confine the fire.
- Shut off power and ventilation, depending on fire conditions.
- Evacuate passengers on ferries to a safe area.
- Consider turning the vessel and opening doors to take advantage of the wind.
- Account for personnel. Perform search and rescue if required and evacuate endangered persons.
- Follow procedures for use of fixed fire suppression systems.
- Conduct postfire activities.

Tanker Fires

Tanker fires most commonly are Class B fires and may include hazardous materials, toxic cargoes, and multiple types and grades of cargoes. Pump room fires are also a possible occurrence. Fires may result from spills, overfills, and ignited drip pans. They may involve manifolds, leaking flanges, and vapor recovery systems. In a tanker fire situation, all crew members may have to wear SCBA and have emergency escape breathing devices (EEBDs). Follow specific emergency procedures according to vessel type for any release of a hazardous gas.

Specific tanker emergency situations may involve internal tank fires, ullage/vent fires, deck fires, and gas and cryogenics (LPG, LNG, and anhydrous ammonia) fires and releases. Mixing of cargoes and products may result in chemical reactions and explosions. Deck tank fires present the threat of a boiling liquid expanding vapor explosion (BLEVE). Tanker fires are usually the result of a collision or an accident with the cargo handling equipment. Depending on the size, area involved, and situation encountered, crew members may take all or most of the actions listed in the following sections after sounding the alarm and notifying the bridge.

Small Fires

Extinguish a small fire with portable fire extinguisher(s). Take the following actions for small fires in tanker vessels:

- Approach from upwind if on deck.
- Use foam or dry chemical fire extinguishers.
- Secure applicable valves to stop any leaks.
- Contain spills; capture leaking material (bucket/drum under leak).
- Shut down cargo pumps, both on vessel and at terminal if in port.
- Recover/absorb spilled product.

Medium Fires

Medium fires require fire hose teams. Provide a minimum of two fire hose teams: attack team (primary) and backup (secondary). Combination techniques (use of two or more agents consecutively or concurrently) may be effective on Class B fires. Take the following actions for medium fires in tanker vessels:

- Extinguish or control the fire using one or two monitors. Sweep burning liquid off the deck with foam monitors to minimize the size of a fire before applying foam.
- Use foam for spill fires (Figure 12.11).
- Use foam and dry chemical combination attack for spraying liquid and spill fires.

- Foam the fire area until the fuel pressure drops, then put out the residual fire with dry chemical. Cool and maintain foam blanket (do not dilute with water) to prevent reignition.

- Use the adjustable nozzle method, which is similar to the galley deep fryer technique No. 1 given earlier:

  — Protect exposures.

  — Start approaching a spraying, three-dimensional fire from a distance behind a wide fog/spray stream using water or foam.

  — Start narrowing spray pattern while approaching to contain and control the fire.

  — Surround immediate fire area with spray cone pattern.

  — Discharge dry chemical from portable extinguisher(s) into spray cone until fire is extinguished.

  — Continue cooling as needed to prevent reignition.

  — Control spilled product on board at fishplate and scuppers.

  — Apply foam at point of collection of runoff.

Large Fires

Large fires are beyond the capability of fire hose teams to control and may include fires inside tanks. These large fires may result from tank explosions, collisions, and cargo hose or loading arm failure. They can be catastrophic if the fixed fire suppression system runs out of agent or fails to extinguish the fire. Use of an inert gas system (IGS) gives protection of exposure tanks and spaces, but probably cannot extinguish a fire because the system does not generate gas fast enough or in sufficient quantities, especially if the involved tank is breached. Take the following actions for large fires in tanker vessels:

- Use hose streams, foam monitors, and fixed fire suppression systems to protect the house (superstructure) and other exposures until the fire burns out (Figure 12.12).

- Turn the vessel to take advantage of the wind.

- Apply a foam blanket inside tanks with a fixed fire suppression system or fire hoses. Press up the foam blanket by filling the tank with cargo or water.

- Use foam conservation and good application techniques because foam has to get to the surface of burning liquid. A limited quantity of foam may be available on board. See Chapter 6, Water and Foam Fire Fighting Equipment.

Gas Carrier Fires

Gas carrier fires are unique to the product they carry: butane, propane, LPG, LNG, etc. They present hazards associated with cryogenic releases (that is, large

Figure 12.11 Foam is the most commonly used extinguishing agent on Class B fires and spills on tankers. *Courtesy of Firefighter Luke Carpenter, Seattle (WA) Fire Department.*

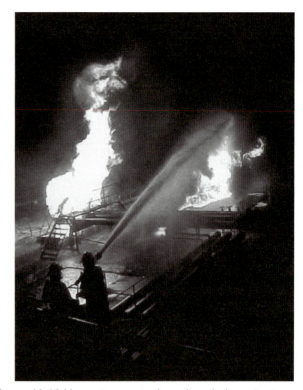

Figure 12.12 Hose streams and monitors help protect exposures on large tanker fires. *Courtesy of Firefighter Aaron Hedrick, Seattle (WA) Fire Department.*

vapor clouds) that can be catastrophic. Void spaces must always be inerted. Inerting spaces is not a fire-fighting technique, but it is critical to the success of fire fighting operations. Emergency situations with gas carriers usually occur during product transfer. For hazardous materials and LNG fires, apply dry chemical from a fixed fire suppression system. For further information, refer to IFSTA's **Hazardous Materials for First Responders** manual and the Fire Protection Publications book, *Hazardous Materials: Managing the Incident.*

WARNING

Never apply water to an LNG spill or fire. Water applied directly into a spill of LNG will cause rapid vaporization causing more fuel to be available.

Passenger Vessel Fires

Passenger vessel fires present a large life safety problem. Otherwise, the types of fires encountered have already been addressed in earlier sections of this chapter. Outside of the engine room, machinery spaces, and galley and holds, passenger vessel fires are likely to involve accommodation spaces, although on a large scale. Passenger vessels also present the possibility of fire spread through multiple elevator shafts and open stairwells. Vertical zoning may allow for horizontal evacuation to protect passengers in place as is done shoreside in a high-rise building.

Management of people/crowd control is the largest and most important responsibility. Crew members must prepare to face the situation calmly and confidently to set an example for the passengers. Panicked crew members cause the passengers to panic; confident crew members instill confidence in passengers and elicit cooperation from them. Terminals should preplan for the possibility of having a large volume of people coming into port during an emergency. Refer to *Standards of Training, Certification and Watchkeeping for Seafarers (STCW)* 1978 Convention amended 1995 (known as *STCW 95*) for additional information on crowd control.

Crowd Control Procedures

In any crowd or group of people, perhaps 15 percent will take action on their own accord. Use these people to work with vessel crew members, otherwise they may take independent action. The majority of people will respond to directions, although some will be incapable of either acting or taking directions. A few may object; neutralize these people by assigning them tasks. Most disputes over command rulings (about 97 percent) are resolved verbally — physical force is a last resort. When addressing a crowd, it is important to follow the following guidelines:

♦ *Stand facing the crowd with a firm stance.*

♦ *Speak clearly, but only a little louder than normal (if you have to shout, any more than a few words will be inaudible).*

♦ *Gesture or direct with open palms — do not point or beckon by crooking the fingers (such*

gestures are insulting or have pejorative meanings in some cultures).

♦ *Remain in clear sight of other officers or crew members.*

♦ *Stay in contact with the bridge or Command officers.*

♦ *Do not stand in doorways unless intending to block passage.*

♦ *Keep people informed of events if possible.*

Catering staff members in particular have an important role to play during emergencies in moving large numbers of people quickly and safely through the vessel to the exits or boat decks. Such evolutions are part of normal business aboard cruise vessels, and very little additional work is needed to ensure safe crowd control in emergencies. However, to ensure effectiveness, practice evacuation procedures.

Upon discovering a fire, *immediately* secure ventilation. Shut it down first, then evaluate. If it is determined that positive pressure is good for clearing the area or smoke removal, manage the ventilation system to produce the desired results.

Miscellaneous/Small Craft Fires

The discussion throughout this chapter has been on fires on large commercial vessels, but many smaller craft exist: fishing vessels, harbor patrol boats, crew launches, water taxis, tug and tow boats, pleasure craft, and small passenger vessels. Many have fixed fire suppression systems, fuel shutoffs, and other features that have been discussed. On smaller vessels, aluminum, fiberglass, or wood construction makes the chance of the vessel burning to the waterline a distinct possibility (Figure 12.13).

The principles of fire attack and fire prevention are valid for small craft as well as for larger vessels. The forces involved may be smaller, with fewer crew members to respond to an emergency, but the concepts do not change, only the application. Vessel stability and free surface effect are still concerns, and understanding fire behavior helps a person to attack a fire successfully.

General safety and personnel accountability are also as important on small vessels as they are on larger ones. For example, in port the concept of personnel accountability is easily practiced, especially if only four people are aboard. Simply devise a board or other means of identifying who is ashore and who is aboard. Accountability ensures that no one will risk his or her life to rescue someone who has gone ashore without telling anyone.

Many people believe that without SCBA availability, they can do nothing to fight a fire. However, harbor patrols, tugs, or other vessels with only a small fire hose or a few portable extinguishers have extinguished many small craft fires. While not a perfect description, the following example illustrates a basic knowledge of fire behavior, size-up, and principles of extinguishment while maintaining personal safety.

Yacht Fire Example

Fire is discovered in the cabin of a 30-foot (9 m) yacht with smoke showing. During size-up, one might note the spare outboard engine on the stern (the main engine is inboard). This is a clue to watch for gasoline. A cylinder of cooking gas is another exposure. Lifting the forward hatch and inserting a fire hose on full fog pattern might achieve extinguishment. Withdraw the hose, close the hatch, and wait. The steam may extinguish the fire.

Figure 12.13 A fishing vessel can burn quickly because many are made of lightweight materials. *Courtesy of Captain John F. Lewis.*

Chapter 13

Shipboard Damage Control

This chapter covers aspects of shipboard damage control (including vessel stability and dewatering concerns) that are pertinent to fire fighting operations and other emergency response actions. *Damage control* deals with remedial measures taken to restore or maintain seaworthiness when the integrity of the hull has been breached, whether by collision, grounding, explosion, or other causes. Important aspects of damage control are anticipating and managing the movement of large amounts of water on board a vessel during an emergency. Damage control deals in great length with measures taken to shore weakened bulkheads, decks, hatches, or deckheads and to prevent ingress of water and further breaching should a vessel encounter adverse weather.

Damage control is also taking control of a situation to keep the vessel from suffering greater damage—in other words, a quick fix. Damage control need not be sophisticated or complicated, nor is specialized equipment needed. Many solutions to damage control problems are done simply and quickly. Some examples of corrective measures illustrate the concept. If the standard compass is damaged, use the gyrocompass, or vice versa. If both are damaged, use a lifeboat compass. If the bridge is destroyed by fire, it is possible to steer the vessel from the emergency steering position using the trick wheel (Figure 13.1). Single-screw vessels have been successfully steered to safety by using weights hung from derricks on both port and starboard sides when all steering power and rudder were lost. The drag caused by lowering a weight to produce a turn affected a crude but ultimately successful steering mechanism.

Following a successful fire fighting operation or other emergency response, inspect the vessel and make repairs. For example, strengthen bulkheads (either above or below the main deck), cover broken windows or portholes that are not fitted with deadlights, rig emergency steering, or repair broken piping. Water need not come directly from outside the

Figure 13.1 A vessel bridge destroyed by fire. *Courtesy of Captain Eric Beetham.*

vessel to cause problems. Uncontrolled water from fire fighting or liquid released from tanks due to broken piping may be sufficient to cause a loss of vessel stability due to free surface effect (FSE) (water moving freely in a confined area). Remedial measures to counteract this effect are discussed briefly in this chapter, but detailed procedures are beyond the scope of this book.

This chapter gives an overview of vessel stability factors, some general principles of free surface effect, emergency response damage control operations, dewatering techniques, inspecting/repairing guidelines, cross-flooding, and advance planning by estimating water quantities and stability. The information given is by no means all-inclusive. More detailed information on these topics is found in U.S. naval publications, British Admiralty publications, nautical textbooks, and many other textbooks.

Vessel Stability

Stability is the measure of a vessel's ability to return to an upright position when heeled by an external force. Stability is dynamic and ever changing. For example, changes occur when cargo is loaded and unloaded and ballast is added or discharged. Stability depends upon the design of the vessel, whether it is empty or loaded, how much additional weight is placed on board, and where additional weight is located.

Figure 13.2 This vessel is listing. *Courtesy of Jon Swain, Texas A & M University System.*

Situations where stability may be a critical factor include the following:

- *Emergency operations (fire fighting, flooding, etc.)* — Uncontained water or liquids may reduce a vessel's stability because of free surface effect (see Gravity/Buoyancy and Free Surface Effect sections).

- *Collision or grounding* — Compartments may be breached (opened) or compromised. Any hole in the vessel's hull, above or below the waterline, affects stability and buoyancy. Depending on a compartment and its contents and the percentage of its space that can be occupied by water (permeability), stability may be reduced. The vessel may list or alter its trim (see List section) (Figure 13.2).

The ability of a vessel to remain in a stable position is controlled by the interaction of two opposing forces, gravity and buoyancy. A shift of gravity can cause list (tilt), which reduces vessel stability. The following sections contain gravity and buoyancy information and describe how to calculate the draft of a vessel that has a list and identify the critical angle of list.

Gravity/Buoyancy

The *center of gravity (G)* is the point where the weight of a vessel and its cargo are assumed to be concentrated. This point is determined by calculations performed during loading and unloading of fuel, water, and cargo. The center of gravity remains in a constant position until the weight distribution on the vessel changes, and then it moves. It may shift fore or aft, up and down, or athwartships (side to side) as cargo is loaded or unloaded or as fuel and water are consumed during a voyage.

The *center of buoyancy (B)* is the center of the immersed volume of the vessel. *Buoyancy* is a force equal to the weight of water displaced by the vessel. Buoyancy is an upward force perpendicular to the surface of the water and is assumed to be located at the center of buoyancy. The center of buoyancy moves laterally, longitudinally, and vertically as a vessel heels, trims, and changes draft (depth of a vessel below the hull's waterline).

The true measure of a vessel's initial stability is the *metacentric height* or *GM* of the vessel. *G* is the vertical location of the center of gravity. *M* is the vertical location of the intersection of the force of buoyancy with the vessel's centerline at a small angle of list. This intersection is the *metacentre* (or *metacenter*) and labeled *M*. The distance between *G* and *M* is called the *metacentric height*. The distance is measured in feet or meters and is positive when *M* is above *G* (Figures 13.3 a and b). A rise in the center of gravity will cause a reduction of GM. For example, the addition of weight at a high point within the vessel causes the center of gravity to rise. When a vessel heels, the center of buoyancy shifts laterally to the lower side. Free surface effect may reduce GM still further. If sufficient water is added to bring the center of gravity (the downward force) outside the center of buoyancy (the upward force), the vessel will capsize. In this case, the GM is reduced to zero or even to a negative figure (a capsizing condition).

List

As a vessel is inclined, the righting arm will increase to a certain angle and then decrease. The angle of heel at which the righting arm starts to decrease is known as the *critical angle of list*. It is important to avoid a situation where the critical angle of list is reached. This angle is normally quite large, but if GM is reduced due to free surface effect caused by fire fighting activities or other emergency actions, then the vessel may easily heel over and capsize. Listing increases the draft. The increase in draft is found by the following formula:

Increase in draft due to list = (Beam/2) × Sine (angle of list)

For example, a vessel with a beam of 92 feet (28 m) has a list of 8 degrees. What is the increase in draft? The answer is as follows:

(92 × 0.5) × Sine 8 = 46 × 0.1392 = 6.4 feet (2 m)

This figure is a considerable increase and may affect a vessel's ability to move (steer) or be moved in port because the increased draft may prohibit a choice of berths. Furthermore, the vessel may lie aground either directly due to the list or as the tide falls. This list may cause the vessel to capsize as it contacts the bottom. At sea, even a small list as a vessel rolls may cause hot porthole glass to break when it contacts cold seawater, thus allowing water to enter. This water increases the list, and progressive flooding may

cause the loss of the vessel. In port, mooring lines must be monitored to prevent list. Lines that become tight must be loosened to allow the vessel to rise and fall with the tide or changes caused by loading/unloading cargo or fire fighting operations.

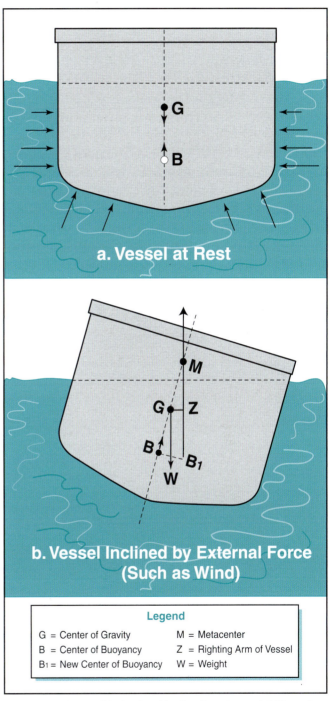

a. Vessel at Rest

b. Vessel Inclined by External Force (Such as Wind)

Legend

G = Center of Gravity
B = Center of Buoyancy
B₁ = New Center of Buoyancy
M = Metacenter
Z = Righting Arm of Vessel
W = Weight

Figures 13.3 a and b Angle of inclination charts. (a) Vessel at rest. The downward force at *G* is equal to the upward force at *B*. (b) Vessel inclined by external force such as wind. The center of buoyancy *(B)* has moved *(B₁)*. The horizontal distance between the vertical lines through *G* and *B₁* (called the *righting arm* or *GZ*) increases. The two parallel forces acting at points *G* and *Z* tend to turn the vessel upright again.

◆ Free Surface Effect

Free surface effect is the tendency of a liquid to remain level in a compartment as a vessel inclines or heels, which allows the liquid to move unimpeded from side to side. The free surface effect of loose liquids anywhere in a vessel impairs stability by decreasing the metacentric height (see Vessel Stability section) (Figure 13.4).

Cruise vessels may be particularly vulnerable to free surface effect when crew members fight fires on upper decks. A series of individually flooded cabins is unlikely to cause a free-surface-effect problem, but the situation could change if an entire section of accommodations burns and floods. Sections containing 12 to 20 cabins are thought of as being separate compartments, but once the Class C or Class B bulkhead divisions have burned through, they actually become one large compartment. The fire may be contained, but a sizable compartment, perhaps running athwartships, may now contain a considerable amount of loose, free water.

In any emergency situation, it is important to know how much water is free and how much water is critical to vessel stability. It is possible to estimate water quantities using several formulas (see Estimating Water Quantities section). Water removal solutions may then be implemented as outlined later in this chapter (see Dewatering section).

Ro/ros and ferries are also susceptible to free surface effect, especially on the vehicle decks. The most effective means of preventing free surface effect in these situations is to install a perforated deck supported by longitudinal divisions. This perforated deck then allows water to drain below its surface where the longitudinal supports act as baffles to reduce the free surface effect. This solution (proposed in a *Safety at Sea* article by Professors Carl T. F. Ross and David Jordan of the University of Portsmouth, UK) is not popular because of the noise during loading and unloading of vehicles, the trip hazards the perforations (holes) present to passengers, and high cost.

In responding to vehicle fires on ro/ros or ferries, fixed fire suppression sprinkler systems may be used to control or extinguish the fire on an involved deck. Generally the freeing ports and scuppers are capable of dewatering as fast as the water is applied (Figure 13.5). Vessel stability may be reduced if ports are blocked, if scuppers are plugged, if the vessel has less reserve stability than usual due to the cargo load configuration, or if the wind and weather are adverse.

◆ Emergency Response Actions

During emergency response actions or fire fighting operations, it is just as essential to remove water as it is to supply it. Flow rates from fire hoses, ruptured piping, or other sources must be monitored so that quantities of water sufficient to endanger the vessel's stability do not accumulate. In some instances, it may be better to either contain water within a fully flooded area (fluid holding tank) until it can be moved lower in

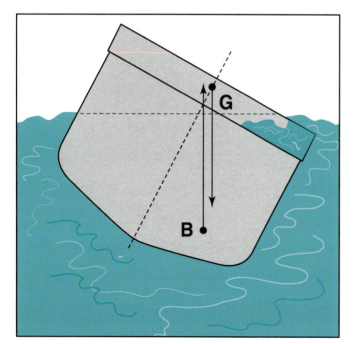

Figure 13.4 Free surface effect. Water (or weight) added high in the vessel affects stability and can cause it to capsize.

Figure 13.5 Scuppers aid the dewatering operation. *Courtesy of Captain John F. Lewis.*

the vessel or remove it entirely than to allow it to flow freely during fire fighting or emergency response activities.

Several factors affect the outcome of an emergency: the condition of vessel systems (such as pumps and piping), the vessel's stability, the quantities of liquids in tanks, and whether watertight doors and fire doors are open or closed. A vessel's officers must have knowledge of all these factors. Measurements (soundings) of liquids are taken daily of all compartments, including void spaces or cofferdams, so weights on board may be calculated and stability figures determined continually. When a fire or other emergency occurs, the vessel's preemergency stability must be known. All altered stability figures are calculated from these baseline stability numbers. Little time is available to perform calculations during an emergency.

Methods of water removal may be planned in advance (see Prefire Planning section). Along with removing water (both during and after a fire or other emergency), inspection and repair of damages after a fire or other emergency are high priorities for vessel safety. Cross-flooding techniques can also influence vessel stability.

Dewatering

The general principle related to removing fire fighting water is to facilitate the flow either overboard or to the lowest possible point in the ship where it can be pumped out (*get it off or get it low*). Use existing drains and scuppers, fixed and portable pumps, or eductors. Few vessels carry portable pumps, but most have the ability to remove liquid from almost any compartment by fixed pumps or eductors. Eductors (permanently installed or portable) work with the fire main system (Figure 13.6). If water is drained to the bilge, it can be removed from the vessel with bilge pumps. Ensure scuppers are open and free of debris so that liquid does not accumulate on deck.

Any tanks located below the vessel's center of gravity that are partially full (slack tanks) should be filled to maximize vessel stability (GM), reduce free surface effect, and add weight. Any tanks located above the vessel's center of gravity that are partially full should be emptied. Trim and list may be altered to the vessel's advantage by addition or removal of ballast. If it is decided to leave a compartment flooded, it may be

advisable to shore adjacent bulkheads to strengthen them because of the pressure created by the water in the compartment.

It may be necessary to cut holes where necessary (in decks or bulkheads, for example) to release water either during or after emergency response actions. These holes must be closed later to prevent ingress of water through the listing or rolling of the vessel.

Inspections/Repairs

After a fire has been completely extinguished or other emergency situations resolved, the damage from fires, fire fighting operations, or other emergency situations must be repaired so that the vessel is returned to a seaworthy state as soon as possible. Shoring of

Figure 13.6 An eductor used in dewatering operations. *Courtesy of R. Wright/Maryland Fire and Rescue Institute/ United States Coast Guard.*

damaged or weakened bulkheads, decks, etc. may be necessary (Figures 13.7 a and b). Damage to the hull may require control of oil pollution. If arson is suspected, it may be necessary to close compartments, leaving them untouched until investigators can examine them. Take photographs, if possible, to aid an investigation. However, the safety of the vessel takes priority over other actions.

Complete a thorough overhaul and examination of all areas below deck. Access may be difficult if ladders are weakened or missing. Lash portable ladders in place before using them. Lighting may have to be rigged if wiring is damaged. The hazards from hidden debris are great, and flashlights are often inadequate in burned or damaged spaces below deck. If the fire occurred in a cabin below the main deck, porthole glass may be broken because of fire or cold water striking hot glass. Fasten deadlights where they are available. Proper ventilation is essential in these areas, but use self-contained breathing apparatus (SCBA) even after overhaul because hazardous gases may linger for days.

Any openings through the hull must be secured and strengthened or shored if damaged. Patching metal structures by welding similar sized plates and framing in place is preferred so that a semipermanent repair is achieved. The weather at sea may not be calm when a fire or other emergency occurs, so prepare for the

worst-case possibility until port is reached. Inspect and test watertight doors to ensure correct operation. Whenever possible, leave watertight doors closed until the vessel reaches port.

Above the main deck, inspect fire doors. If any are inoperable, install fire curtains (if available) or close off the involved section. Cover any broken windows or portholes. Steel is the preferred material, but an adequate covering may be made from ¾-inch or 1-inch plywood, some threaded stud bars, and 4- × 4-inch or 2- × 4-inch strongbacks. Lighting will be required inside the space if its use is essential.

If a fire was in a hold or pump room, examine all sides (as far as practicable) for structural weakness or damage. Cracked plating, distorted or broken frames, and ill-fitting hatches may lead to flooding and a catastrophic breakup of the vessel in bad-weather conditions.

Repair damaged piping by replacing it or by using banded patches or pipe clamps (Figure 13.8). Soft patches can be made on low-pressure pipes but *not* on fuel lines or high-pressure lines (Figure 13.9). In

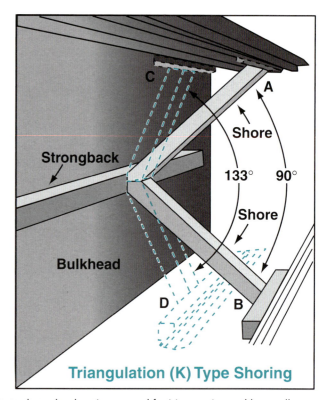

Direct Compression Shoring

Triangulation (K) Type Shoring

Figures 13.7 a and b Types of shoring structures. (a) Direct compression: simple, strong, and fast to erect; used in small compartments and for deck stiffening and hatch shoring. (b) Triangulation or K: used for bulkheads, doors, and patch support (90-degree angles or less); normally supported by deck and deckhead permanently mounted structures such as I beams, frames, and hatches.

some cases, rubber gaskets held in place by jacks or cribbing may stop pipe leaks.

Cooling and lubricating systems exposed to fire may appear undamaged, but when they are brought to working pressure, problems may occur. Damage may lead to the release of oil overboard, and engine intakes may suck in oil or debris and cause blockages or explosions.

If it is necessary to remove debris while still at sea, provide a well-secured bin or suitable receptacle. The days of throwing debris overboard are gone.

Cross-Flooding Techniques

Some vessels — usually large passenger vessels — have cross-flooding equipment or flume tanks that counteract listing and vessel movement for both vessel stability and the comfort of passengers. At sea, water is pumped rapidly from one side to another to counteract the rolling of the vessel. If available on a vessel, the equipment may give an advantage in fire fighting operations by counteracting list caused by loose fire fighting water. Another application is to use this equipment to make a vessel list in order to contain a hazardous liquid spill. However, crew members need to understand the implications of using this equipment before attempting to use it to correct or create a list.

Flume tanks are dynamic, roll-reduction devices that help only when the vessel is rolling at sea. They may worsen the stability situation if the vessel develops a static list. It may then be necessary to drain them or close their normally open cross-connections to minimize free surface effect. Flume tanks operate as a "tuned" system, allowing water to slosh from side to side but with the movement of water lagging behind the rolling motion of the vessel so that the weight of the water tends to counteract the roll. In port or in calm water, this system creates a free surface effect and makes a list worse. Disable the system in port by closing the cross-connections or by either emptying or completely filling the flume tanks.

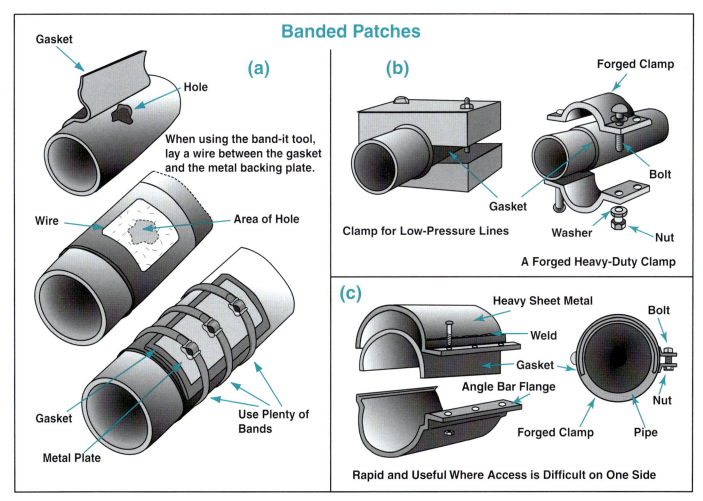

Figure 13.8 Banded patches or clamps can repair damaged piping.

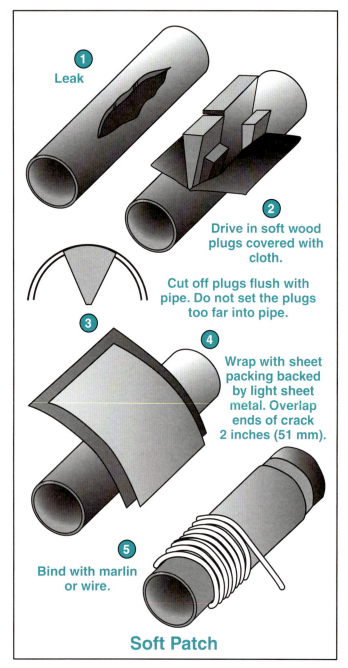

1. Leak

2. Drive in soft wood plugs covered with cloth.

Cut off plugs flush with pipe. Do not set the plugs too far into pipe.

3.

4. Wrap with sheet packing backed by light sheet metal. Overlap ends of crack 2 inches (51 mm).

5. Bind with marlin or wire.

Soft Patch

Figure 13.9 Soft patch: Do not use on high-pressure lines or fuel lines.

WARNING

Emptying or filling the flume tanks to correct a list that has already developed changes both static stability and hull stresses. The master should only use this technique with full knowledge and an analysis of the effects and with the advice of a naval architect.

In anything but the simplest case, cross-flooding increases free surface effect and may degrade stability. The simplest case is when one off-center compartment is damaged, open to the sea, and flooded, causing the ship to develop a list because the corresponding compartment on the other side of the vessel's centerline is still buoyant. Flooding the other compartment *might* correct the list, but it may also create new problems by reducing overall buoyancy and/or increasing bending stress on the hull.

WARNING

Never attempt to correct a list by shifting weights or counterflooding (flooding a hold or space) without fully understanding what is causing the list. If stability is marginal, the movement of free surface liquid within the vessel from one side to the other due to a weight shift or the addition of free surface liquid due to counterflooding may capsize the vessel.

◆ Prefire Planning

The potential for excess fire fighting water becoming a hazard is rarely planned for in advance, not because of a lack of awareness, but simply because it may not be recognized as being capable of preplanning. Fire fighting is only part of the emergency response to a fire. Success in extinguishment must not endanger the vessel in other ways. Seafarers must plan and drill for realistic scenarios. Vessels have been lost when relatively minor fires have been extinguished using excessive amounts of water.

Plan for eventualities. Estimate the amount of water to be used for a given scenario and its potential effect on the vessel's stability. Plan ways to remove water for every scenario. Make stability estimates for various situations and conditions.

Estimating Water Quantities

When planning emergency responses to fire situations and damage control operations, consider vessel stability. Little time is available during a response to make calculations, but the amount of water flowing may be easily estimated by using the following guidelines:

- For every 1½-inch (38 mm) fire hose operating, assume 125 gallons per minute (gpm) (473 L/min) and 0.46 ton (0.47 tonnes) of water per minute.

- For every 2½-inch (64 mm) fire hose operating, assume 250 gpm (946 L/min) and 0.93 ton (0.94 tonnes) of water added per minute.

- Every 100 gpm (379 L/min) of water flow adds 22 tons of seawater per hour.

- Every 100 L/min (26 gpm) of water flow adds 6 tonnes of seawater per hour.

Retained water may affect stability by causing free surface effect, which may be calculated in advance. Calculate the moment of inertia *(i)* (resistance to acceleration) to estimate the consequences of free surface effect of a given compartment. The formula for calculating the moment of inertia is as follows:

$$i = (LB^3)/12$$

where i = *Moment of inertia in feet[4] (meters[4])*

L = *Forward-to-aft dimension of the compartment in feet (meters)*

B = *Athwartships dimension of the compartment in feet (meters)*

The moment of inertia may be calculated for any compartment. Then the reduction of GM in feet or meters may be found by using the following formula:

$$(i/V) \times (d_1/d_2) \times (1/n^2) = reduction\ of\ GM\ (feet\ or\ meters)$$

where i = *Moment of inertia for compartment (feet[4] or meters[4])*

V = *Volume of displacement (vessel) (feet[3] or meters[3])*

d_1 = *Density of liquid in the compartment*

d_2 = *Density of the water in which the vessel is floating*

n = *Number of longitudinal bulkheads into which the compartment is subdivided*

Making Stability Estimates

Use the vessel's loading computer (specific computer used to calculate weight and stability) to obtain stability figures. Make estimated stability calculations during prefire planning. For example: A fire in the aft storeroom could require *x gpm (L/min)* for *y minutes* to extinguish, giving a quantity of retained water (free surface) weighing *w tonnes*. Assume the fire fighting water stays in the storeroom and cannot escape. The potential reduction of *GM* may be as high as *z meters (feet)*. This reduction is not a problem when a vessel is fully loaded but could compromise stability in light-ship conditions, especially during adverse wind and weather. Therefore, if a fire occurs in this storeroom while the vessel is light, either use a portable pump to remove the water or ballast the appropriate tank to compensate. An engineering solution could be implemented also such as installing a drain (remotely operated) in the storeroom for fire fighting use only.

Stability calculations made during prefire planning can also estimate the effects of flooding a hold or space (counterflooding). Flooding a large space such as a hold can have a detrimental effect on a vessel's stability. This flooding technique has been done successfully but may be a hazardous operation. The following considerations need to be assessed to ensure vessel stability: permeability, shear forces, and bending stresses. All of these considerations may be estimated in advance.

Prepare a stability chart for compartments in the hull and superstructure to estimate the effects of fire fighting water (Figure 13.10). The compartment name block can be color-coded as follows:

- ***Green color code*** — Compartments located below the normal range of the vessel's vertical center of gravity (added weight will lower the center of gravity)

- ***Red color code*** — Compartments located above the normal range of the vessel's vertical center of gravity (added weight will raise the center of gravity)

The example on page 315 shows how to make the calculations for a storeroom and prepare the stability calculation chart. The example uses English System units, but the formulas and calculations also work with the International System of Units.

Stability Calculations

Compartment Name and Location	Amount of Flooding	Free Surface GM Reduction	Water Weight (tonnes)	Water Vertical Center of Gravity	Change in Draft	Change in Trim	Change in GM
	25%						
	50%						
	100%						
	25%						
	50%						
	100%						
	25%						
	50%						
	100%						
	25%						
	50%						
	100%						
	25%						
	50%						
	100%						
	25%						
	50%						
	100%						
	25%						
	50%						
	100%						
	25%						
	50%						
	100%						
	25%						
	50%						
	100%						
	25%						
	50%						
	100%						

Figure 13.10 Use this blank stability calculation form to estimate the effects of fire fighting water for compartments in the hull and superstructure.

Stability Calculation Examples

Preplan the effects of fire fighting on a storeroom on a vessel displacing 40,000 tons. The storeroom has dimensions of 20 feet longitudinally, 30 feet athwarthships, and 10 feet high:

$$volume\ displaced\ by\ the\ ship = 40{,}000 \times 35\ ft^3/ton = 1{,}140{,}000\ ft^3$$

$$moment\ of\ inertia\ of\ the\ compartment = (20 \times 30^3)/12 = 45{,}000\ ft^4$$

$$free\ surface\ correction = 45{,}000/1{,}140{,}000 = 0.04\ feet$$

♦ *25 percent flooding:*

If the storeroom is 25 percent flooded, the volume of water is

$$0.25 \times 10 \times 20 \times 30 = 1{,}500\ cubic\ feet$$

1 ton of freshwater occupies 36 cubic feet, so if the storeroom is 25 percent flooded, the weight of water is

$$1{,}500/36 = 41.7\ tons$$

The water depth in the storeroom will be

$$0.25 \times 10\ feet = 2.5\ feet$$

and the center of gravity of the flooding water will be

$$0.5 \times 2.5 = 1.25\ feet$$

above the deck.

♦ *50 percent flooding:*

If the storeroom is 50 percent flooded, the volume of water is

$$0.5 \times 10 \times 20 \times 30 = 3{,}000\ cubic\ feet$$

1 ton of freshwater occupies 36 cubic feet, so if the storeroom is 50 percent flooded, the weight of water is

$$3{,}000/36 = 83.3\ tons$$

The water depth in the storeroom will be

$$0.5 \times 10\ feet = 5\ feet$$

and the center of gravity of the flooding water will be

$$0.5 \times 5 = 2.50\ feet$$

above the deck.

♦ *100 percent flooding:*

If the storeroom is 100 percent flooded, the volume of water is

$$10 \times 20 \times 30 = 6{,}000\ cubic\ feet$$

1 ton of freshwater occupies 36 cubic feet, so if the storeroom is 100 percent flooded, the weight of water is

$$6{,}000/36 = 166.7\ tons$$

The water depth in the storeroom will be 10 feet, and the center of gravity of the flooding water will be

$$0.5 \times 10 = 5\ feet$$

above the deck.

The stability chart entries for the storeroom example are shown in Figure 13.11. The weight and vertical center data can be entered into the vessel's loading computer to obtain the change in GM, draft, and trim. Note that when the compartment is full, there is no free surface effect.

With this type of preplan information, a crew member can count the number and size of fire hoses operating into a compartment, estimate how may tons (tonnes) of water are being added per hour, and see the effect on GM. Alternatively, the crew member can look into a compartment, estimate the percentage of flooding, and look up the effect on GM. These estimates minimize the amount of calculation and also the chances of making errors in calculations under the stress of fire conditions.

Making these calculations during planning may seem troublesome, but in practice only a few compartments would likely retain sufficient water to cause a problem, so the actual number of calculations would not be great. The advantages of making stability estimates during prefire planning are as follows:

- The information is always available, whether at sea or in port.

- A prefire plan with stability considerations and current data is invaluable if a fire department/fire brigade is ever called to fight a fire on board.

- Focusing on all aspects of a response makes a vessel's personnel more aware of potential hazards.

Stability Calculations

Compartment Name and Location	Amount of Flooding	Free Surface GM Reduction	Water Weight (tonnes)	Water Vertical Center of Gravity	Change in Draft	Change in Trim	Change in GM
Storeroom on Third Deck Fr 220-240	25%	0.4	42	1.25			
	50%	0.4	83	2.5			
	100%	0.0	167	5.0			

Figure 13.11 Stability calculation form with storeroom example information.

Chapter 14

Fire and Emergency Training and Drills

Training is the process of achieving proficiency through instruction and hands-on practice in the operation of equipment and systems that will be used in the performance of assigned duties. A *drill* is the process of training or teaching that requires those trained in a skill to gain experience by repeating an exercise again and again. Fire and emergency training has three major components: (1) knowledge sessions, (2) skill sessions, and (3) drills.

A knowledge session could be a classroom session in which crew members are oriented to proper confined-space rescue procedures for example. A skill session would then consist of a practical exercise on the fantail and in the after-steering space where crew members practice the use of confined-space rescue procedures. A fire and emergency drill gives crew members an opportunity to practice skills in a realistic setting. For example, a simulated rescue of a crew member from the cargo pump room could be part of a weekly fire and emergency drill schedule. Full-scale drills come after crew members have learned what to do; otherwise, they serve no purpose except to reinforce bad habits.

Fire and emergency drills are often regulated and differ from training in that they require a mandated response covered by shipping company, national, and international regulations. Even while meeting these requirements, some crews have been known to practice only routine "fire-and-abandon-ship" drills. These are the drills in which the same crew members report to the same fire stations, deploy the same fire hoses, squirt water over the same railings, and then report to the same lifeboat stations time after time. These types of drills are of little or no value in preparing a crew to fight fire or handle other emergencies. Beneficial drills must be practical, realistic, varied, challenging, safe, and designed to accurately portray various emergencies that may occur on board a vessel. Drills need not be lengthy to be successful.

If both response and command functions are practiced regularly, officers and crew members will be capable of responding to emergencies effectively and efficiently. A relationship between planning and practicing exists. Figure 14.1 illustrates the planning/practicing cycle. The pre-incident planning (see Chapter 10, Prefire Planning) assesses the hazards and plans for possible actions to take in emergencies. The training program develops knowledge and skills, and crew members practice these until they are competent. Command functions are developed through command and communications exercises.

The fire and emergency drill is where all the elements of planning and practicing come together. A scenario (or potential fire/emergency) is initiated. During the drill, vessel personnel use the appropriate plans and exercise their skills and knowledge. Thus, the plans are tested, and the performances of crew members are assessed.

A critique of the drill is essential. In reviewing the drill, the elements that work well are identified, and the areas where change and improvements are needed are noted. The pre-incident plans are then reviewed and altered if necessary. The training program is

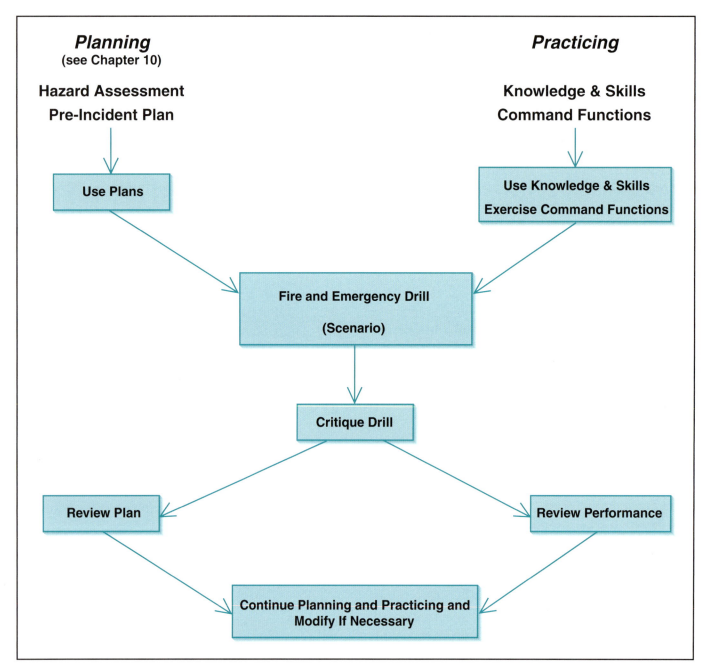

Planning
(see Chapter 10)

Practicing

Hazard Assessment
Pre-Incident Plan

Knowledge & Skills
Command Functions

Use Plans

Use Knowledge & Skills
Exercise Command Functions

Fire and Emergency Drill
(Scenario)

Critique Drill

Review Plan

Review Performance

Continue Planning and Practicing and
Modify If Necessary

Figure 14.1 Planning and practicing training process.

reviewed based on the performance of crew members. The planning/practicing cycle is repeated (using different scenarios) with continual improvement of planning and practicing.

This chapter explains the best practices for training shipboard personnel to fight vessel fires and handle other emergencies by describing fire and emergency training and drill requirements, lesson plan development, how to plan and prepare for fire and emergency training and drills, and how to deliver this training. Training for the command role through communica-

tions exercises is also described. The requirements for new crew member safety orientation sessions and how to evaluate a crew's fire and emergency response knowledge and performance are also given. The training program is also evaluated to ensure that the design of teaching/learning activities remains effective. By following and practicing the recommendations in this chapter, successful training instruction is possible. For further details on instruction, see IFSTA's **Fire and Emergency Services Instructor.**

New Crew Member Safety Orientation

Upon joining a vessel, it is essential that new crew members be correctly instructed in the fire and emergency duties on board (also see Fire Prevention Program section of Chapter 2, Onboard Fire Causes and Prevention). The safety officer conducts a safety orientation at the beginning of each voyage. The "fire and boat" drill held shortly after leaving port will be little more than a confusing experience for new crew members who have not been properly instructed.

New crew members must first find their assigned fire and emergency locations on the muster list. A second document, the station bill, gives the duties to be performed in the event of emergency situations such as fire, man overboard, or abandon ship. In addition, some vessels list other emergency situations such as fuel spill, piracy, collision, or grounding. These situations may vary depending on the vessel and trade (business). Alarm signals are listed on the station bill and must be explained and understood. For example, some vessels sound the whistle and bells for 10 to 15 seconds followed by one or more short blasts: One blast denotes fire in the forecastle or No. 1 hold, two blasts means fire in No. 2 or No. 3 holds, and five blasts means fire in the engine room.

The safety officer ensures that new crew members know the locations of fire and lifesaving equipment, especially in their work areas (for example, galley, engine room, etc.). Identify the locations of fire doors and watertight doors, and demonstrate their operation. Have crew members locate the nearest portable fire extinguishers in their work and accommodations areas and also the nearest emergency escape routes. Explain that emergency escape routes lead to the boat deck; note alternative routes in case one way is blocked.

New crew members need to be aware of the hazards of the cargo, including any hazardous materials or dangerous goods on board. Post material safety data sheets (MSDSs) for crew members to read and identify the locations of the products described (Figure 14.2). Require observance of no-smoking regulations, and stress that good housekeeping routines must be practiced. Finally, encourage new crew members to ask questions and report any unusual circumstances.

 ## Fire and Emergency Training and Drill Requirements

No matter how sophisticated the navigational equipment, how automated the engine room, or how computerized the cargo operations, none of these factors are of much help if officers and crew are not proficient in responding to fire and emergency situations. Mariners need to know the best practices associated with preventing and suppressing fires on board and how to handle emergencies quickly. To effectively handle emergencies, crew members must be properly trained to function as a team. Experience and competence are then gained through practice and drills. Coordination and cooperation are important for team success. Training, practice, and drills are essential because without them little coordination or cooperation is achieved. Instruction is customized to fit with operations on board the vessel. Do not confuse it with the formal training courses offered ashore. The aim on board is to foster competency through instruction and frequent, objective practice.

Although the training of fire and emergency teams may be delegated to the chief officer, the master reviews and approves the plans for proposed lessons, practice, and drills. Every crew member has a responsibility for safety on board. Vessel leaders, whether officers, petty officers, or other crew members, have additional responsibilities to foster a safety culture. Fire and emergency response training topics should not be seen as apart from day-to-day operations, but as part of the safe and effective operation of the vessel. Skill evolutions give opportunities for practice after training, and drills develop confidence and competency.

Responsibilities of Crew Members and Leaders

Training can be performed in one of two ways: *formal* or *informal*. Formal training follows a set (sometimes rigid) routine and is focused on an instructor's leadership. Crew members may play passive roles and follow instructions. Normal procedures require crew members and emergency teams to attend periodic instructional sessions that deal with a variety of emergencies that could occur on board the vessel. They must also learn how to operate fire and emergency equipment safely.

Material Safety Data Sheet

May be used to comply with
OSHA's Hazard Communication Standard,
29 CFR 1910.1200. Standard must be
consulted for specific requirements.

U.S. Department of Labor

Occupational Safety and Health Administration
(Non-Mandatory Form)
Form Approved
OMB No. 1218-0072

IDENTITY (As Used on Label and List)

Note: Blank spaces are not permitted. If any item is not applicable, or no information is available, the space must be marked to indicate that.

Section I

Manufacturer's Name	Emergency Telephone Number
Address (Number, Street, City, State, and ZIP Code)	Telephone Number for Information
	Date Prepared
	Signature of Preparer (optional)

Section II — Hazardous Ingredients/Identity Information

Hazardous Components (Specific Chemical Identity; Common Name(s))	OSHA PEL	ACGIH TLV	Other Limits Recommended	% (optional)

Section III — Physical/Chemical Characteristics

Boiling Point		Specific Gravity (H$_2$O = 1)	
Vapor Pressure (mm Hg.)		Melting Point	
Vapor Density (AIR = 1)		Evaporation Rate (Butyl Acetate = 1)	

Solubility in Water

Appearance and Odor

Section IV — Fire and Explosion Hazard Data

Flash Point (Method Used)		Flammable Limits	LEL	UEL

Extinguishing Media

Special Fire Fighting Procedures

Unusual Fire and Explosion Hazards

(Reproduce locally)   OSHA 174, Sept. 1985

Figure 14.2 Material safety data sheets (MSDSs) identify the locations of hazardous products or dangerous goods on board. This sample form shows the categories of information required.

Section V — Reactivity Data

Stability	Unstable		Conditions to Avoid
	Stable		

Incompatibility (*Materials to Avoid*)

Hazardous Decomposition or Byproducts

Hazardous Polymerization	May Occur		Conditions to Avoid
	Will Not Occur		

Section VI — Health Hazard Data

Route(s) of Entry:	Inhalation?	Skin?	Ingestion?

Health Hazards (*Acute and Chronic*)

Carcinogenicity:	NTP?	IARC Monographs?	OSHA Regulated?

Signs and Symptoms of Exposure

Medical Conditions
Generally Aggravated by Exposure

Emergency and First Aid Procedures

Section VII — Precautions for Safe Handling and Use

Steps to Be Taken in Case Material Is Released or Spilled

Waste Disposal Method

Precautions to Be Taken in Handling and Storing

Other Precautions

Section VIII — Control Measures

Respiratory Protection (*Specify Type*)

Ventilation	Local Exhaust		Special	
	Mechanical (*General*)		Other	
Protective Gloves			Eye Protection	

Other Protective Clothing or Equipment

Work/Hygienic Practices

Page 2

☆ U S G P O 1986-491-529/45775

Figure 14.2 Continued

Informal training develops through the interaction between crew members and instructors. Both groups play active roles. Informal training can be a very effective teaching tool. The informal atmosphere gives more people a chance to listen and speak. A full exchange of information and ideas can lead to a better understanding of the responsibilities of crew members, relative to their specific skills, and hence a safer vessel. Informal discussions allow people to learn from each other, maintain interest, and establish good attitudes toward safety.

An individual (often the first mate or chief officer) may be responsible for onboard training, but it is recommended that the actual training be delegated to other instructors. Usually deck and engineering officers are the primary trainers. A vessel's officers share the task of training. A competent officer must be capable of training other members of the vessel's crew. The most qualified persons should teach most classes, regardless of their ranks. The objective is to have the most qualified individuals give the instruction on given topics. If an individual has specific expertise and responsibilities in maintaining lifesaving or fire fighting equipment (for example, starting the emergency generator or emergency fire pump), let that individual instruct colleagues in the correct use of that equipment. Thus, all crew members learn from the best resource available.

Fire and emergency drills are mandatory, but skill practice evolutions are essential in order for these drills to be realistic and effective (Figure 14.3). Crew members must become competent in basic skills such as advancing fire hose and the use of self-contained breathing apparatus (SCBA) before they can coordinate their efforts in a full-scale fire and emergency drill. Likewise, officers must exercise command and communications functions in small lessons before testing their abilities in coordinating the many facets of a required fire and emergency

drill, which include fire fighting, use of survival craft, first-aid support, and evacuation. It is important to coach junior officers in leadership skills. They should take an active part in onboard instruction and follow the seafaring tradition of learning the job at the next level while teaching to those at the level below. Postpone large-scale simulations until officers and crew show competence in the basic skills and have a clear understanding of related concepts.

One technique is to present different problems and situations during instructional sessions and discuss possible solutions until satisfactory options are found. Have crew members manipulate the necessary tools and equipment for given situations. The regularly scheduled fire and emergency drills then become demonstrations of efficiency rather than learning sessions.

Realistic training means crew members will be ready to respond to any emergency. An easy excuse — *"but we'll do it differently when it is a real emergency"* — undermines the readiness of a crew to respond during an actual emergency. Do not give a crew member the same task at every drill (such as holding a nozzle over the side or hosing down decks to test the fire stream). During a real emergency alarm, that task might not have relevance to the emergency situation.

Figure 14.3 Skill practice evolutions are essential in order for drills to be realistic and effective. *Courtesy of R. Wright/Maryland Fire and Rescue Institute/United States Coast Guard.*

Keep accurate, up-to-date training records. For example, enter drills performed into logbooks. Records provide documentation of required training completion. These records also identify training areas that have been emphasized as well as areas that require more attention.

Fire and Emergency Response Training Topics

Some of the fire and emergency response topics that could be covered in training are listed below. These are not exhaustive lists; other topics may be needed for any given vessel.

Basic Skill Evolutions

Skill evolutions are for practice. Some of the best practices for fire fighting and safety training are as follows:

• *Portable fire extinguishers*

— *Recommended practice:* Discharge a percentage of onboard cartridge-operated (or other easily

rechargeable) portable fire extinguishers each month (Figure 14.4). Recharge after use, therefore combining maintenance and training. Follow recommended procedures.

Figure 14.4 Practice discharging a percentage of onboard portable fire extinguishers each month.

 Basic Fire Fighting Skills and Knowledge

♦ *Theory of fire*
♦ *Classes of fire*
♦ *Fire behavior*
♦ *Personal protective equipment (PPE)*
♦ *SCBA and other safety equipment*
♦ *Portable and semiportable fire extinguishers*
♦ *Search and rescue operations*
♦ *Advancing fire hose functions*
  — *Opening/closing doors*
  — *Ventilation techniques*
  — *Stream patterns*
♦ *Fixed fire suppression systems*

 Strategy and Tactics

♦ *Size-up*
♦ *Offensive vs. defensive attacks*
♦ *Direct attack*

♦ *Indirect attack*
♦ *Boundary cooling*
♦ *Ventilation planning*
♦ *Fixed fire suppression systems*
♦ *Dewatering techniques*
♦ *Overhaul*
♦ *Fire cause determination*
♦ *Salvage (loss control)*

 Command Functions

♦ *Communications*
  — *Internal*
  — *Distress*
♦ *Accountability*
♦ *Planning*
♦ *Logistics*
♦ *Interface*
  — *Other vessels*
  — *Shore*

- *Hoses and nozzles*

  —*Recommended practice:* Lay uncharged fire hose along decks and up and down interior and exterior ladders into machinery spaces and accommodation areas. Flow water and operate nozzles where practical.

- *SCBA*

  —*Recommended practice:* Use SCBA with air flowing during both fire and emergency training practice and drills. Use in both open and confined spaces. If compressed air is not available to refill cylinders, wear the complete SCBA assembly while breathing atmospheric air in safe areas.

- *Fire pumps*

  —*Recommended practice:* Prepare (line up the valves) and start the fire pumps. All engineering personnel and as many crew members as possible should be capable of doing this. Start the emergency fire pump as a precaution. Use the emergency fire pump if the main pump is inoperative or ineffective.

- *Ventilation techniques*

  —*Recommended practice:* Practice closing watertight doors and ventilation dampers and operating emergency fan shutdowns to secure spaces. Practice placing fans/blowers for assisted ventilation procedures.

- *Search and rescue operations*

  —*Recommended practice:* Always wear appropriate personal protective equipment when entering confined spaces. Practice the buddy system. Practice searching various parts of the vessel for crew members posing as missing persons. Locate and remove them to a safe location.

- *Lifeboats and life rafts*

  —*Recommended practice:* Become familiar with the use and operation of lifeboats and life rafts. Practice boarding, launching, operating, and recovery of survival crafts and using hydrostatic releases.

- *Portable emergency communication equipment*

  —*Recommended practice:* Become familiar with the operation of the Emergency Position Indicating Radio Beacon (EPIRB), two-way communications devices, and portable very high frequency (VHF) radios (Figure 14.5). Practice operating the handheld two-way radios and the lifeboat radio.

- *Foam application equipment*

  —*Recommended practice:* Prepare and operate foam application equipment.

- *Personal floatation devices (PFDs)*

  —*Recommended practice:* Practice donning and doffing PFDs (also known as life jackets or life vests), including the use of attached emergency lights and whistles (Figure 14.6).

- *Immersion suit*

  —*Recommended practice:* Don and use an immersion suit. Practice floating and swimming while wearing the suit.

- *Firefighter's personal protective clothing*

  —*Recommended practice:* Demonstrate proper donning and wearing of the firefighter's personal protective clothing: hood, pants, boots, coat, helmet, gloves, and eye protection.

Figure 14.5 Know the operating principles of the Emergency Position Indicating Radio Beacon (EPIRB). *Courtesy of R. Wright/Maryland Fire and Rescue Institute/United States Coast Guard.*

Fire and Emergency Drill Topics

Cover all aspects of fire fighting and survival craft use in these drills in addition to topics such as response to oil spills, man overboard emergencies, abandon ship procedures, and other emergency response procedures.

◆ Lesson Plan Development

A *lesson plan* is a written outline of units of instruction. It maps out a plan to cover information and skills and is a step-by-step guide for any type of presentation. It clearly states what an instructor will accomplish with crew members during a particular session. A lesson plan is part of a systematic approach to training. A lesson plan may cover any training activity from a 15-minute practical evolution to a 3-hour block of instruction. Plans may be handwritten, typed, or keyed into a computer. They may be prepared on board a vessel or provided by the vessel's management team ashore. Preparation and use of lesson plans are vital to the success of any ongoing systematic shipboard fire and emergency training program. Lesson plans are valuable to the shipboard trainer for the following reasons:

- Teaching and learning are easier with sequential, orderly instruction.

- A standard or basis is formed for practical evaluations.

- The specific vessel, equipment inventory, and crew are considered when preparing lesson plans.

- Periodic updating keeps lesson plans current.

- Uniform instruction is ensured when lesson plans are used by different instructors.

- Documentation of training records is easier.

Three areas or learning domains apply to shipboard training: affective, cognitive, and psychomotor. All three areas must be addressed when teaching any subject. The domains are not independent areas of learning. Rather, they are interrelated areas in which learning occurs. When combined in instructional methods, they enable crew members to perform a behavior or a job skill. An easy way to visualize these three learning domains is to arrange three circles in an overlapping pattern and include three images representing each one: heart (affective), head (cognitive), and hand (psychomotor) (Figure 14.7). The three domains are briefly described as follows:

- *Cognitive domain* — Deals with what information a crew member needs to have about a topic; for example, crew members gain understanding about a behavior.

- *Psychomotor domain* — Deals with the physical skills a crew member must be able to do for task

Figure 14.6 Demonstrate the use of personal floatation devices (PFDs). *Courtesy of Captain John F. Lewis.*

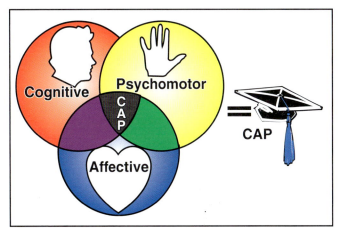

Figure 14.7 For effective learning, the three domains of learning (cognitive, psychomotor, and affective) must overlap or interrelate with one another.

completion; for example, crew members perform the skills associated with the behavior.

- **Affective domain** — Deals with a crew member's attitudes, beliefs, feelings, and values about a given subject; for example, crew members develop a willingness to perform the behavior correctly and safely.

The following sections explain the component parts of the lesson plan and the general formats commonly used. Examples, questions, and activities are included in lesson plans to ensure participation by crew members. By reviewing the components of a lesson plan, an instructor knows the lesson details and logistics and how to prepare appropriately for teaching ahead of time. A lesson plan template and sample SCBA lesson are included.

Lesson Plan Components

Lesson plan developers typically separate lessons into *manipulative* (skill) and *cognitive* (information or technical) lessons. Lesson plan components are described in three major areas: (1) *instructor preparation*, (2) *four-step instructional process* (preparation, presentation, application, and evaluation), and (3) *crew member reinforcement*. These areas are outlined in the following sections.

Instructor Preparation

Preparation gives the instructor confidence and credibility. The following lesson plan components guide the instructor's preparation for presenting the lesson:

- **Job or topic** — Short descriptive title of the information covered

- **Time frame** — Estimated time it takes to teach a lesson; allows for variations in class size and experience level of crew members

- **Level of instruction** — Job performance requirements or desired learning domain/level (cognitive, psychomotor, or affective)

- **Learning objectives** — Minimum acceptable behaviors crew members must learn and be able to demonstrate upon completion of instruction; learning objectives can be behavioral, enabling, or performance types:
  - *Behavioral:* Measurable statement that describes behavior the learner is expected to exhibit as a result of instruction

  - *Enabling:* Specific objective that learners must attain as a component of completing a behavioral objective

  - *Performance:* Explicitly worded statement that specifies learners' performances, the conditions in which they will perform, and the criteria they will meet

- **Equipment and materials needed** — Types, quantity, and availability of materials (includes equipment, handouts, videotapes, etc.)

- **Resources and references** — Names of assistants and logistical support personnel, textbooks, references, learning activities, unique training locations, etc.

Four-Step Instructional Process

The four-step instructional process (preparation, presentation, application, and evaluation) prepares the crew member for learning, involves them in active participation in the instruction, and gives a way to measure their understanding (Figure 14.8). A number of ways are available to familiarize crew members with fire and emergency equipment and to teach the required skills. An instructor works through all four steps whenever a new topic is discussed or a new piece of equipment is operated. A half dozen or so *key points* (basic steps in an operation), in the proper order, may have to be learned for an operation to be successful (Figure 14.9).

 Four-Step Instructional Process

Step 1: Preparation — *Find out how much crew members know about the subject or the equipment under discussion. Arouse interest by relating personal experiences. Motivate crew members to answer the question "Why do I need to know this?" Relate the topic to aspects of their jobs. Encourage discussion. State the learning objectives. Prepare crew members to listen for key points by briefly stating the main topics that will be covered.*

Step 2: Presentation — *Explain, illustrate, and demonstrate the operation. Present the information in an orderly, sequential* *outline. Use the most effective types of instructional materials to discuss or illustrate each part of the operation. Emphasize and illustrate key points.*

Step 3: Application — *Provide opportunities for activities, work groups, skill practice, etc. Have crew members actually handle the tool or equipment. Have them explain its operation. Make sure key points are repeated. Most of the learning takes place during this step.*

Step 4: Evaluation — *Have the crew member demonstrate what has been learned through a written or performance test. Evaluate the use of tools and the operation of equipment. If the lesson is not concerned with manual operation, a verbal or written description could be sufficient. The training session is not over until crew members have demonstrated that the lesson objectives have been met.*

How the Steps Work in Practice

In an actual teaching situation, the four steps in the instructional process tend to blend together. For example, demonstrations or explanations in the presentation step may be repeated during the application step. The same written test may be used during pretest and evaluation. The only real difference between the application and evaluation steps is supervision.

The steps of *Preparation, Presentation/Delivery, Application,* and *Evaluation* become logical and natural parts of the written lesson plan. With practice, the instructor simply does what is needed — without giving much thought to which step is involved.

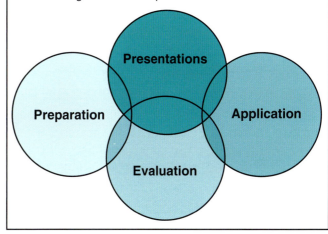

Figure 14.8 *(above)* The four steps of the instructional process (preparation, presentation, application, and evaluation) overlap in a lesson.

Figure 14.9 Four-step instructional method: (1) preparation, (2) presentation, (3) application, and (4) evaluation.

Crew Member Reinforcement

It is not enough to prepare and present instruction. Instructors must give crew members opportunities to review, remember, and reinforce the instruction. The following lesson plan components reinforce lesson information:

- *Summary* — Restate or reemphasize key points. Ask crew members to recall information or list steps. Safety precautions are always key points.

- *Assignment* — List optional work or activities to perform outside class. Relative and practical learning assignments enable crew members to practice and apply what they have learned.

Lesson Plan Format

A lesson plan format allows for flexibility. Many ways to arrange a lesson plan are possible. The lesson plan template shown in Figure 14.10 illustrates the features recommended for successful planning and execution of a lesson. The typical lesson plan has six components: Introduction, Instructor Preparation, Main Body (presentation and application steps), Testing or Evaluation, Summary, and Assignments. These parts require the following features:

- *Introduction* — An introductory section guides the developer when designing the lesson and the instructor when teaching it. This part of the lesson plan states the learning objectives so that crew members know what they are to do in the lesson. The following elements are generally listed: lesson title/topic or job, learning objectives, time frame, and level of instruction (learning domain).

- *Instructor preparation* — This section includes information that will tell instructors, at a glance, the purpose of the lesson and what they need to teach it. The instructor provides a motivational statement that gets the crew members' attention and interest. The section contains reminders or instructions such as list of materials needed, notes about equipment and materials, references, course goals, and possibly a mission statement. The following elements are generally listed: equipment and materials needed, resources, references, motivational statement, enabling/performance objectives, and overview (key points).

- *Main body* — This section is actually the outline based on the lesson overview and includes the *pre-*

sentation and *application* steps. The instructor introduces each topic (taken from the list of overview points), uses appropriate instructional methods to discuss or demonstrate, uses instructional materials effectively, and relates known or familiar information and tasks to unknown or new and unfamiliar information and tasks. *Application* activities work together with the presentation. An instructor who presents effectively not only *tells* information but *shows* how it is applied. The application activities provide opportunities for crew members to practice applying information or skills in a simulated work environment.

- *Testing or evaluation activities* — Various types of evaluation instruments can also be used at points during the lessons or course to measure knowledge and skill. Tests and evaluations can be one or more types such as written or performance, oral, diagnostic, and comprehensive.

- *Summary* — In this section, instructors bring closure to a lesson. They can review key points with crew members and check their understanding of the topics discussed, demonstrated, applied, and practiced.

- *Assignments* — Some kind of assignment can be given so crew members can think about or process information further. Readings, research, or projects that reinforce the information introduced may be helpful.

Sample SCBA Lesson

Although this lesson deals specifically with SCBA, the applicability of the method to other topics should be obvious. Prior to the lesson, assemble the following materials: at least one SCBA complete with harness, cylinder, spare cylinder, and carrying case and handouts depicting the complete and assembled SCBA unit used on board the vessel. Have sufficient resource materials available for crew members.

The following four steps tie the specific knowledge and skills required to learn the use of an SCBA on board a vessel to the four-step instructional process described earlier:

Step 1: *Preparation* — Tell crew members the goal of the lesson is to learn the safe operation of SCBA. Give the learning objectives and key

Lesson Plan Template

1. *Introduction*

    Job or Topic:

    Time Frame:

    Level of Instruction:

    Learning Objectives:

2. *Instructor Preparation*

    Equipment and Materials Needed:

    Resources:

    References:

    Motivational Statement:

    Enabling/Performance Objectives:

    Overview (Key Ideas/Skills):

3. *Main Body: Presentation and Application*

    Content Outline   Instructor's Notes/Application Activities

    1.

    2.

    3.

    4.

    5.

    Etc.

4. *Summary*

    Overview:

    Key Points:

5. *Testing or Evaluation*

    Written Exams:

    Performance Evaluations:

    Instructor Evaluations:

6. *Assignment*

    Reading:

    Video:

    Other:

Figure 14.10 Sample lesson plan template: a guide for the marine instructor.

points of the lesson and describe how the objectives are to be measured. Determine what the crew members already know about SCBA: What is it? What is its purpose? Where might it be used? Encourage discussion by asking if anyone has had experience with SCBA in school, aboard this vessel, or on a previous vessel. Throughout the discussion, try to arouse interest and get crew members involved.

Step 2: *Presentation*—Show the SCBA to crew members, describe its construction, and explain its operation. Use the diagram to explain the airflow through the device. Use the resource materials to illustrate the various steps of donning and doffing the unit. Demonstrate donning/doffing the SCBA, and explain the procedures. Note where the operating instructions for the unit are kept (crew safety training manuals, SCBA storage case, emergency gear locker, etc.). Emphasize the key points such as the following:

— Integration of unit components by proper connections

— Proper donning of the unit, including positive-pressure testing for a well-sealed facepiece

— Value of the buddy system; use when entering and moving about a darkened space

— Low-pressure alarms; leaving the contaminated atmosphere

— Replacing and refilling cylinders and storing the units

Step 3: *Application*—Allow each crew member to examine the SCBA. Ask selected individuals to explain the key steps in donning and doffing the unit. Encourage crew members to ask questions and take part in a discussion on the construction and use of the unit.

Step 4: *Evaluation (demonstration)*—Require each crew member to don and doff the SCBA, starting with removal of the unit from its container, using each of the donning/doffing steps. Correct any mistakes as they are made.

 ## Instructor Planning and Preparation

Comprehensive onboard training is attainable, but not without effort. It requires planning, preparation, and using the best available onboard resources. Perhaps the most difficult onboard emergency training is fire fighting. Few seafarers fully understand the many principles involved. However, without the motivation of smoke and flame, onboard fire fighting training may seem dull or unrealistic just because a little imagination and adequate preparation by the instructor are lacking. Even as little as 30 minutes of realistic and relevant training per week brings benefits. Variety in training includes using lectures, discussions, illustrations, demonstrations, mentoring (coaching), role-play with scenarios, brainstorming, and safety videos as well as practical skill evolutions. Often, a planned discussion around the mess room table results in more learning than a confusing drill where crew members are unsure of the concept or their roles.

Even the most experienced teacher prepares carefully for each lesson. The amount of planning and preparation depends on the subject matter and the type of lesson. However, the instructor's planning and preparation should include the following areas:

- *Designing program* — Select lesson topic, develop objectives, list important information to be presented, determine key points, design a lesson plan, and create evaluation instruments. The basic design may be complete if the lesson has been given before.

- *Scheduling* — Schedule training regularly at times when most personnel can attend. Many sessions may not require more than an hour at a time, but schedule breaks in longer training sessions.

- *Notification* — Publicize training schedules and topics (bulletin boards, word of mouth, etc.). Acquire a list of the crew members who can attend the sessions.

- *Class/practice locations* — Conduct classroom training in the quietest place available on the vessel; select an appropriate area for practice. Secure permission to use specific areas such as the engine room. Arrange to have an adequate number of chairs and tables available, and arrange seating to mini-

Figure 14.11 An instructor ensures that enough equipment is available for all members of the class. *Courtesy of R. Wright/ Maryland Fire and Rescue Institute/United States Coast Guard.*

mize visual distractions. Examples: Use a tiered arrangement with audiovisual equipment; put a demonstration area in the middle of a semicircular grouping.

- *Materials/equipment* — Assemble everything that will be needed for the session/practice: tools or equipment, chalkboard or other marker board or easel pad, other appropriate teaching aids (audiovisuals, handouts, etc.), and pencils and notebooks for crew members. Have large enough quantities to train the number of crew members expected (Figure 14.11).

- *Assistants* — Choose assistants (crew members or other instructors) to assist in demonstrating skills. Recognize the experience and pride that seasoned crew members have. Encourage their participation, and highlight their advanced skills. Discuss questions about lessons with other instructors.

◆ Delivery of Fire and Emergency Training

Instructors cannot just appear in the classroom and teach; neither can crew members just arrive for a class and learn. Both must prepare for the training experience. Both must have expectations and anticipations of what they want to accomplish and how to accomplish it. Effective instructors take steps to ensure that the learning experience is worthwhile, relevant, and interesting. Taking time to follow these steps moti-

vates crew members to think, question, and get involved in the learning experience.

An important aspect of teaching is how instructors communicate. Instructors must consider how they actually present themselves to crew members. A professional appearance demonstrated through well-prepared presentation methods and a positive attitude and demeanor add to instructor credibility. An effective instructor focuses on several important areas: preparing to teach, presentation techniques in both class and practice sessions, evaluation techniques, and the lesson closing.

Preparing to Teach

Prior to the actual training session, review the material to be presented. Read the lesson plan several times until thoroughly familiar with it, and review it immediately prior to conducting the lesson. Add supplemental reference material to the lesson plan folder, and add pertinent information on note pages. Underline key instructional points. Research references in the vessel's technical library if necessary. Bring reference materials to class sessions for crew members to use as on-the-spot resources.

It is important to check equipment and training aids in advance. Practice using any new devices that may be introduced, and preview audiovisual materials. Ensure that audiovisual equipment is set up properly and in good operating order. Have spare bulbs available during class sessions. Prepare visual charts or overhead transparencies for display. Have slides, videotapes, and the like ready for showing before class sessions begin.

Before the lesson starts, assess the learning environment. Arrive early to prepare and answer the following questions: Will it be difficult to see notes or aids? Is it cold, hot, humid, or stuffy? Can the lighting be adjusted? Is there enough seating? Is equipment in the proper position for training? Check room lighting and temperature before class sessions. Test audiovisual equipment for sound levels, tracking, and focus adjustments. Ensure projection is large enough for everyone to see (Figure 14.12).

Presentation Techniques

Approach the class with a cooperative attitude: Be pleasant, professional, and enthusiastic. A competent instructor demonstrates mastery of the teaching/

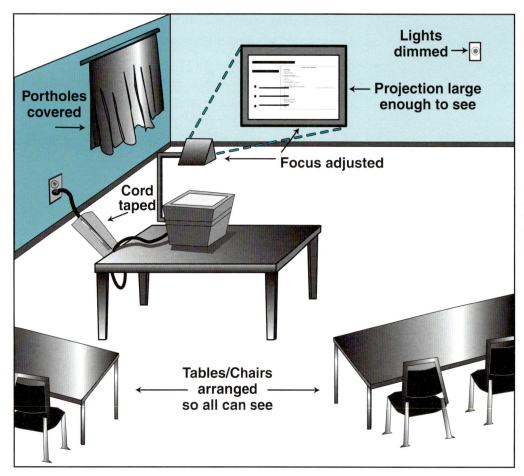

Portholes covered →

Lights dimmed →

Projection large enough to see ←

Focus adjusted ←

Cord taped ←

Tables/Chairs arranged so all can see

Figure 14.12 Arrange and test audiovisual equipment before class begins to avoid distractions. Make sure everyone can see and hear the presentation.

learning environment by demonstrating a dynamic personality when leading instructional activities. It is important to start on time. Ease into the instruction by first warming up the group. Learn the names of crew members, and use their first names as often as possible. Give the lesson's objective, and treat the subject seriously by explaining the reason for the lesson and why it is important to know it. List the knowledge and skills that crew members will review or acquire. Make sure crew members are aware of training policies. For example, arriving late or leaving early may disqualify a crew member for attendance credit. Enforce policies against eating, drinking, or smoking in class.

Make comments interesting, and encourage participation. Do not read the lesson plan to crew members. Some crew members may have limited skills in the instructor's native language. Speak slowly (but not loudly) when this is the case. Paraphrase information and use different terminology or describe topics in different ways so all can understand. Praise crew members for work well done or significant comments. Praise goes a long way toward encouraging a person to actively participate. Make training enjoyable; smile.

When demonstrating skills, practice safety first at all times. Carefully explain each step in any technique. Show a skill step-by-step before expecting anyone to perform it independently. Demonstrate techniques in slow motion. Have crew members work in small groups of four to six members when practicing. Place one well-experienced person with each group. Keep crew members focused on tasks.

Summarize key points throughout the session. Ask crew members to give key points for the class. Encourage discussion of lesson material. Adjust pace and vocabulary to the level of the group. Be available during breaks to discuss lesson material. Use audiovisual aids as an adjunct to the lesson, not as a replacement for dialog.

Questions are useful teaching and learning instruments. They stimulate and maintain communication, arouse curiosity, and provoke interest that triggers content-related questions from crew members. Defer questions that might be better addressed later in the lesson, but do not forget them. If unsure of an answer, promise to get the information later, and then do it; do not bluff.

25. Maintain appropriate pacing and keep the lesson moving. Do not get bogged down in relating particular fire and emergency incidents or *sea stories*. Avoid spending too much time on any particular part of the lesson. Get to the point as quickly as possible, and check to see if the point comes across to crew members. Ask for comments and questions, and then move on to the next point. Follow the lesson plan, but be flexible. Assess and reassess lesson progress, and make adjustments in style and methods as needed.

Two self-evaluation methods help instructors identify areas that need improvement: (1) Practice delivery of a lesson before its presentation, and (2) videotape a presentation, and review it for distracting actions and speech patterns. This self-evaluation can help eliminate annoying habits and identify tendencies that distract from delivery of the message. The following techniques can also improve an instructor's delivery:

26. • Speak clearly and distinctly; clearly pronounce each word. Do not mumble or slur words together.

• Use expressive voice inflection and add emphasis to words. Do not speak in a monotone voice.

• Govern speaking pace or speed. Begin slowly with new information, and gradually pick up speed as information becomes familiar.

• Pause periodically so that crew members can catch up with their thoughts and ask questions.

• Use correct grammar. Avoid slang and expletives.

• Relax, and speak in a conversational tone.

• Make eye contact, which enforces the feeling that the instructor is interested in crew members and concerned that they understand.

• Use appropriate gestures.

• Avoid distracting mannerisms such as pacing, jingling keys or change, or repeating worn phrases such as *"you know"* or *"uh."*

• Remember to use the golden words *"please"* and *"thank you."*

Skills Practice Techniques

In order for lesson material to be learned well by crew members, time must be allowed for some type of practice. As crew members use and reuse a particular concept or technique, it becomes more and more a lasting piece of knowledge or skill. Repetition helps form lasting habits. As crew members become proficient in certain skills, they can speed up their performance to the point of competing against other members of the crew when performing fire and emergency drills.

The following techniques for the instructor add interest and variety to the skills practice sessions:

27. • Have each member of each team rotate roles in the various exercises. For example, each member can perform as rescuer, patient, and observer in a rescue simulation or as nozzle operator and backup in a fire hose evolution.

• Encourage the key person in each group to provide a positive model for other crew members.

• Move about and observe the various groups practicing their skills. Correct improper techniques as they occur in a constructive, pleasant manner. Criticism should be directed at the work, not at the crew member. For example, *"There seems to be a problem with the tightness of this harness; try putting it on like this, and it will work better."* A questioning process also allows persons to stop, think, and recall the information they are attempting to apply. For example, *"Before you go further, do you remember . . . ?"*

• Stress patience on the part of crew members.

• Have observers take notes for classroom critiques and review. Be good-natured about any personal complaints. Accept recommendations for change graciously, and thank the person who offers advice.

Evaluation Techniques

The purpose of training is to improve knowledge and performance. The purpose of testing is to evaluate knowledge and performance. Ongoing evaluation of students and instructors is valuable inasmuch as it encourages self-awareness and promotes continued efforts at self-improvement. Crew members can help evaluate each other while doing practical exercises. To ensure accurate and effective evaluation, use the following guidelines:

• Ask questions of crew members based on objectives listed in each lesson.

- Use observers' written comments to help evaluate performance.

- Be nonjudgmental in comments about less-than-adequate performance. Review material again if necessary.

- Use instructor-made quizzes and activities to keep crew members interested and motivated.

Lesson Closing

To reinforce the importance of the lesson and the essential information and skills, close the lesson in a formal manner after any testing is completed. Use the following guidelines:

- Summarize the main steps and key points.

- Allow crew members to critique each session by completing a brief evaluation form.

- Provide a preview of coming attractions; make assignments.

- Have a definite ending or wrap-up with closing comments.

- End on time.

◆ Planning Fire and Emergency Drills

After basic knowledge and skill training and practice have been conducted successfully on a regular basis for a period of time, instructors can start conducting multifaceted training events and fire and emergency drills. Regular training and practice sessions give crew members proper instruction in their respective duties. Drills can also allow for *cross-training:* the training of personnel to fill each other's roles in time of need. Backups give the vessel's fire and emergency organization greater capabilities.

Planning, managing, and conducting fire and emergency drills (including both entire company and emergency team drills) require conscientious effort. Both shipping company fire and boat regulations and *International Convention for the Safety of Life at Sea (SOLAS)* regulations require that fire and emergency drills be held weekly on all vessels at sea and in port. The following points help ensure success:

- Simulate actual conditions as closely as possible.

- Require crew members to perform as though it is an actual fire or other emergency.

- Make adequate arrangements for removing water used to extinguish simulated fires.

- Provide sufficient refills or replacements for portable fire extinguishers.

- Conclude drills with a review and discussion.

- Maintain written records of drills in logbooks. Retain the records for future reference and to provide documentation of their required completion.

Generally, the chief officer and other officers down the chain of command supervise fire and boat drills, with the master in command on the bridge and the chief engineer in charge of the engine room. The chief mate or second engineer is usually the officer designated to be in charge of a fire fighting operation. This person is not on an attack team but directs the use of the necessary personnel and equipment. For the purpose of drills, it is recommended to have other officers supervise from time to time in case the designated persons are ashore or unavailable when a fire or emergency occurs. In emergencies, promotions may be sudden and unexpected. Leadership is a key element in effectively managing emergency drills. Officers conducting drills must give clear, concise orders.

Conduct fire and emergency drills with as much realism as possible so that crew members will know their duties in actual emergency situations. Realistic simulations motivate and challenge crew members to do their best. Fire fighting exercises particularly need to simulate actual fire conditions. Just operating the fire pump and discharging a stream of water overboard is not an adequate drill.

Ideally, conduct simulations of fire outbreaks in high-risk areas (galley, engine room, paint locker, and any other area where flammable materials are stored) (Figure 14.13). Change the location of each simulated fire on each occasion so crew members become familiar with the vessel and its target hazards. Obtain smoke generators to create more realistic conditions. Use them also for drills with SCBA. For some SCBA drills, black out or otherwise obscure the facepieces (cover with paper or cloth, for example) to simulate limited visibility conditions.

Practice using the fire control plans for each area during fire suppression drills. Some topics to cover include following a routine to familiarize crew members with the positions of ventilation louvers and flaps, shutoff valves, emergency stop devices for machinery and the location and operation of fire fighting equipment. If possible, discharge at least one portable fire extinguisher at each fire drill. Good radio communication skills must also be practiced and demonstrated during these drills (Figure 14.14).

◆ Evaluation of Crew Members and Training Programs

Regardless of the approach, there must be a process to measure that learning took place. An objective is specified, the crew members perform, and their success or failure in meeting the objective forms the basis for review of the training program. Feedback is required from crew members, instructors, and observers. The evaluation process collects information for the purpose of making decisions. Based on the objectives and results of tests, instructors can determine whether the results indicate accomplishment of the objectives that were set at the beginning of the program. When the design of teaching/learning activities is no longer effective, the process must change. Instructors must always be alert to ways of improving or updating instruction. The purposes of evaluation include the following:

- Determine whether objectives and test results match.

- Discover weaknesses in learning and instruction.

- Diagnose causes of learning problems or weaknesses.

- Establish guidance or recommendations for further study.

Basic performance skills and knowledge of crew members may be evaluated either formally or informally. Often, the informal approach is preferred because it is less threatening and more motivational. Enthusiasm on the part of the instructor is essential. The formal and informal approaches have different characteristics and are briefly described as follows:

- *Formal*

  — *Written tests:* Test knowledge through multiple choice, true-false, matching, short answer/completion, or essay tests

Figure 14.13 Conduct simulations of fire outbreaks in high-risk areas like an engine room. *Courtesy of Firefighter Aaron Hedrick, Seattle (WA) Fire Department.*

Figure 14.14 Practice communications skills when conducting drills. *Courtesy of R. Wright/Maryland Fire and Rescue Institute/United States Coast Guard.*

— *Performance tests:* Test skill abilities using detailed checklist (Figure 14.15)

- *Informal*

  — *Oral quizzes:* Test communication ability; also test persons who cannot read at the level of a written test

  — *Observation:* Occurs throughout instruction; uses peers, supervisors, or third parties

All shipboard training programs can be improved by continual (1) validation, (2) review, (3) analysis, (4) critiquing, and (5) incorporation. Each time a real or simulated fire and emergency situation occurs, the functional *validity* of a training program is tested. *Review* a vessel's training program at least once a year. As mentioned earlier, training must occur on a regular basis. *Analysis* of a training program's quality is important to ensure that it meets its mission requirements or designated tasks. *Critiquing* after drills and exercises is important to assess the quality of a training program. Drills, classes, and practice evolutions should be complete and comprehensive. *Incorporate* lessons learned from drills, classes, and practice evolutions into the existing training program.

◆ Training for the Command Role

The command process is more than issuing orders. It is the interplay of individuals working together for a common end. The leader is the focus; communications are the lifelines. Information, situation assessments and updates, available resources, and knowledge of operational gains or losses are all vital components of a successful command process.

Well-executed routines apply to command functions as much as they apply to other fire and emergency response skills. Evaluate the command process so that no member of the team is in doubt as to the actions required. Team building begins with meetings, assignment of duties, review, and testing of processes. This process testing can be done as a table-top exercise where the command process is applied to simulated fire and emergency situations.

The way that vessel officers command in everyday situations is close to the way they will respond in emergencies (*"As you train, so shall you respond"*). The parts of a successful command process that may

be evaluated during onboard training for emergencies are (1) orderly presentation and recording of data, (2) regular reporting and updating, and (3) use of common terminology. The ability to properly demonstrate specific tasks gives a foundation upon which to build. Emergency command competency is achieved after experience, conscious review, and improvement based on lessons learned.

Command and Communications Exercise

A functional exercise for the assessment of command and communications functions can be designed using the vessel's fire control plan. Planning an exercise based upon a fire-in-the-galley scenario is an example. An instructor plans the exercise and writes several outcomes. The exercise is written so that a quick response using correct procedures may achieve a quick conclusion, but failure to take proper actions may cause escalation of the incident and lead to disaster. For example, failure to check for extension of a fire in areas through which the galley stove vent extends will allow an extension of fire into those areas during the exercise. Identification of this potential by participants is generally sufficient justification to close off that avenue of disaster. Fairness in judging participants' decisions is essential.

The goal is to give participants enough detail to keep the exercise interesting, without delving too much into strategy and tactics. Plan two or three possible outcomes, and write events, based on previous actions, that may be given to a team as the exercise progresses. For example, if the galley stove is electric, then a water attack without de-energizing the circuit might reasonably cause injury to the fire attack team. On the other hand, if the team leader identified de-energizing as an action to be taken and ordered an attack with carbon dioxide (CO_2) portable fire extinguishers, then successful extinguishment is a reasonable outcome. The intent of the exercise is to assess and improve command decision-making capabilities, not to prove the participants are incompetent in any way.

During this exercise, four positions of command are assessed: officer of the watch (OOW), master, engineer on duty, and fire team leader. The vessel's fire control plan is laid out in a room or posted on a

Personal Protective Equipment
Performance Test

Choose the following problems or set up similar situations in which the candidate will have to use collectively those psychomotor skills performed independently.

When creating your own performance test problems, make the problem as realistic as possible by presenting it within a framework that simulates real-life situations that will be encountered by the candidate after completing training.

To pass the performance test, the candidate should be able to determine required processes in a given situation and to perform those processes to a competency level of at least 2.

Competency Rating Scale

3 — Skilled — Meets all evaluation criteria and standards; performs task independently on first attempt; requires no additional practice or training.

2 — Moderately skilled — Meets all evaluation criteria and standards; performs task independently; additional practice is recommended.

1 — Unskilled — Is unable to perform the task; additional training required.

N/A — Not assigned — Task is not required or has not been performed.

✔ **Evaluator's Note:** Formulate and inform the candidate of the standards for this performance test (time allowed and number of attempts, if applicable). Observe the candidate perform the task, note whether the performance meets each of the evaluation criteria, and then use the rating scale above to assign an overall competency rating. If the candidate is unable to perform any step of this task, have the candidate review the materials and try again.

Student's Name _____ Date _____

Performance Task 1   ☐ Assigned   ☐ N/A

Don and doff protective clothing.

Performance Criteria	Yes	No
Donning		
Trouser legs over boots	☐	☐
Hood on properly (no exposed hair)	☐	☐
Coat completely fastened	☐	☐
Coat collar up and fastened	☐	☐
Helmet secured under chin	☐	☐
All articles of protective clothing donned within 1 minute	☐	☐
Doffing		
Gloves in coat pocket	☐	☐
Helmet stored with front facing back of storage unit	☐	☐
Hood in coat pocket	☐	☐
Coat on storage hook with collar up and front completely unfastened	☐	☐
Boots and trousers doffed as one	☐	☐
Other		
1. _____	☐	☐
2. _____	☐	☐

Competency Rating _____

Figure 14.15 Sample performance test rating sheet for one task of personal protective equipment requirements.

bulkhead (Figure 14.16). The situation is not revealed to the participants. Each officer is given a portable radio (to add realism and reinforce correct communication procedures) and a command position. The OOW is first on scene. She enters the room and is given a simulated situation and instructions. For example: *"You are on the bridge. It is 1000 hours. The fire panel shows fire in the galley. Proceed."* As each participant is contacted by the OOW, assessments are made. If required, the instructor may make further information known to the team as outlined previously, but additional facts are kept to a minimum so maximum concentration is on the command process. A log of communications is kept to form the basis for debriefing.

The entire exercise may take 10 to 20 minutes and may be repeated with others assuming the lead roles. It may be beneficial to have other officers observing. If participants are nervous and personnel are available, two may be paired to each role with one participant acting as a mentor (adviser) to the other. If time is available, an exercise can be conducted in "start/stop" mode with periodic time-outs to discuss how the situation is progressing. This guided exercise slows down the tendency to rush toward a resolution without thinking through all the important considerations.

It is surprising how realistically participants behave during such exercises. Improved communication procedures with clear and unambiguous transfer of command and information and a strong sense of teamwork are some benefits of such exercises. Training for the command role may be achieved using these scenarios without conducting time-consuming and costly full-scale drills.

Exercise Review and Evaluation

In addition to scenarios and portable radios, the command and communications exercise requires evaluation checklists. Sample checklists for the role of officer of the watch and master are shown in Figures 14.17 and 14.18. Use these samples to design the checklists for engineer on duty and fire team leader. The checklists given are comprehensive but not all-inclusive. Not all items on a list may be assessed during each exercise. The purpose of a checklist is to help focus the participants on the best process for each situation. All items on the sample checklists are followed by check boxes for *yes, no,* or *N/A.*

Review of the exercise should be quick, to the point, and focused on the team rather than on an individual. Thus, participants do not feel threatened, but they do learn. They are left with no doubt as to the desirable action to take. Exercises of this type may be conducted with greater frequency than regular drills until the desired level of competency is attained.

Figure 14.16 Using the fire control plan for a command and communications exercise. *Courtesy of R. Wright/Maryland Fire and Rescue Institute/United States Coast Guard.*

Officer of the Watch
Exercise Evaluation Checklist

Task	Task Accomplished		
	Yes	No	N/A
Assess situation.			
Call master.			
Sound general alarm.			
Contact engineer on watch.			
Put crew member on wheel.			
Add second steering motor (backup or emergency motor to control rudder).			
Prepare emergency message.			
Give situation update.			
Transfer command.			
Assume emergency position as navigator.			
Keep master informed as to vessel's position and traffic considerations.			
Maneuver vessel as directed by master.			

Figure 14.17 Sample exercise evaluation checklist for the officer of the watch's role.

Master
Exercise Evaluation Checklist

Task	Task Accomplished		
	Yes	No	N/A
Seek situation update.			
Assume command.			
Communicate with fire team leader.			
Communicate with boat team leader.			
Communicate with chief engineer.			
Order transmittal of emergency message.			
Seek situation updates from fire team leader, boat team leader, chief engineer, and others.			
Order fire attack.			
Maneuver vessel if required for best ventilation.			

Figure 14.18 Sample exercise evaluation checklist for the master's role.

Chapter 15

In-Port Fire Fighting and Interface with Shore-Based Firefighters

Active trading vessels spend most of their time at sea, and most of a crew's fire training is done at sea, based on the reasonable assumption that the only fire fighting resources available are those on board. Yet, most vessel fires (approximately 65 percent) occur in port. This chapter examines the additional factors that may affect a fire fighting response when in port. The vessel's crew is responsible for the *initial* fire fighting efforts on board. An extensive fire will almost certainly require assistance from other sources, and the mariner needs to prepare for the complexities that occur when outside help is needed and shoreside firefighters arrive on the scene.

Some port municipal fire departments (called *fire brigades* in Europe) are well trained and equipped to respond to vessel fires. Some fire departments/brigades may be supplemented by commercial or industrial fire fighting organizations. Marine terminal fire teams are often trained specifically to fight the type of fires that could occur on their properties. But the mariner should also understand that worldwide some fire departments/brigades are not trained or equipped for vessel fires. They may not even plan for vessel fires. Shoreside fire fighting personnel may not be willing or prepared to come aboard a vessel in an emergency. Even if they are assigned to a marine facility with a pier for deep draft vessels, they may only be trained and responsible for fire suppression operations ashore. The master has a responsibility to know what resources are available in the various ports and take appropriate actions for the different situations.

Do not assume that a fire department/brigade can or will assume responsibility for onboard fire fighting in port. Some firefighters assume that the vessel's crew will fight a vessel fire while shoreside personnel only provide support. Misconceptions by shoreside firefighters about a vessel's crew fire fighting capabilities may be compounded by the vessel's crew members' misconceptions of shoreside firefighters' capabilities. Either community may underestimate or overestimate the shipboard fire fighting capabilities of each other.

In some cases, fire department/brigade personnel have no training in marine fire fighting, while crew members have little understanding of the strategy and tactics used by land-based firefighters. For an example of this situation, see the case history of the *Ambassador*, Belledune, New Brunswick, December 31, 1994 (see Appendix A, Case Histories). A factor contributing to the *Ambassador* fire was that the vessel's officers were not aware that shoreside firefighters were not trained in shipboard fire fighting and allowed ventilation of the fire site prematurely.

Some fire fighting strategies that work well on land may be hazardous on board. For example, a common ventilation practice shoreside is to open a structure and remove smoke and heat. Then an interior fire attack is made with fire hoses. However, common practice on a vessel uses vessel construction features to close a compartment and confine the fire. This action begins to smother the fire, and if extension through conduction is prevented by boundary cooling, the fire may be extinguished without spreading. If the fire is in a space equipped with a fixed fire suppression system, the extinguishing agent is introduced and boundaries are cooled with fire hoses. Shoreside ventilation techniques may be useful on board a vessel, but they must be used with caution and after consultation with a vessel's officers.

Shoreside fire departments/brigades should seek information from vessel personnel before committing their resources. Reliance on a vessel's fire control plan will not give sufficient information. Fire department/brigade officers are reluctant to respond to fire on board without clear information, guidance, and assistance from vessel personnel.

Any effective emergency response by crew members or shoreside personnel depends on the following factors:

- Timely notification of an emergency
- Speed of emergency response
- Knowledge of the location of personnel
- Accurate knowledge of the location of a fire or other emergency
- Knowledge of vessel systems and layout of the vessel
- Sufficient resources to extinguish a fire or to contain it until it burns out

This chapter discusses both fire department/brigade and vessel emergency organizations and the priority of response actions when a local fire department/brigade arrives at a fire incident on board a vessel. Working together with the local fire department/brigade is extremely important for crew members, and several aspects of this teamwork are described: response tasks, access, vessel stability, strategy and tactics, incident management systems, and joint training exercises and tours.

 ## Fire Department/Brigade Organization

To acquaint the mariner with the rank structure of fire departments/brigades, examples are given in a chart. These are the most often used ranks, but others may be used by some departments/brigades.

When fire departments/brigades refer to their chiefs, they are not talking about a position in the engineering or deck department. In North America, an engineer is a firefighter assigned the duties of driv-

ing apparatus to the fire scene, operating the apparatus for pumping capability, or operating an aerial device at the scene (Figure 15.1).

United Kingdom	North America
• Chief Officers	• Chief Officers
• District Officer	• Battalion or District Chief
• Station Officer	• Captain
• Sub Officer	• Lieutenant
• Leading Firefighter	• Sergeant
• Firefighter	• Firefighter

In North America, a fire department's responsibility is to perform the actions needed to satisfy the goals and objectives it has identified: life safety, incident stabilization, and property conservation. Actions are governed by rules, regulations, or policies that define how the department plans to operate. The standard operating unit of a fire department is the company, a group of firefighters assigned to a particular piece of fire apparatus or to a particular station. The following four basic organizational principles allow a firefighter to operate effectively as a team member:

- **Unity of command** — A person can report to only one supervisor directly, but indirectly everyone reports to the fire chief through the chain of command.

- **Span of control** — The number of personnel one individual can effectively manage ranges from three to seven.

- **Division of labor** — Large jobs are divided into small jobs that are assigned to specific individuals (prevents duplication of effort).

- **Discipline** — The boundaries or limits for expected performance are set and enforced through written rules that define how the department plans to operate.

In the United Kingdom, fire brigades are administered on a county basis and vary in size from London, the largest, with 100 fire stations (all professionally staffed) down to the Isle of Wight with 5 stations (only one professionally staffed). Each county has a headquarters with a control center. Large counties are divided into divisions of 10 to 15 stations. Department

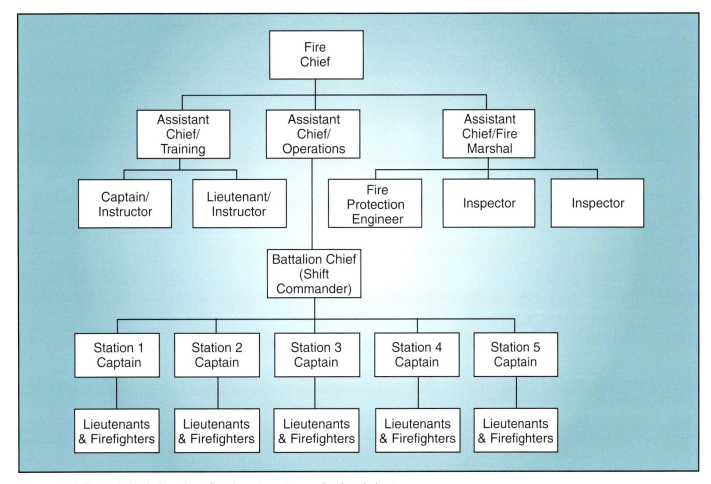

Figure 15.1 Typical North American fire department organizational chart.

organizational units include fire safety, training, vehicle workshops, and technical/research. Figure 15.2 is a typical example.

 Vessel Emergency Organization

An emergency team or squad is organized for fire fighting and damage control operations on board a vessel. The station bill and muster list show assignments, duties, and responsibilities of crew members. Always maintain minimum staffing requirements appropriate to the vessel when in port so sufficient officers and crew remain aboard to effect an emergency response, guide fire department/brigade personnel, or move the vessel if needed. For example, the vessel may need to be moved to a safer or more accessible location.

Some vessels have personnel boards that show who is aboard and who is ashore when vessels are in port. Likewise, most fire departments/brigades also maintain some form of personnel accountability system during incidents. A personnel board need not be complex, for example, cup hooks located beside labels that show each crew member's name hold washers (tags) painted white on one side and black on the other to denote presence or absence. When a person is aboard, the white side of the tag shows; when a person goes ashore, the tag is turned so the black side is visible. If an evacuation is ordered, crew members aboard turn their tags to the black side and leave. If everyone is ashore after evacuation, only black tags remain. Any white tags left would denote missing persons. It is helpful to give a copy of the crew list to dock security personnel so that they may identify who is aboard and who is ashore.

A weathertight container, tube, or enclosure ("fire wallet") containing the fire control plan is placed at the head of the gangway outside the deckhouse for use by fire department/brigade personnel. The enclosure should be red and have a mark of a red ship silhouette on a white background. Copies are stored on the port and starboard sides of the deckhouse at the main deck level (Figure 15.3). Fire control plans

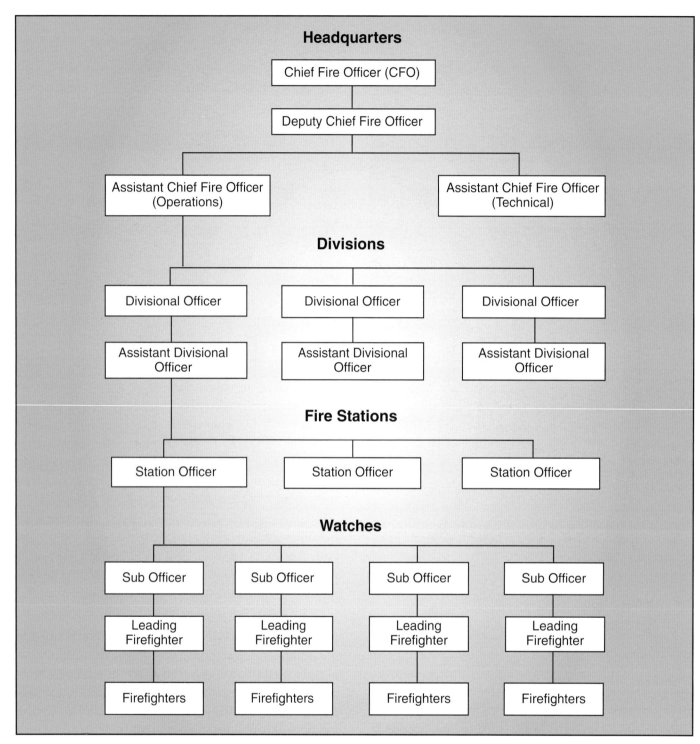

Figure 15.2 Typical United Kingdom fire brigade structure.

are required by the *International Convention for the Safety of Life at Sea (SOLAS)* to be in the official language of the vessel's flag state and also translated into either English or French.

Fire control plans indicate the locations of fire stations, fire doors, portable and semiportable fire extinguishers, and control stations for fire detection and suppression systems. They also show the locations of fire-resistant bulkheads along with their fire ratings and give the means of access and/or escape from different compartments. Ideally, in addition to the fire control plan, these containers or enclosures include copies of the following items:

• Current list of crew members

Figure 15.3 Fire control plans can be in several locations. *Courtesy of R. Wright/Maryland Fire and Rescue Institute/ United States Coast Guard.*

- Cargo manifest
- General arrangement plans for each deck, indicating the following:
  - Side elevation
  - Midship cross section
  - Electrical system controls
  - Fuel system controls
  - Fans and ventilators
  - Ventilation system (dampers and fans) controls
- Prefire plans (booklet containing plans of action in the event of fire) (see Chapter 10, Prefire Planning)
- Information on fixed fire suppression systems, automatic sprinkler systems, foam systems, fire detection systems, alarm systems, and ventilation systems
- Other pertinent information

After an alarm is sounded on board a vessel in port, the appropriate authorities on shore are notified (Figure 15.4). All crew members need to know how to

Figure 15.4 Notify on-shore authorities after an alarm is sounded. *Courtesy of Jon Swain, Texas A & M University System.*

contact shoreside help. A variety of ways are possible: global maritime distress and safety system (GMDSS) radio, ship's whistle, VHF radio (Channel 16), telephone, or runner. Cellular telephones may be on board, or the nearest telephone may be on the pier. The emergency number to call varies. For example, a person dials *9-1-1* in North America, *9-9-9* in Britain, *1-1-2* in Europe, or *1-1-9* in Japan. In some cases, an outside line may be needed first, for example, a person may need to dial *(9) 9-1-1*. Special emergency telephones are available in some ports (for example, Rotterdam port personnel place an emergency telephone on board all vessels).

Emergency services may be needed for reasons other than a fire such as medical emergencies (heart attacks, injuries, etc.) or a release or spill of hazardous materials. When in port, a vessel's crew members must be prepared to work with personnel from other agencies as well as fire departments/brigades.

◆ Arrival of the Local Fire Department/Brigade on Scene

In an ideal situation, any fire that occurs on board is extinguished before shore resources arrive on the scene. For the sake of illustration, it is assumed that the best fire fighting practices are taking place on board but that the fire is still uncontrolled when fire department/brigade personnel arrive. Know how long it normally takes for a fire department/brigade to respond. It could take from 4 to 12 minutes or longer (Figure 15.5). Ask the fire department/brigade its expected response time upon arrival in port. In many ports, a 10-minute response time is not unusual, and an additional 15 or 20 minutes may be needed to connect a fire engine to a continuous water supply and bring fire hoses on board.

Mariners must understand how shoreside firefighters operate in order to work well with them and assist them properly. The fire fighting abilities of the vessel's crew are unknown to local firefighters. Depending on their experiences, local firefighters may see the vessel's personnel as either help or hindrance. Because vessel personnel are the experts on maintaining vessel stability, they need to provide stability information to help local firefighters. The best option is for both groups to work together in a unified response. Some information that is helpful to the fire officer in the initial stages of an incident include the following:

- Account of what caused the incident
- Reports of injuries, deaths, missing persons, or where people are trapped
- Specific location of fire/incident
- Action taken (hatches or doors closed, ventilation turned off, etc.)
- Type of cargo or machinery in incident area or adjacent areas
- Other resources requested
- Actions planned by the vessel's crew

Figure 15.5 It may take some time for the fire apparatus to arrive. *Courtesy of Firefighter Luke Carpenter, Seattle (WA) Fire Department.*

- Assistance required from the fire department/brigade

A description of the priority response actions used by most fire departments/brigades help crew members understand shoreside operations and how they can support these operations. The response actions and components of the *RECEO* model: rescue, exposures, confine, extinguish, and overhaul are described in the sections that follow.

Priority of Response Actions

The RECEO model was used in Chapter 12, Fire Fighting Strategies and Tactical Procedures, as a way to establish tactical priorities. Most fire departments/brigades also work on a similar priority model. The *RECEO* model as it applies to a marine incident is listed, and then each item is briefly discussed in the sections that follow.

- *Rescue (life safety)* — Rescue those who are endangered.
- *Exposures* — Protect vessel areas, other vessels, or structures that are threatened.
- *Confine* — Contain the fire, and prevent its spread (may include ventilation).
- *Extinguish* — Control and extinguish the fire.
- *Overhaul* — Check for and extinguish hot spots (prevent reigntion/reflash); investigate the fire cause; conduct salvage/loss control operations.

Rescue

Vessel officers must know with certainty who is aboard and where they are so that the names of missing persons may be ascertained in an emergency. Maintain a personnel board at the top of the gangway so a single glance shows this information. Report the outcome of any search and the probable locations of any missing persons to the fire officer. When searching for missing persons, devise and practice a system of cabin searches so time is not spent searching the same spaces several times.

Exposures

In port, an onboard fire can affect other vessels or port facilities. Environmental exposures (air, water, and ground) are also possible. Knowing the location of a fire is essential, whether it is determined by alarm panels or visual signs. Information given to the first responding fire officer includes the following:

- Fire doors are open or closed
- Fire is contained or increasing
- Fire is likely to spread (to where) or not likely to spread

The fire officer assesses other exposures such as other vessels, sheds, warehouses, and the pier or dock (Figures 15.6 a and b). The master and fire officer together with the captain of the port/harbourmaster continue to assess threats to the vessel and exposures. For example, they may decide to move the vessel to a safer location. If moving the vessel is needed, vessel personnel (chief officer and engineer on watch) provide the following information:

- Emergency towing wires are installed or not available

- Vessel has power or is a dead ship
- Main engines are available or out of service
- Length of time it will take to move the vessel

Confine

Marine firefighters usually confine an interior fire to smother it. Land-based firefighters open a path to release hot gases and smoke so that they may enter and extinguish the fire. The marine option is not generally available in land structures. However, boundary cooling must be used on a vessel to limit the spread of fire through conduction (Figure 15.7).

Figure 15.7 A fireboat can provide boundary cooling to exposures. *Courtesy of R. Wright/Maryland Fire and Rescue Institute/United States Coast Guard.*

Figures 15.6 a and b (a) In some cases the fire department may need to board from the water side. *Courtesy of Captain John F. Lewis.* (b) Some exposures can be assessed from the fireboat. *Courtesy of Captain John F. Lewis.*

Regardless of which strategy is used, fire department/brigade personnel need to know the location and status (open/closed or on/off) of several vessel damage-control components. While some of this information is available from the fire control plan and supporting documentation, the on-scene officer in charge provides other information such as what tactical procedure is currently being employed. These components include the following:

- Watertight and/or fire doors
- Fans and ventilators
- Fuel and electrical systems
- Fixed fire suppression systems (plus type and operating instructions)

Extinguish

Crew members must be prepared to work with shore-based firefighters to extinguish the fire. Vessel personnel may need to guide firefighters to check the status of ventilators or fixed systems or even to enter the vessel should an aggressive interior attack be attempted (Figure 15.8). Land-based firefighters may not be familiar with vessel construction features or terminology. Features such as steep ladders, door thresholds, and low deckheads may hinder them until they become familiar with the vessel. Crew members who are proficient in the use of self-contained breathing apparatus (SCBA) greatly facilitate a speedy and safe response by guiding firefighters aboard a vessel.

Overhaul

Overhaul is a hazardous part of a fire response. Firefighters may not understand the degree to which heat may linger in a vessel. Personal protective equipment and SCBA should be worn when entering burned spaces, even after ventilation. Crew members may assist in securing openings and hatches and guiding investigators who may be unfamiliar with a vessel's construction.

Crew Members and Firefighters Working Together

Vessel officers should become familiar with the fire departments/brigades in the ports they visit in order to better relate to each other in emergency situations. It is recommended that a vessel's officers encourage visits from fire departments/brigades and other emergency service agencies. The objective is to assist personnel from these departments and agencies in relating not only to the vessel's systems and layout but also to the officers, crew, and passengers. Cultural differences and language barriers may also be minimized through structured tours (Figures 15.9 a–c).

As mentioned earlier, not all port fire departments/brigades are experienced or even trained in marine fire fighting. It is important to learn from each other so that in an emergency, a unified response may be successful. Response to shipboard fires is a team effort. Crew members and firefighters can share tasks such as closing vents, conducting boundary cooling, and making aggressive fire attacks. Vessel crews can work with shoreside fire teams by guiding them inside the vessel and assisting them in the following ways:

- Assist with the logistics of getting people and equipment on board (access).
- Maintain vessel stability and location by monitoring mooring lines (tightening or slacking as needed to prevent excessive movement), current, tide, and

Figure 15.8 Vessel personnel should guide firefighters to necessary systems. *Courtesy of R. Wright/Maryland Fire and Rescue Institute/United States Coast Guard.*

Figures 15.9 a–c Firefighters should tour vessels before an incident. (a) *Courtesy of Captain John F. Lewis.* (b and c) *Courtesy of R. Wright/Maryland Fire and Rescue Institute/ United States Coast Guard.*

weather conditions that shore-based teams may not recognize as significant.

- Offer advice on best strategy and tactics as part of the Command function.

Access

Firefighters may place ladders or aerials against the vessel for additional access. Personnel or equipment may be ferried on board using the stores crane or dock cranes. Staging areas or places to lay equipment may be required (Figure 15.10). To assist firefighters, the vessel's crew can perform the following tasks:

- Tend mooring lines.
- Have heaving lines ready to hoist fire hoses.
- Carry equipment.
- Operate cranes.

Vessel Stability

Fire department/brigade personnel may or may not be familiar with the concepts of vessel stability. They may not know that their actions may cause a vessel to list, sink, or capsize. Most fire officers do not have training in calculating a vessel's stability and free surface effect.

Vessel personnel should gather the vessel's current stability figures, stability booklet, and fire flow figures (amount of water being pumped on board). Free surface effect should be calculated and monitored by a vessel officer and reported to the Command Post. This information is used to make decisions on tactical

Figure 15.10 A staging area may be onboard the vessel. *Courtesy of Robert E. "Smokey" Rumens.*

operation by the master and fire officer in the Unified Command process (see Incident Management System section). Vessel pump systems, ballast capabilities or other systems or conditions that can affect stability in a positive or negative way should also be reported to Command.

In some situations (for example, when a release or spill of a hazardous material has occurred), it may be advantageous to change the vessel's list or trim to make the substance flow to where it can be contained. Ballasting or deballasting of peak tanks and double bottoms may also effect the desired result, but firefighters may be unaware of this possibility or unable to achieve it.

Strategy and Tactics

The goal at a fire incident is to safely and quickly extinguish the fire. One strategy might be to directly attack the fire with the tactic of extending fire hoses from the poop deck at the aft end of the upper deck. A means to achieve this tactic might be to take fire hoses from the pier through the fairleads using heaving lines instead of advancing the hose up the gangway and along the deck. In this instance, a marine perspective offers a quicker way of achieving a desired end.

 ## Incident Management System

Many land-based fire/rescue service agencies work together using the concept of the Incident Management System (IMS). While the terminology of IMS may be unfamiliar to mariners, the incident management concept is well understood: command and control, delegation, and communication. The purpose of IMS is to provide for a systematic development of a complete, functional Command organization designed to allow for single or multiagency use, which increases the effectiveness of Command and firefighter safety. A marine hierarchy fits the concept well, although methods and terminology may differ. Crew members should become familiar with the IMS and learn how to work within it. For more information, see the series of *Model Procedures Guides* prepared by the National Fire Service (Incident Management System Consortium Model Procedures Committee) and published by Fire Protection Publications.

Key features of the system include the following:

- Systematic development of a complete, functional organization is achieved. Major functions include Command, operations, planning, logistics, and finance/administration.

- Multiagency adoption in federal, state/provincial, and local fire agencies is allowed. Organizational terminology is acceptable to all levels of government.

- Basic, everyday operating system is provided for all incidents within an agency. Transition to large and/or multiagency operations requires minimum adjustment.

- Organization builds from the ground up; initial management of major functions is the responsibility of one or two persons. Functional units handle the most important incident activities. As the incident grows in size/complexity, additional persons are assigned to functional unit management.

- Jurisdictional authority of involved agencies is not compromised; each agency is assumed to have full Command authority within its jurisdiction at all times. Assisting agencies normally function under the direction of the Incident Commander of the jurisdiction where the incident occurred.

- Multijurisdictional incidents are normally managed under a Unified Command management structure involving a single Command Post and a single incident action plan (strategic goals, tactical objectives, and support requirements for the incident).

- The System is staffed and operated by qualified personnel from any agency and could involve personnel from a variety of agencies working in many different parts of the organization.

- The System expands and contracts organizationally according to the needs of the incident; organizational structure is never larger than is required.

Command is responsible for overall management of an incident. The Command function within IMS is conducted in two general ways: Single Command and Unified Command.

Single Command

Within a jurisdiction where an incident occurs and where no overlap of jurisdictional boundaries is involved, a single Incident Commander is designated by the jurisdictional agency to have overall management responsibility for the incident. The Incident Commander prepares incident objectives that become the foundation for subsequent action planning and establishes overall management strategy for the incident. The Incident Commander is responsible for ensuring that all functional area actions are directed toward accomplishing the strategy. The Incident Commander approves the final action plan and all requests for ordering and releasing of primary resources. The Incident Commander may have a deputy (with the same qualifications) who works directly with the Incident Commander, performs a relief role, or performs specific assigned tasks. Figure 15.11 depicts an incident with Single Command authority on board a vessel.

Unified Command

A Unified Command structure is used when the incident is totally contained within a single jurisdiction but also when more than one agency shares management responsibility because of the nature of the incident or the kinds of resources needed. The individuals designated by their agencies must jointly determine objectives, strategy, and priorities. A maritime example of a Unified Command structure is given in Figure 15.12.

 ## Joint Training Exercises

Joint training exercises are seldom performed, but the benefits are considerable. Such joint exercises leave personnel of the vessel and participating agencies better prepared should an emergency occur in that port. Both the strong and weak points of each agency are identified. A plan of action can then evolve to make a better response the next time a real emergency occurs. The primary disadvantage is cost — cost of time and money. Thus this type of training occurs infrequently.

In some ports, vessel personnel, fire officers, and government agency personnel meet and plan large-scale exercises for a response to a simulated emergency. These simulations involve multiple levels of management of the vessel owners, vessel operator, and vessel crew. The fire service provides marine-trained firefighters, emergency medical services personnel, hazardous materials technicians, Command Post staff, etc. Coast guard forces, environmental protection agencies, occupational safety and health organizations, port authorities, and law enforcement agencies participate and play their projected roles.

It is not necessary to have everyone mobilized at a great cost of time and money just to train together. Vessel officers can meet with fire department/brigade officers and participate in tabletop incident management exercises. This type of exercise can be conducted using a port model complete with miniature vessels and emergency response vehicles. In addition, portable radios, projected images, paper documentation, and computer databases can be used (Figure 15.13).

Another effective method for joint training is to have the vessel crew practice exercises with fire department/brigade personnel following a structured tour of a vessel during which the fire officers complete a pre-emergency survey. This tour would include visits to all major areas of the vessel and discussions with crew members. The fire departments/brigades would then invite the vessel's crew to the fire station for a similar tour. At least one major U.S. shipping company routinely conducts fire department/brigade tours worldwide (Figure 15.14).

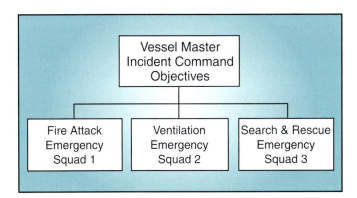

Figure 15.11 Single Command structure onboard a vessel.

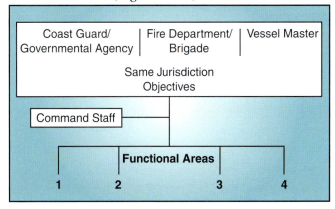

Figure 15.12 Unified Command structure for a multiagency maritime incident.

Figure 15.14 Firefighters and crew members should visit each other's facilities and practice exercises. *Courtesy of R. Wright/Maryland Fire and Rescue Institute/United States Coast Guard.*

Figure 15.13 Tabletop models can be used when conducting joint training exercises. *Courtesy of R. Wright/Maryland Fire and Rescue Institute/United States Coast Guard.*

Case Histories

Alaska Spirit

Overview

About 0200 on May 27, 1995, the U.S. fish-processing vessel *Alaska Spirit* caught fire and burned while moored alongside a dock at the Seward Marine Industrial Center, Seward, Alaska. The fire took nine hours to extinguish and resulted in $3 million in damage. The master of the vessel was found dead adjacent to his quarters, indicating that he had been awakened by the fire alarm but was overcome by smoke.

When the fire was discovered on the No. 2 deck in the area of the aft port door and the wheelhouse, the crew attempted to extinguish the fire using the shipboard fire fighting equipment. The crew used shipboard hoses connected to the ship's fire pump along with the general service and seawater service pumps. Despite their efforts, the fire continued to grow in intensity and was finally extinguished by land-based firefighters who were able to eventually gain access to the immediate fire area by wearing proper protective equipment and self-contained breathing apparatus (SCBA).

Scene Circumstances

The ship's fire detection system consisted of 27 heat detectors located in the deckhead throughout the vessel. The majority of these detectors were located in the area of the crew members' rooms. Only three smoke detectors were on the ship, and they were located in the freezer holds. No automatic fire sprinklers were in place on the *Alaska Spirit*. Had there been sprinklers in place at the time of the fire, they may have contained it to the area of origin.

The *Alaska Spirit* had two types of fire hose. One type was a standard 1½-inch (38 mm) fire hose with standard couplings. The second type was a nonstandard Armitex type that required adapters to connect it to the hydrants on the ship. The crew choose to use the second type of hose, and the adapters, which were stored in the chief engineer's room, had to be retrieved before fire-fighting operations could take place.

Two SCBAs were available on board the ship for use by the crew during fire-fighting operations. None of the crew used these SCBA during the fire and as a result was unable to make an effective attack on the fire.

It was reported that the *Alaska Spirit* had conducted a fire drill on the previous fishing trip. This drill was conducted without the crew ever laying out or charging the ship's fire hoses from any of the hydrants or donning the SCBA or personal protective equipment. As a result, the crew was not familiar with the equipment, and the fire-fighting system was not tested.

Conclusions

The U.S. National Transportation Safety Board (NTSB) investigation determined that a cook pot that was left on in the assistant fish master's room ignited bedding materials and started the fire. Factors contributing to the spread of the fire included combustible wooden bulkhead and deckhead finishes and combustible foam insulation on all the interior hull and deck surfaces. The NTSB concluded that the rapid growth and spread of the fire were due to the high fuel load and lack of noncombustible construction. Additionally, had smoke detectors been in place instead of heat detectors, the crew would have received an earlier alarm of fire, possibly allowing the master adequate time to escape. Due to the delayed alarm, the fire was able to gain significant headway, and the master was not able to escape prior to succumbing to the smoke and toxic gases produced. Finally, fire-fighting efforts were hampered by the use of nonstandard hose and the failure of the crew to use personal protective equipment and SCBA. The NTSB determined that the lack of realistic fire fighting drills led to the unpreparedness of the crew and compromised the ship's fire fighting system.

Reference

Fire On Board the U.S. Fish Processing Vessel Alaska Spirit, Seward, Alaska, May 27, 1995, National Transportation Safety Board, Marine Accident Report MBR-96/01, Washington, D.C., June 11, 1996.

 ## MV Ambassador

Overview

On December 31, 1994 the *MV Ambassador*, a self-unloading bulk carrier was unloading a cargo of rock phosphate at Belledune, New Brunswick, Canada. A fire occurred in the conveyor tunnel resulting in heavy damage to the conveyors and accommodations. The vessel's crew and several shore-based fire departments combined to bring the fire under control before it was finally extinguished 28 hours later. There was no damage to harbor installations, no serious injuries, and no reported pollution as a result of the fire.

Scene Circumstances

Unloading operations were ceased at approximately 0115, and the conveyor belts were shut off at 0205 in order to allow the dust in the tunnels to settle. Operations were stopped frequently because of spilled cargo and dust. At 0220, a heat sensor in the

area of the transfer belt indicated the presence of a fire, and smoke was seen issuing from the loop belt casing. A section of one of the conveyor belts ignited, probably because the belt was in contact with an overheated roller. The roller probably overheated because of a bearing failure or from being jammed with refuse. The temperature at the time was -13°C (9°F) with a wind chill of -28°C (-18°F).

When the fire was discovered, the sprinkler system in the loop belt casing was activated, and all mechanical ventilation in the tunnel was stopped. The sprinkler system was ineffective in controlling the fire. Ship personnel were unable to enter the area to determine the location of the fire and extinguish it due to the heavy volumes of dense smoke. Three fire hoses were directed into the loop belt casing in an attempt to douse the fire. Two of these three fire hoses were run from the engine room, through the accommodations, and to the fire area because all but one of the deck hydrants were frozen. Local emergency services were notified, and three fire engines along with 115 fire-fighting personnel from seven different emergency response agencies arrived to assist in fire control efforts.

Fire-fighting personnel attempted to gain access to the fire area but were turned back by intense heat and heavy smoke conditions. The No. 3 hold, which was empty, was flooded with seawater in an attempt to quench the fire. This strategy allowed the burning longitudinal belt to be submersed for approximately 90 percent of its length. At 1100 the fire was thought to be under control. The vessel's personnel believed the best course of action was to keep the area sealed and continue cooling efforts, while shore-based firefighters believed that hot spots should be extinguished immediately. Vessel personnel deferred to the land-based firefighters, and ventilation fans were restarted at 1330 to evacuate smoke and toxic gasses. At 1430, just prior to personnel entering the area, smoke again began to issue from the tunnel. The fire began to burn again with even greater intensity. Heat was spread via conduction through bulkheads, and fires were ignited in the accommodations. Approximately 28 hours after the initial alarm of fire, the fire was fully extinguished.

Conclusions

Flooding as a method of extinguishment should not be chosen in haste. The *Ambassador's* stability information gave no guidance as to the effects of flooding the tunnel area on the stability of the ship. The decision to flood the tunnel that contained the burning conveyor system was made without direct knowledge that the vessel was capable of withstanding such an operation. It turned out that the lack of motion in the sheltered conditions that existed in the harbor contributed to the success of this method of extinguishment. This situation may have incurred a different result had the vessel been in a location where it would have subjected to motion from waves or wind.

Once the fire had reached a point where it was first thought to be under control, the vessel's officers, who had been trained in shipboard fire fighting techniques, wanted to keep the fire area sealed and continue the application of water. They deferred, however, to the land-based firefighters (who had no shipboard fire fighting training) who believed that the area should be opened so that ventilation and overhaul operations could start. The vessel's personnel should have first determined whether or not the land-based fire officers had experience in fighting shipboard fires. After determining that they did not, the course of action recommended by the vessel's personnel should have been followed. This procedure may have prevented or lessened the effects of the second flare-up of fire.

The extreme cold weather significantly hampered fire-fighting efforts. It was noted that the crew of the *Ambassador* had readied the vessel's fire main for cold weather conditions. The extreme cold that existed at the time of the incident rendered the fire main system virtually useless. Only one outside deck hydrant could be used by fire fighting personnel because all of the others were frozen.

It was found during the investigation by the Transportation Safety Board of Canada that the crew had not drilled in fire fighting methods during weekly emergency musters. This lack of training resulted in the crew being unaware of the condition of the fire fighting equipment. Some of the vessel's fire fighting equipment was found to be substan-

dard. In addition, the program of routine testing and maintenance of safety equipment proved ineffective. It was further noted that although there was a fixed fire suppression sprinkler system in place in the loop belt casing, it was not capable of controlling a fire of this magnitude.

Reference

Fire in the Cargo-Handling Conveyor System of the Self-Unloading Bulk Carrier Ambassador, Belledune, New Brunswick, Canada, December 31, 1994, Marine Occurrence Report, Report No. M94M0057, Transportation Safety Board of Canada, Hull, Quebec, Canada.

MV Carnival Ecstasy

Overview

At approximately 1710 on July 20, 1998, the Carnival Cruise Lines ship *Ecstasy* caught fire 2 miles (3.2 km) off the coast of Miami, Florida. The fire was extinguished by 2109 that evening, and the vessel returned to its berth in Miami at 0220 the next morning. All passengers were safety disembarked before 0600. No deaths or serious injuries occurred, although 54 people were treated for smoke inhalation and 20 were taken to hospitals. Estimated damage to the vessel was $30 million, and the vessel was returned to service in September, 1998.

Scene Circumstances

The fire was discovered by two mechanics who were preparing to weld in the laundry room on Deck 2 at the aft of the ship. The mechanics had not yet begun to weld but had turned on their welding equipment. One of the electrodes came into contact with the metal on the mangle, a type of machine used to iron linens. The electrical arc made sparks that started a small fire underneath the mangle. The mechanics tried to extinguish the fire with water and extinguishers. Then they noticed that fire had spread into the vent opening over the machine. The fire spread upward through the vent to Deck 4 (mooring deck). The vessel lost propulsion and steering as a result of the fire. The closing of fire doors and the activation of sprinkler systems confined the fire to the rear of the vessel.

At 1730, an announcement was made requesting passengers to stay away from the stern of the vessel. No call was made to the U.S. Coast Guard at that time, but the Coast Guard received an initial report of a fire from Miami Beach watchstanders who observed black smoke coming from the stern of the cruise vessel. The Coast Guard established communications at 1732 with the *Ecstasy* master who confirmed that a fire was onboard, that it was under control, and that no Coast Guard assistance was required. At 1740 the first Coast Guard unit arrived on the scene and observed a significant increase in black smoke billowing from the stern of the *Ecstasy*. At 1800 the *Ecstasy* requested fire-fighting assistance. Firefighters aboard the vessel were able to extinguish the fire with help from responding tugboats.

All decks on the *Ecstasy* were protected by sprinkler systems with the exception of Deck 4, the mooring deck. It was not protected by sprinkler systems because it was considered an open deck, and *The International Convention for the Safety of Life at Sea (SOLAS)* does not require sprinkler systems on open decks.

Conclusions

The official NTSB report has not been released as of this date. It is highly likely, however, that improper hot-work practices will be blamed for this particular incident. The mechanics did not have a hot-work permit. A fire watch was not posted, but at the time the fire occurred, they were not actually welding. The major issues are the origin and propagation of the fire, the adequacy of shipboard fire protection, the adequacy of passenger safety procedures aboard the vessel (including shipboard emergency and contingency planning), and Carnival's management oversight.

Reference

Wolf, Alisa. "Smoke on the Water," *NFPA Journal*, January/February 1999, pp. 38-41, Vol. 93, No. 1.

SS Hanseatic

Overview

About 0730 on September 7, 1966, a fire started in the diesel generator room on the B deck level of the German passenger vessel *SS Hanseatic* while it was docked at Pier 84, North River, New York. Fuel from a leaking line on the No. 4 diesel generator ignited and burned. Then fire extended through an intake ventilator through seven decks. The New York Fire Department brought the fire under control at approximately 1430 the same day. However, as a result of the fire, the vessel suffered damage estimated at $1 million with no reportable injuries.

When the fire was discovered at 0730 in the diesel generator room, three crew members attempted to extinguish the flames with the portable dry chemical and foam fire extinguishers immediately available in the area. After the alarm sounded in the engineer's office, the chief engineer proceeded below and with other crew members attempted to extinguish the fire with semiportable foam fire extinguishers from the No. 2 boiler room. While shutting down fuel tanks at 0745, all power was lost on the vessel. The New York Fire Department arrived at 0750 and proceeded to fight the fire. During the first hour, firefighters realized that the compartment was equipped with a fixed carbon dioxide (CO_2) fire suppression system; however, it failed to extinguish the fire upon activation.

Scene Circumstances

The CO_2 fire suppression system on the *SS Hanseatic* contained 3,300 pounds (1,497 kg) of CO_2 gas and was connected to the forward holds, No. 1 boiler room, No. 2 boiler room, the diesel generator room, and the after holds. Because the seat of this fire was several feet (meters) above the bilge, this fire suppression system was not initially used to fight the fire because it was believed to be a bilge flooding system. All mechanical and electrical equipment in the diesel generator room were exposed to extreme heat and flame and suffered considerable damage.

A sprinkler system was installed in the accommodation and service areas, however, it activated in the starboard passageway only, and discharge ceased when the gravity storage tank that supplied the system exhausted its supply. Furthermore, the automatic pump was unable to start due to the power failure.

Conclusions

From the reports given by both firefighters and crew members, the cause of the fire abroad the *SS Hanseatic* was determined to have originated in the No. 4 diesel generator when diesel fuel sprayed onto the engine head and exhaust manifold. Also, due to the extensive fire damage, the cause of the fuel line failure could not be determined. The primary fire spread vertically through the ship but was held within the steel bulkheads while secondary fires were on all decks by direct conduction of the heat from the primary fire through the heated steel bulkheads. The primary fire was hot enough to melt power cables passing through the generator room thereby causing the power failure that de-energized two of the fire pumps on the vessel. Furthermore, lack of adequate training for crew members, delayed reporting of the fire to the master, and inadequate fire-fighting equipment caused the fire to rage out of control. It was the opinion of the Marine Board Investigation, U.S. Coast Guard, that the crew could not have successfully combated the fire had it occurred while the vessel was at sea.

Reference

"Case Histories of Shipboard Fires," *Marine Fire Prevention, Firefighting and Fire Safety*, U.S. Department of Transportation, Maritime Administration, Robert J. Brady Company.

near Harbor Island in Seattle. The fire was contained two hours after it started but continued to smolder for several days before final extinguishment. The source of the fire was difficult to reach. The fire was located near the stern of the ship in a hold containing about 50 cargo containers. The *Manulani's* cargo was mostly household goods. One firefighter was taken to a hospital for heat exhaustion, but there were no other injuries. The *Manulani's* crew attempted to fight the fire, but was forced from the hold by smoke after a container exploded.

Scene Circumstances

More than 100 Seattle firefighters as well as U.S. Coast Guard personnel battled the blaze and were able to contain the fire using water and foam, seal the area, and then extinguish it with CO_2. Sixteen storage containers above deck were damaged from smoke and heat.

The fire was located in the stern area in a cargo hold below the deck. A liquefied petroleum gas (LPG) tank was located in the fire area, but firefighters kept it cool with streams of water. A welding torch and a failure to remove or protect combustibles from the hot-work area may have started the fire.

Conclusions

The official NTSB report has not been released at this time.

References

Cavanagh, Sean. "Fire Breaks Out Below Container Ship's Deck," *The Seattle Times*, Sunday, August 31, 1997.

CV Manulani

Overview

About 1000 hours on Saturday, August 30, 1997, the container vessel *Manulani* caught fire while docked

Scandia/North Cape

Overview

At 1320 on January 18, 1996, the U.S. tug *Scandia* caught fire and burned while pushing an unmanned

U.S. tank barge, the *North Cape*, in Block Island Sound off the coast of Rhode Island. The fire was not extinguished, and the crew abandoned the vessel at 1357 amid 10-foot (3 m) waves and 25-knot winds. The tug and the barge both ran aground because the crew was unsuccessful in its attempts to release the anchor of the barge. The barge spilled 828,000 gallons (3 134 311 L) of home heating oil. Damage to the tug was estimated at $1.5 million. Estimated damage to the barge was $3.6 million. The oil spill was the largest pollution incident in Rhode Island history and led to the closing of local fisheries. No crew members were injured.

Scene Circumstances

When the fire was first discovered in the space above the engine room (the fiddley) at approximately 1320, two crew members tried unsuccessfully to fight it with handheld carbon dioxide fire extinguishers. The crew members were unable to enter the fiddley and activate the engine room's semiportable CO_2 extinguisher system due to the intense smoke and heat generated by the fire. The *Scandia* did not carry SCBAs or firefighter protective clothing. The crew members were not trained in marine firefighting, nor were they required to be.

Conclusions

The NTSB investigation determined that the fire was probably caused when a container of grease for the towline spilled onto the hot engine. Inadequate oversight of maintenance and operations aboard the vessels permitted the fire to become catastrophic and eliminated any realistic possibility of stopping the grounding of the vessels. If crew members had been trained in fire fighting and had access to protective clothing and SCBAs, the fire could have been extinguished and the oil spill avoided. Contributing to the accident was the lack of adequate U.S. Coast Guard and industry standards addressing towing vessel safety.

Reference

Fire Aboard the Tug Scandia and the Subsequent Grounding of the Tug and the Tank Barge North Cape on Moonstone Beach, South Kingston, Rhode Island, January 19, 1996, National Transportation Safety Board, Marine Accident Report No. MAR-98-03, Washington D.C., July 14, 1998.

SS Sea Witch/ SS Esso Brussels

Overview

On June 27, 1973, the container vessel *SS Sea Witch* lost steering control, struck, and penetrated the anchored Belgian tankship *SS Esso Brussels* in New York harbor near the Verrazano Bridge. Approximately 31,000 barrels of oil were released, and the resulting fire destroyed both vessels. The two vessels, locked together, drifted under the bridge and grounded in Gravesend Bay. Fireboats attacked the fires on the vessels and rescued survivors. Traffic was stopped on the bridge, which suffered some damage. Search and overhaul operations on the *Esso Brussels* continued for two days. Operations on the *Sea Witch* were interrupted by a sudden list of the vessel caused by the movement of fire fighting water below decks. The fires in the containers and holds on the *Sea Witch* burned for 2 weeks. Damage to the vessels and cargo amounted to $23 million. A total of 16 crew members died, including the masters of both vessels.

The *Sea Witch* collided with the *Esso Brussels* when it lost control of its steering system caused by a failure of the universal coupling connection in the shaft between the hydraulic rotary power receiver units and the differential gear mechanism in the steering engine room. Ten separate instances were documented on the *Sea Witch* where the steering gearing and equipment had been repaired or modified due to loss of control or sluggish control.

Scene Circumstances

The *Sea Witch* was equipped with 1½-inch (38 mm) interior and 2½-inch (64 mm) exterior fire hoses. The crew of the *Sea Witch* was only able to use the interior hoses because the exterior hoses were inaccessible due to the intense heat and smoke generated by the fire. Emergency gear lockers were also inaccessible. Crew members used the interior hoses with straight-stream nozzles to cool exterior doors. Approximately 10 minutes before the order to abandon ship was given, the fire pump shut down because too many nozzles were in use or left open. This situation showed a lack of crew training and deprived the crew of fire fighting capability before rescue.

The *Esso Brussels's* fire fighting equipment included deck fire hydrants and steam smothering fire suppression systems to the individual cargo tanks. Neither the deck fire main nor the steam smothering systems was activated before the crew abandoned the vessel. Engineering personnel did start the emergency generator before leaving the vessel. The bulk of the oil spilled from the *Esso Brussels* was consumed in the fire. Most of the casualties from the *Esso Brussels* occurred when a lifeboat could not start its engine and was pulled into the flames next to the vessel.

Conclusions

The NTSB determined that the accident could have been avoided if the *Sea Witch* had had a redundant steering system that was functioning at the time of the accident. Also a lack of adequate and timely action by the crew to control the vessel after the failure occurred contributed to the accident. The cause of the loss of steering was the deficient design of the system. The NTSB also said that combination fire nozzles with shutoff capability at the nozzle would have been more effective in this situation and would have prevented the fire pump shutdown.

Reference

SS C.V. Sea Witch-SS Esso Brussels Collision and Fire, New York Harbor, June 2, 1973, National Transportation Safety Board, Marine Accident Report No. MAR-75-06, Washington D.C., December 17, 1975.

Surf City

Overview

On February 20, 1990, the 760-foot-long (232 m) U.S. tankship *Surf City* loaded with naphtha and automotive diesel oil exploded and burned in the Persian Gulf. The fire burned for two weeks, and 196,985 barrels of the 606,215 barrels of cargo were lost. The damage resulting from this accident was $31.53 million. The accident resulted in the deaths of the master and the chief mate.

Scene Circumstances

Crew member Eugene Williams was in his sixth week of service aboard the U.S. tankship *Surf City* when an explosion and fire occurred. Williams stood the 0400 to 0800 bridge watch with the first mate. Upon completion of their tours, he and the first mate set up a water-driven fan to blow air into the No. 5 starboard ballast wing tank. Williams recalled later that they did not test the electrically grounded water hose for continuity or measure for oxygen or flammable vapor content. Williams then retired to his stateroom. While lying in his bunk, he looked out the window and saw the master and the chief mate at the entrance to the tank.

On the deck the master and the chief mate were investigating the source of a strong naphtha odor. The helmsman and the boatswain both saw the master and the chief mate peer into the No. 4S ballast tank and immediately jerk their heads back. Vapors were visible emanating from the access door. The chief mate donned a breathing apparatus and entered the ballast tank access trunk with a gas analyzer. The master was standing by. The chief mate emerged from the tank after about 5 minutes "panting for air." The master and the chief mate them used a mirror to look into the tank. The chief mate was preparing to reenter the tank when a violent explosion occurred.

Seaman Williams heard and felt the explosion at 1015. He looked out his window to see a fireball and column of smoke. He went immediately to his muster station, the port lifeboat. Due to the intensity of the flames on the starboard side, the starboard lifeboat could not be reached. Several crew members attempted to begin fire suppression operations. The intensity of the fire and radiant heat made this task impossible All crew members then proceeded to the port lifeboat on the A deck. The forward and aft gripes were released and the boatswain began to lower the 20 crew members in the life boat. The lifeboat snagged because the aft tricing pendent did not release. When a crew member released the pendent, the boat dropped the rest of the way, and an ordinary seaman was thrown overboard. Once the boat was in the water the motor was started by the third mate. The boat returned to pick up the ordinary seaman who had fallen overboard and the boatswain who had jumped into the water. They also retrieved the radio operator and the second officer, both of whom had been unaccounted for at the time of the lowering of the lifeboat but had subsequently arrived and jumped overboard.

The main engine had been shut down, so the vessel was moving slowly. The lifeboat was pulled into the flaming oil on the water by the current. Seaman Williams reported that the heat was intense, and crew members crouched as low as possible in the lifeboat. The crew was in the lifeboat for 30 minutes prior to being picked up by the *USS Simpson*, a U.S. Navy Guided Missile Frigate escort vessel.

A second tank also exploded later. Firefighters on several fire fighting tugs fought the fire but were unable to extinguish it. The vessel burned for 12 days, and then two-thirds of the cargo was recovered.

Conclusions

The National Transportation Safety board determined that the explosion was probably the result of a noninstantaneous combustion of naphtha vapor and air (commonly called deflagration) in the No. 4S ballast tank. The NTSB could not identify a specific ignition source but believed it to be within the ballast tank, most likely a spark from some type of metal-to-metal contact. The fan used to blow air into the ballast tank was not believed to be a factor in the explosion.

Reference

Explosion and Fire on the U.S. Tank Ship Surf City, Persian Gulf, February 22, 1990, National Transportation Safety Board, Marine Accident Report No. PB92-916302, Washington D.C., March 31, 1992.

Universe Explorer

Overview

The Panamanian passenger ship *Universe Explorer* was en route from Juneau, Alaska, to Glacier Bay, Alaska, with 1,006 people aboard. A fire started in the main laundry room early on July 27, 1996. Dense smoke and heat spread upward to the crew accommodations deck. Five crew members died, and 55 crew members and 1 passenger sustained injuries, with a total of 69 people being transported to area hospitals. The estimated damage to the vessel was $1.5 million.

Scene Circumstances

Origin of the fire was not determined, but factors contributing to the loss of life and injuries were the lack of sprinkler systems, lack of automatic local-sounding fire alarms, and the rapid spread of smoke through open doors into the crew berthing area. A lack of effective oversight by New Commodore Cruise Lines allowed the physical conditions and operating procedures to exist that compromised the fire safety of the *Universe Explorer*.

Conclusions

The NTSB determined that the major safety issues were the adequacy of shipboard communications, emergency procedures, and oversight. A prime safety issue was the adequacy of fire prevention, detection, and control measures.

Reference

Fire on Board the Panamanian Passenger Ship Universe Explorer in the Lynn Canal Near Juneau, Alaska, July 27, 1996, National Transportation Safety Board, Marine Accident Report No. MAR-98/02, Washington D.C., April 14, 1998.

HMAS Westralia

Overview

The Royal Australian Navy's supply ship *HMAS Westralia* sailed from Fleet Base West for the Western Australia Exercise Area on May 5, 1998, at 0900. Fire broke out in the main machinery space (engine room) at 1035. After unsuccessful attempts to extinguish the fire with fire hoses, the carbon dioxide fire suppression system was activated at 1101. Only half of the carbon dioxide cylinders activated. Fire teams reentered the space after 15 minutes and controlled the fire within an hour. Four crew members were found dead in the space when the fire was finally declared extinguished at 1232. Eleven crew members were treated for injuries and others suffered mild smoke inhalation.

Scene Circumstances

Before the fire started, a significant fuel leak was noticed in the area of the No. 9 cylinder on the inboard side of the port main engine. The port main engine was shut down so repairs could be made. Thirty minutes later fire was reported on the outboard side of the starboard main engine. The fire was caused from fuel spraying under pressure from a hole in a flexible fuel hose on the starboard main engine coming into contact with a hot machinery component. Four people escaped from the engine room with three injured, but several crew members were missing.

The starboard main engine was shut down and electrical power to the engine room was isolated. The fire was intense, causing rapid smoke buildup

and extreme heat. A contributing factor to the size of the fire was fuel from the earlier leak. Despite some heroic but unsuccessful fire fighting efforts, the engine room was fully involved. A recommendation to release the carbon dioxide fire suppression system was made to the commanding officer, but it was not accepted initially because of the missing persons. Other unsuccessful attempts were made to enter and fight the fire. The carbon dioxide system was finally remotely activated about 20 minutes later when it was clear no one could survive the fire conditions, but some cylinders had to be manually discharged 7 minutes later. Fire teams reentered the space 15 minutes later to attack the fire again. Foam was also pumped into the space through the funnel.

The vessel's situation was communicated to Fleet Base West by mobile telephone at 1045. Maritime Headquarters West notified other naval vessels in the area shortly after. The *Stirling* arrived at 1143 along with a helicopter from the *Success* and provided medical personnel and fire fighting equipment. Other naval vessels arrived at 1220 along with the civilian tug *Wambiri* from the Port of Fremantle.

Conclusions

An investigation by the Naval Board of Inquiry could not say whether the carbon dioxide extinguished the fire or not. The decision to reenter the space to fight the fire after only 15 minutes was premature and showed a lack of understanding of the way in which a fire suppression system using oxygen depletion works. This lack of understanding increased the risk to the vessel and the fire hose teams.

The Board concluded that all crew members contributed to successfully overcoming the engine room fire and providing medical care to those injured. The requirement for boundary cooling was well understood, and fire-fighting teams performed competently. Ventilation control was not completely suitable for the situation. The standard operating procedure was to close both the supply and exhaust ventilation to the engine room in case of fire. Although appropriate as preparation for use of the carbon dioxide system, it prevented heat and hot gases from escaping, thus increasing the dangers faced by crew members reentering the space. Training was inadequate for some crew members on emergency systems and damage control. Escape drills using respiratory devices had not been practiced regularly.

The four deaths were attributed to acute carbon monoxide poisoning resulting from smoke inhalation. It was concluded that all died within 10 minutes of the fire, well before the carbon dioxide system was activated.

The *Westralia* did not make any general emergency or urgency broadcast to alert other shipping and civilian authorities, although the Port of Fremantle was only 7 nautical miles away. The Command team did not know that the port had significant fire fighting resources on almost immediate call. The *Westralia* received well-coordinated support from six naval vessels within 70 minutes.

Reference

"HMAS Westralia Fire Report," *Fire Australia Journal*, August, 1999. Official Journal of the Fire Protection Association Australia and the Institution of Fire Engineers Australia, Incorporated. The report is an extract of the Naval Board of Inquiry report into the fire onboard the Royal Australian Navy's supply ship, *HMAS Westralia*.

References and Supplemental Readings

 ## Books

American Merchant Seamans Manual
H. Keever
Seiler, Cornell, 1994

The Captain's Guide to Life Raft Survival
Captain Michael Cargal
Sheridan House, Inc.
Dobbs Ferry, NY, 1990

Fire Fighting on Ships
Edward W. Reanney
Brown, Son, and Ferguson, Limited
Glasgow, 1969

Firefighting Strategy and Leadership, 2nd Ed.
Charles V. Walsh and Leonard G. Marks
McGraw-Hill Book Company, NY, 1977

*International Safety Guide for Oil Tankers and
   Terminals (ISGOTT), 4th Ed.*
Whitherby & Co., Ltd.
32/36 Aylesbury Street
London, EC1R OET, England

*Marine Fire Prevention, Fire Fighting, and Fire
   Safety*
U.S. Department of Commerce
Maritime Administration
Robert J. Brady Company
Bowie, Maryland, 1979

Survival Guide for the Mariner
Robert J. Meuen
Cornell Maritime Press, Inc.
Centerville, MD

An Introduction to Marine Firefighting (Workbooks)
B. L. Hansen & Associates

 ## Classification Societies

American Bureau of Shipping (ABS)
Bureau Veritas (BV)
Det Norske Veritas (DNV)
Lloyd's Register of Shipping (LR)

 ## Organizations and Their Publications/Standards/Programs

American Petroleum Institute
   Central Committee on Transportation by Water
   Committee on Tank Vessels

Canadian Coast Guard
   Training Program in Marine Emergency Duties

Fire Protection Publications (FPP)
   www.fireprograms.okstate.edu/index.ssi
   *Hazardous Materials Emergencies Involving
      Intermodal Containers*
   *Hazardous Materials: Managing the Incident
   Incident Management System Model Procedures
      Guides*

International Fire Service Training Association
  (IFSTA)
   www.ifsta.org
   Essentials of Fire Fighting
   Fire and Emergency Services Instructor

Fire and Life Safety Educator
Fire Inspection and Code Enforcement
Fire Service Orientation and Terminology
Fire Service Ventilation
Hazardous Materials for First Responders
Introduction to Fire Origin and Cause
Principles of Foam Fire Fighting
Private Fire Protection and Detection
Self-Contained Breathing Apparatus

International Association for Safety and Survival
 Training
   www.karoo.net/iasst/index.htm

International Maritime Organization (IMO)
   www.imo.org
   *Code for the Construction and Equipment of Mo-
    bile Offshore Drilling Units*
   *Code of Safe Practice for Solid Bulk Cargoes (BC
    Code)*
   IMO Model Courses
   *International Convention for the Safety of Life at
    Sea 1974 (SOLAS) and amendments*
   *International Maritime Dangerous Goods
    (IMDG) Code*
   International Safety Management (ISM) Code

International Phonetic Association
   www2.arts.gla.ac.uk/IPA/fullchart.html
   Department of Linguistics, University of Victoria
   Victoria, British Columbia, Canada

Marine Chemist Association, Inc.

National Cargo Bureau, Inc.

National Fire Protection Association (NFPA)
   www.nfpa.org
   *Fire Protection Guide to Hazardous Materials,
    12ᵗʰ edition, 1997*
   Fire Protection Handbook, 18ᵗʰ edition, 1997
   NFPA 10 Standard for Portable Fire Extinguishers
   NFPA 11 Standard for Low-Expansion Foam
   *NFPA 12 Standard on Carbon Dioxide Extinguish-
    ing Systems*
   *NFPA 13 Standard for the Installation of Sprinkler
    Systems*
   *NFPA 30A Automotive and Marine Service Station
    Code*

*NFPA 52 Compressed Natural Gas (CNG) Vehicu-
 lar Fuel Systems Code*
NFPA 70 National Electrical Code®
NFPA 72 National Fire Alarm Code®
NFPA 101® Life Safety Code®
*NFPA 301 Code for Safety to Life from Fire on
 Merchant Vessels*
*NFPA 302 Fire Protection Standard for Pleasure
 and Commercial Motor Craft*
*NFPA 303 Fire Protection Standard for Marinas
 and Boatyards*
*NFPA 306 Standard for the Control of Gas Haz-
 ards on Vessels*
*NFPA 307 Standard for the Construction and Fire
 Protection of Marine Terminals, Piers, and
 Wharves*
*NFPA 312 Standard for Fire Protection of Vessels
 During Construction, Repair, and Lay-Up*
*NFPA 471 Recommended Practice for the Classifi-
 cation of Flammable Liquids, Gases, or Vapors
 and of Hazardous (Classified) Locations for Elec-
 trical Installations in Chemical Process Areas*
*NFPA 472 Professional Competence of Responders
 to Hazardous Materials Incidents*
*NFPA 704 Standard System for the Identification
 of the Hazards of Materials for Emergency Re-
 sponse*
*NFPA 750 Standard on Water Mist Fire Protection
 Systems*
*NFPA 921 Guide for Fire and Explosion Investiga-
 tions*
*NFPA 1405 Guide for Land-Based Fire Fighters
 Who Respond to Marine Vessel Fires*
*NFPA 1962 Standard for the Care, Use, and Service
 Testing of Fire Hose Including Couplings and
 Nozzles*
*NFPA 1971 Standard on Protective Ensemble for
 Structural Fire Fighting*
*NFPA 1982 Standard on Personal Alert Safety Sys-
 tems (PASS)*
*NFPA 2001 Standard on Clean Agent Fire Extin-
 guishing Systems*

Transportation Safety Board of Canada
   http://www.bst-tsb.gc.ca
   Technical Reports

U.S. Coast Guard
  www.uscg.mil
  Proceedings Magazine
  Marine Safety Newsletter
  Fire Fighting Manual for Tank Vessels
  Marine Safety Manual
  Advisory Committee on Hazardous Materials
  Regulations/Marine Courses

U.S. Department of the Navy
  Naval Ship's Technical Manual
  Shipboard Firefighting

U.S. Department of Transportation
  Emergency Response Guidebook
  Chemical Data Guide for Bulk Shipment by Water
  CHRIS, A Condensed Guide to Chemical Hazards

U.S. National Transportation Safety Board
  www.ntsb.gov
  Technical Reports

◆ Codes/Regulations/Reports

American National Standards Institute
  ANSI Z88.5 *Practices for Respiratory Protection for the Fire Service*

Standards of Training, Certification, and
  Watchkeeping for Seafarers (STCW 95)
  www.dnv.com/stcw/Revl/default.htm

Transport Dangerous Goods (TDG), Canada

U.S. Code of Federal Regulations (CFR)
  www.access.gpo.gov/nara/cfr/cfr-table-
    search.html
  Title 33 Navigation and Navigable Waters, Parts
    125-199
  Title 46 Shipping, Parts 1-40
  Title 46 Shipping, Part 95.10-5 Fire Protection
    Equipment
  Title 46 Shipping, Parts 140-155
  Title 46 Shipping, Part 161 Electrical Equipment
  Title 49 Transportation

NBSIR 85-3223 *Data Sources for Parameters Used in
  Predictive Modeling of Fire Growth and Smoke
  Spread*

NBS Monograph 173, *Fire Behavior of Upholstered
  Furniture*

Underwriters Laboratories Inc.®
  www.ul.com
  *UL 521, Standard for Heat Detectors for Fire Pro-
    tective Signaling Systems*

Underwriters' Laboratories of Canada
  www.ul.ca.index.htm

Training Facilities/Courses

Delgado Community College
New Orleans, LA
Basic and Advanced Fire Fighting/Industrial
  Safety Courses

Fire Service College
Moreton-In-Marsh, England
Merchant Navy/Marine Fire Fighting Courses

Justice Institute
Fire and Safety Training Center
Maple Ridge, British Columbia, Canada
Marine Emergency Duties/Shipboard Fire Fighting
  Courses

Lamar University
Beaumont, TX
Basic and Advanced Marine Fire Fighting Courses

Louisiana State University
Baton Rouge, LA
Petroleum and Marine Fire Safety Courses

Marine Emergency Technology
Oakland, CA
Maritime Fire Fighting and Disaster Planning Work-
  shops

Marine Institute, School of Maritime Studies
Offshore Safety and Survival Center
Memorial University of Newfoundland
St. John's, Newfoundland, Canada
Marine Emergency Duties/B2 Fire Fighting/Re-
  sponse to Shipboard Fires Courses

Maritime Institute of Technology and Graduate
   Studies
Linthicum Heights, MD
Marine Fire Fighting for Mariners (Basic/Advanced)
   Courses

Maryland Fire and Rescue Institute
College Park, MD
Marine Fire Fighting for Land-Based Firefighters
   Course

Northeast Maritime Institute
New Bedford, MA
Basic/Advanced Fire Fighting/STCW Safety Train-
   ing Courses

State of Alaska
Fire Service Training
Department of Public Safety
Juneau, Alaska
Basic/Advanced Marine Shipboard Fire Fighting
   Courses

Texas A&M Center for Marine Training and Safety
Galveston, TX

Texas A&M University System
College Station, TX
Marine Fire Fighting and Emergency Training
   (Basic/Advanced) Courses

 ## Miscellaneous Web Sites

www.fire.org.uk/marine/home.htm
www.resolvemarinegroup.com (click on services,
   then fire training)
www.gulfpub.com/videos/offshore/basic-marine.
   html
www.seaportinfo.com/firefigh.html
www.safetycenter.navy.mil
http://marine6.hypermart.net
www.ci.tampa.fl.us/fire/inside.htm
www.geocities.com/pentagon/9221
www.kelvinhughes.co.uk/publicat/index.htm
http://mypage.direct.ca/s/seafire

Glossary

A

Abeam — Directly off the side of a vessel; in a direction at right angles to the middle of the vessel's length. An object is said to be abeam when it is to the side of a vessel.

Aboard — In or on a vessel; opposite of ashore.

Accommodation Ladder — Vessel's own gangway (usually one on each side) fitted with means of raising and lowering; also a set of steps or ladder used for getting from one deck to another.

Accommodation Spaces — Areas of a vessel (cabins) designed for living. They are subdivided into officer, crew, and passenger accommodations.

Aft (After) — Direction towards the back end or stern of a vessel; term used relative to some other part of a vessel indicating the direction toward the stern.

Afterpeak — Area in the hull at the extreme rear end of a vessel; usually used for storage. *Also see* Forepeak.

Aground — Vessel resting wholly or partly on the ground instead of being entirely supported by the water. If done intentionally, a vessel is said to "take the ground"; if by accident, it is said to have "run aground."

Ahead — In front of a vessel; may indicate direction (an object may lie ahead) or to indicate movement (proceed at "full speed ahead").

Amidships — Center of a vessel's length, halfway between the bow and the stern.

Anchorage — Designated areas, identified on navigational charts, where ships may safely anchor.

Anchor Light — Light a vessel carries when at anchor; must be visible for 2 miles (3.22 km) at night in every direction. Vessels over 150 feet (45.7 m) must carry two lights visible for 3 miles (4.83 km).

Angle of Loll — Angle at which an imbalanced vessel is leaning and to which the vessel will stabilize. *Also see* Loll and List.

Ashore — Leaving a vessel and stepping on land; opposite of aboard.

Athwartship — Direction from side to side. To move across a vessel is to move athwartships.

B

Backstay — Line made of rope or wire supporting a mast (vertical pole); extends from the top of the mast to the stern.

Ballast — Additional weight placed low in the vessel's hull to improve its stability; may be steel, concrete, or water. *Also see* Ballasting, Ballast Tank, and Trim.

Ballasting — Process of filling empty tanks with seawater to increase a vessel's stability. *Also see* Ballast, Ballast Tank, and Trim.

Ballast Tank — Watertight compartment that holds liquid ballast. *Also see* Trimming Tank.

Barge — Long, large vessel (usually flat-bottomed, self-propelled, or towed or pushed by another vessel) used for transporting goods on inland waterways. *Also see* Lighter.

Barrel — Measure of liquid volume used in the marine industry; for petroleum, 1 barrel = 42 U.S. gallons (159 liters).

Batten — (1) Thin iron bar used to hold down the coverings of hatches on merchant vessels. (2) Strip of wood used to keep cargo away from the hull of a vessel or to prevent it from shifting. (3) Galley range support bar.

Beam — Width of a vessel measured at the widest point.

Below — Anywhere on board below the level of the upper deck; downstairs.

Berth — (1) Mooring or docking a vessel alongside a pier, wharf, or bulkhead. (2) Sleeping space. *Also see* Berthing Area and Mooring (3).

Berthing Area—(1) Space at a wharf or pier for docking a vessel; place where a vessel comes to rest. (2) Bed or bunk space on a vessel. *Also see* Berth and Mooring (3).

Bilge — Lowest inner part of a vessel's hull; flat part of bottom of vessel.

Bilge Pump—Small pump located in the bilge used to remove internal water.

Bill of Lading — Document that establishes the terms of a contract between a shipper and a transportation company; serves as a document of title, a contract of carriage, and a receipt for goods.

Bitts—Single or twin set of upright wood or steel posts located on deck along the sides of a vessel used for securing mooring lines. *Also see* Bollard.

Boat — Small craft capable of being carried onboard a vessel.

Boat Deck — Uppermost deck on which lifeboats and other lifesaving appliances are stowed; used as a promenade space on passenger vessels. *Also see* Deck.

Boat Hook — Long pole with distinctive hook at the end used for fending off other boats and retrieving or picking up mooring lines.

Boatswain (Bosun) — Petty officer on a merchant vessel who has charge of the deck crew, hull maintenance, and related work.

Boiler Room — Compartment containing boilers but not containing a station for operating or firing the boilers.

Bollard — Stout vertical post (single or double) on a pier or wharf used for securing a vessel's mooring lines; common along piers where large vessels are moored. *Also see* Bitts.

Boom — (1) Pole rigged for use as a crane on board a vessel. (2) Floating object used to confine materials upon the surface of the water.

Bow — Front end or forward part of a vessel; opposite of the stern.

Bow Thruster — Large propeller mounted in a tunnel located in the forward part of the vessel used to assist the vessel in docking and undocking; reduces the need for assistance from tugs.

Break Bulk Cargo — Loose, noncontainerized cargo commonly packaged in bags, drums, cartons, crates, etc.

Break Bulk Terminal — Shore facility handling cargo shipped in bags, steel drums, cartons, crates, or pallets.

Typical cargoes are rolls of paper, bags of fertilizer, coils of wire, or packages of steel.

Bridge — (1) Control center on modern mechanized vessels; forward part of a vessel's superstructure. (2) Persons in charge of a vessel.

Bulk Cargo — Homogeneous cargo (oil, grain, coal, bricks, lumber, or ore) stowed loose in a hold and not enclosed in any container such as boxes, bales, or bags.

Bulkhead — (1) Upright, vertical partition (wall) dividing a vessel into compartments (rooms); serves to retard the spread of liquids or fire. (2) Vertical row of wood or metal pilings or stone blocks along a shoreline that has been back-filled to protect the shore from erosion or form a berth for a vessel. *Also see* Main Transverse Bulkheads and Main Watertight Subdivision.

Bulk Terminal — Handling area for cargoes (unpackaged commodities carried in holds and tanks of cargo vessels and tankers) that are loaded and unloaded by conveyors, pipelines, or cranes. A *liquid bulk terminal* handles cargoes such as fuel, lubricating oils, and chemicals. *Also see* Dry Bulk Terminal.

Bulwark — Wall built around the edge of a vessel's upper deck.

Buoyancy (B) — (1) Tendency or capacity to remain afloat in a liquid. (2) Upward force of a fluid upon a floating object. *Also see* Center of Buoyancy.

C

Canadian Coast Guard (CCG) — Marine law enforcement and rescue agency in Canada; responsible for the safety, order, and operation of maritime traffic.

Captain — Commander of a vessel. *Also see* Master.

Captain of the Port (COTP)—U.S. Coast Guard person who has broad powers over all vessels in a port area in the United States; equivalent to Harbourmaster in the United Kingdom.

Cargo Manifest—Document that lists all cargo carried on a specific vessel voyage.

Cargo Plan — View of a vessel showing all the storage space available for cargo; shows the amount and type of cargo carried, its destination, and how it will be stowed.

Car Terminal — Facility for loading and unloading vessels specially designed to transport automobiles.

CCG — Abbreviation for Canadian Coast Guard.

Centerline — Imaginary line running the length of a vessel from the point of the bow to the center of the stern; equidistant from the port and starboard sides of a vessel.

Center of Buoyancy — Geometrical center of the underwater volume of a body; considered to be the point through which all forces of buoyancy are acting vertically upwards with a force equal to the weight of a body. *Also see* Buoyancy.

Center of Gravity — Point through which all the weight of a vessel and its contents may be considered as concentrated so that if supported at this point, the vessel would remain in equilibrium in any position. *Also see* Gravity.

Chief Engineer — Senior engineering officer responsible for the satisfactory working and upkeep of the main and auxiliary machinery on board a vessel.

Chief Officer — Deck officer immediately responsible to a vessel's master on board a merchant vessel; officer next in rank to the master; also called *chief mate, first mate,* or *mate.*

Chief Steward — Person in charge of the steward's department; responsible for the comfort and service of passengers on passenger vessels; obtains and regulates the issue of provisions and stores and is in charge of the inspection and proper storage of provisions.

Chock — (1) Cast metal ring mounted to the deck edge to control a mooring line or prevent chafing of the line; closed chock requires one end of the mooring line to pass through the center of the chock; open chock allows the line to be dropped in from the top. (2) Piece of wood or other material placed at the side of cargo to prevent rolling or moving sideways. *Also see* Fairlead.

Cleat — (1) Fitting consisting of two arms fastened on deck around which mooring lines may be secured. (2) Strip of wood or metal to give additional strength, prevent warping, or hold in place.

Coaming — Raised framework around deck or bulkhead openings; used to prevent entry of water.

Cofferdam — Narrow, empty space (void) between compartments or tanks of a vessel that prevents leakage between them; used to isolate compartments or tanks.

Collision Bulkhead — Stronger-than-normal bulkhead located forward to control flooding in the event of a head-on collision.

Companionway — Interior stair-ladder used to travel from deck to deck (usually enclosed).

Company — Term that embraces the whole crew of a vessel.

Compartment — Interior space (room) of a vessel; numbered from forward to aft with odd numbers on starboard side and even numbers on port side.

Compartmentation — Subdividing of a vessel's hull by transverse watertight bulkheads; may allow a vessel to stay afloat under certain flooding conditions. *Also see* One-Compartment Subdivision.

Containers — Boxes of standardized size used to transport cargo by truck or rail car when transported over land and by cargo vessels at sea; sizes are usually 8 x 8 x 20 feet or 8 x 8 x 40 feet (2.4 m by 2.4 m by 6 m or 2.4 m by 2.4 m by 12.2 m). *Also see* Reefer Containers and Container Terminal.

Container Terminal — Facility for loading and unloading cargoes shipped in containers and their stowage; usually accessible by truck, railroad, and marine transportation. *Also see* Containers and Reefer Containers.

COTP — Abbreviation for U.S. Coast Guard Captain of the Port.

Crew List — Part of a vessel's papers listing the names and nationalities of every member of the crew, capacity in which each member serves, and amount of wages each member receives.

Critical Angle of List — Point at which critical events will occur; not a point that remains constant in all cases; determined by stability calculations made by qualified personnel along with their professional judgment. *Also see* List and Heel.

D

Damage Control Locker — Compartment containing fire fighting/emergency equipment.

Deck — Continuous, horizontal surface (floor) running the length of a vessel; some may not extend the whole length of a vessel but always reaches from one side to the other. *Also see* Boat Deck, Main Deck, Tank Top, Tween Deck, Poop Deck, Upper Deck, and Weather Deck.

Deckhead — See *Overhead.*

Dewatering — Process of removing water from a vessel.

Displacement — Weight or volume of water displaced by a floating vessel of equal weight; weight of the vessel (including its load) is measured in long tons (1 long ton = 2,240 pounds or 1 tonne [1,000 kilograms]).

Dog — Locking levers or bolts and thumbscrews on watertight doors.

Dog the Hatches — Close the doors.

Double Bottom — Top of a series of tanks and void spaces placed along the bottom of a vessel; extra watertight floor within a vessel above the outer watertight hull; void or tank space between the outer hull of a vessel and the floor of a vessel. Also know as the *inner bottom* or *tank top*.

Draft — Vertical distance between the water surface and the lowest point of a vessel; depth of water a vessel needs in order to float. Draft varies with the amount of cargo, fuel, and other loads on board. *Also see* Draft Marks.

Drafting — Act of acquiring water for fire pumps from a static water supply by creating a negative pressure on the vacuum side of the fire pump.

Draft Marks — Numerals on the ends of a vessel indicating the depth of the vessel in the water. *Also see* Draft.

Dry Bulk Terminal — Facility equipped to handle dry goods (such as coal or grain) that are stored in tanks and holds on a vessel. *Also see* Bulk Terminal.

Dry Dock — Enclosed area into which a vessel floats but where water is then removed leaving the vessel dry for repairs, cleaning, or construction.

Dunnage — Loose packing material (usually wood boards and wedges) that is placed around cargo in a vessel's hold to support, protect, or prevent it from moving while the vessel is at sea.

E

Economizer — Assembly of coils in a vessel's stack (chimney) designed to transfer heat rising up the stack to water within the tubes. *Also see* Fiddley and Stack.

Escape Trunk — Vertical, enclosed shaft with a ladder providing an escape path for crew stationed in low areas of a vessel.

Explosionproof Equipment — Encased in a rigidly built container so it withstands an internal explosion and also prevents ignition of a surrounding flammable atmosphere; designed to not provide an ignition source in an explosive atmosphere.

F

Fairlead — Chock or opening, sometimes fitted with a roller device designed to lead a rope or line from one part of a vessel to another (change line direction); also controls lines and minimizes chafing. *Also see* Chock (1).

Fantail — Back part of a vessel that hangs out over the water; stern overhang.

Fast — Term referring to a vessel being securely attached to a wharf or dock.

Fender — Buffer between the side of a vessel and a dock or between two vessels to lessen shock and prevent chafing.

Fiddley — Vertical space extending from the engine room to a vessel's stack (chimney). *Also see* Economizer and Stack.

Fire Alarm Signal — Continuous rapid ringing of a vessel's bell for a period of not less than 10 seconds supplemented by the continuous ringing of the general alarm bells for not less than 10 seconds.

Fire Control Plan — Set of general arrangement plans for each deck that illustrate fire stations, fire-resisting bulkheads, and fire-retarding bulkheads together with particulars of fire detecting systems, manual alarm systems, fire extinguishing systems, fire doors, means of access to different compartments, and ventilating systems (including locations of dampers and fan controls). Plans are stored in a prominently marked weathertight enclosure outside the house for the assistance of land-based fire fighting personnel.

Fire Main System — System that supplies water to all areas of a vessel; composed of fire pumps, piping (main and branch lines), control valves, hose, and nozzles.

Fire Pump — Centrifugal or reciprocating pump that supplies seawater to all fire hose connections.

Fire Station — Location on a vessel with fire fighting water outlet (fire hydrant), valve, fire hose, nozzles, and associated equipment.

Fire Wire — Length of wire rope or chain hung from the bow and stern of a vessel in port to allow the vessel to be towed away from the pier in case of fire; also called *fire warp* or *emergency towing wire*.

Flag State — Nation in which a vessel is registered.

Forecastle (Fo'c's'le or Fok-sul) — Section of the upper deck located at the bow of a vessel; forward section of the main deck; a superstructure at the bow of a ship where maintenance shops, rope lockers, and paint lockers are located.

Forepeak — Watertight compartment at the extreme forward end of a vessel; usually used for storage. *Also see* Afterpeak.

Forward (Fore) — Direction toward the front (bow) of a vessel.

Frames — Structural members of a vessel's framework that attach perpendicularly to the keel to form the ribs of the vessel. *Also see* Keel.

Freeboard — Vertical distance between a vessel's lowest open deck and the water surface; measured near the center of the vessel's length where the deck is closest to the water.

Free Surface Effect — Tendency of a liquid within a compartment to remain level as a vessel moves, which allows the liquid to move unimpeded from side to side. Loose water anywhere in a vessel impairs stability by raising the center of gravity. *Also see* Center of Gravity, Center of Buoyancy, and Stability.

G

Galley — Vessel's kitchen facility.

Gangway — Opening in the railings on the side of a vessel for a ladder or ramp providing access to a vessel from the shore.

Gantry — Overhead cross-girder structure on which a traveling crane is mounted or from which heavy tackle is suspended. Supporting towers at each end of the structure are on wheels.

Gas-Free Certificate — Document stating that an authorized and trained person has evaluated the atmosphere of a space, tank, or container (using approved equipment and methods) and determined that the atmosphere is safe for a specific purpose. Also called *gas certificate* or *certified gas-free*.

Gravity (G) — Force acting to draw an object toward the earth's center; force is equal to the object's weight. *Also see* Center of Gravity.

Gunwale (Gunnel) — Raised edge along the side of a vessel that prevents loose items on deck from falling overboard; sometimes called *fishplate*.

H

Harbourmaster — Person in charge of a port (anchorages, dock spaces, etc.) in the United Kingdom; equivalent to U.S. Coast Guard Captain of the Port in the United States.

Hatch — Opening in the deck of a vessel that leads to a vertical space down through the various decks (hatchway); covered by a hatch cover (hinged or sliding).

Heel — Angle a vessel leans to one side due to wind, waves, or turning of the vessel; measured in degrees. *Also see* Critical Angle of List, Heeling, and List.

Heeling — (1) Tipping or leaning to one side. (2) Causing a vessel to list (continuous lean to one side). *Also see* Heel and List.

Hog — Vertical distance of a vessel's keel at amidships above a vessel's keel at the bow and stern; to bow up in the middle and sag at the ends as a result of improper loading. *Also see* Hogging, Sag, and Sagging.

Hogging — Straining of a vessel that tends to make the bow and stern lower than the middle portion; the middle section has greater buoyancy. *Also see* Hog, Sag, and Sagging.

Hot Work — Any construction, alteration, repair, or shipbreaking operation involving riveting, welding, burning, or similar fire-producing operations.

House — Structure located above the main deck. *Also see* Superstructure.

Hull — Main structural frame or body of a vessel below the weather deck.

I

IMO — Abbreviation for International Maritime Organization.

Inclinometer — Instrument that measures the angle at which a vessel is leaning to one side or the other.

International Convention for the Safety of Life at Sea (SOLAS) — International convention dealing with maritime safety; covers a wide range of measures designed to improve the safety of shipping. The first version was adopted in 1914. Since then, four more versions have been adopted. The present version was adopted in 1974 and became effective in 1980. The Protocol of 1978 and Amendments of 1990 and 1991 have been added since.

International Maritime Organization (IMO) — Specialized agency of the United Nations devoted to maritime affairs. It first met in 1959. Over the years, IMO has developed and promoted the adoption of more than 30 conventions and protocols as well as 700 codes and recommendations dealing with maritime safety. Its main purposes are safer shipping and cleaner oceans.

International Shore Connection — Pipe flange with a standard size and bolt pattern allowing land-based fire department personnel to charge and supply a vessel's fire main.

Intrinsically Safe Equipment—Incapable of releasing sufficient electrical energy to cause the ignition of a flammable atmospheric mixture.

J

Jacob's Ladder — Flexible ladder made of rope or chain but having solid rungs (wood or iron) used for boarding a vessel or scaling the sides of a vessel.

Jettison — To throw objects or cargo overboard in order to lighten a vessel's load in time of distress.

Joiner Construction — Bulkheads that subdivide the ship into compartments but do not contribute to the structural strength of the ship. Also known as *nonstructural bulkheads.*

K

Keel—Principal structural member of a vessel running fore and aft extending from bow to stern; forms the backbone of a vessel to which frames are attached; lowest member of a vessel framework. *Also see* Frames.

Knot — International nautical unit of speed; 1 knot = 6,076 feet or 1 nautical mile per hour (1.15 miles or 1.85 kilometers per hour).

L

Ladder—Any stairway or ladder (often nearly vertical) onboard a vessel.

Lash — Secure or tie anything down or to something else with rope or line.

Lift on/Lift off (Lo/Lo) — Refers to a vessel capable of loading and unloading its own cargo without shoreside crane assistance.

Lighter — Large boat or barge (usually nonpowered) for conveying cargo to and from vessels in harbor, transporting coal or construction materials, transporting garbage, etc. *Also see* Barge.

Line — Length of rope in use on a vessel.

List—Continuous lean or tilt of a vessel to one side due to an imbalance of weight within the vessel. *Also see* Heel, Heeling, Loll, Critical Angle of List, and Angle of Loll.

Load Line — *See* Plimsoll Mark.

Loll — Neutral equilibrium when vessel comes to rest within a range of stability as opposed to a point of stability, that is, instead of being stable when upright, the vessel may be stable within 1 degree to port or starboard sides and thus will lean either port or starboard. *Also see* List and Angle of Loll.

Longitudinal Stability —Ability of a vessel to return to an upright position when forced from its rest condition by pitching. *Also see* Stability and Static (Initial) Stability.

Long Ton—Unit of weight used in the marine industry; 1 long ton = 2,240 pounds or 1 tonne (1,000 kilograms). A short ton = 2,000 pounds or 0.9 tonne (900 kilograms).

Longshoreman — *See* Stevedore.

M

Machinery — Vessel's main and auxiliary engines, pumps, deck winches, steering engine, hoists, etc.

Main Deck — Uppermost continuous deck of a vessel that runs from bow to stern. *See* Deck.

Main Transverse Bulkheads—Watertight bulkheads that subdivide a vessel into watertight compartments. *Also see* Bulkhead (1) and Main Watertight Subdivision.

Main Watertight Subdivision — Space between two main transverse watertight bulkheads. *Also see* Bulkhead (1) and Main Transverse Bulkheads.

Manifest — *See* Cargo Manifest.

Marina — Special harbor with facilities constructed especially for yachts and other pleasure craft.

Maritime Law — Laws relating to commerce and navigation on the high seas and other navigable waters; a court exercising jurisdiction over maritime cases. Also known as *admiralty law.*

Mast — Vertical pole rising from the keel or deck of a vessel supporting sailing rigging; also used for radio antennas and signal flags.

Master — Commander of a merchant vessel. *Also see* Captain.

Mate — *See* Chief Officer.

Mayday — International distress signal broadcast by voice.

Metacenter (M) — Point through which the force of buoyancy works; point of intersection of the vertical through the center of buoyancy of a floating body with the vertical through the new center of buoyancy when the body is displaced. *Also see* Metacentric Height and Center of Buoyancy.

Metacentric Height (GM) — Measure of a vessel's initial stability; a geometric relationship between the center of gravity, the center of buoyancy, and the metacenter; distance of the metacenter above the center of gravity of a floating body. *Also see* Metacenter, Center of Gravity, and Center of Buoyancy.

Mooring — (1) Permanent anchor equipment (attached by a chain to a buoy) to which a vessel may connect a line, wire, or chain, eliminating the need to use the vessel's anchor. (2) Act of securing a vessel. (3) Location where a vessel is berthed. *Also see* Anchorage, Berth, and Berthing Area.

MT — Prefix to the name of a tank vessel powered by diesel machinery.

Muster List — List of crew members/passengers and their duty/emergency stations on a vessel.

MV — Prefix to the name of a vessel powered by diesel machinery

N

Naval Architecture — Branch of knowledge concerned with the design and construction of things that float (vessels, submarines, docks, yachts, etc.)

Navigable — Term for any body of water suitable for navigation by any particular vessel (not necessarily all vessels).

Night Order Book — Written instructions, special orders, or reminders from the captain or master for each officer taking night watch; placed in the chart room before the captain or master retires for the night.

O

Oil Tanker — Tank vessel specially designed for the bulk transport of petroleum products by sea. Also called *tanker.*

One-Compartment Subdivision — Subdivision of a vessel by bulkheads that will result in a vessel remaining afloat with any one compartment flooded under certain conditions. *Also see* Compartmentation.

Outboard — Anything that is on the seaside of a vessel; anything mounted outside the hull.

Overhead — Underside of a deck; ceiling of a vessel's compartment; also known as *deckhead.*

P

Passageway — Any interior walkway, corridor, or hallway in a vessel.

Pier — Platform (usually wood or masonry) extending outward from the shore into the water for use as a landing place for vessels; supported on pilings and open underneath allowing the berthing of vessels alongside. *Also see* Wharf.

Pilot — Person knowledgeable of the local waters who meets vessels and steers them safely into and out of port.

Platform — (1) Horizontal surface extending partway through a vessel; usually in the cargo space. (2) Any flat-topped vessel capable of providing a working area for personnel or vehicles.

Plimsoll Mark — Symbol placed on the sides of a vessel's hull at amidships, indicating the maximum allowable draft of the vessel. Also called *Plimsoll line* or *load line.*

Poop Deck — Partial deck above main deck at stern. *Also see* Deck.

Port — General area of a shore establishment having facilities for the landing, loading/unloading, and maintenance of vessels; harbor with piers.

Portable Pump — Small gasoline-driven pump used in emergencies to deliver water to a fire independent of a vessel's fire main system.

Port Authority — Person(s) entrusted with the duty or power of construction, improving, managing, or maintaining a harbor or port. Also called *harbor authority, harbor board, port trust,* or *port commission.*

Porthole — Circular window in the side of a vessel.

Port of Registry — Port at which a vessel is registered. Also called *home port.*

Port Side — Left-hand side of a vessel as a person faces forward.

Port State — Nation in which a port is located.

Port State Authority — Government agency having authority over port operations.

Positive-Pressure Ventilation — Method of ventilating a space by mechanically blowing air into the space in sufficient volume to create a slight positive pressure within and thereby forcing the contaminated atmosphere out the exit opening. *Also see* Ventilation.

Pump Room — Compartment in tank vessels where the pumping plant for handling cargo is installed. Pumps are placed as low as possible in order to facilitate draining. In oil tankers over 400 feet (12.2 m) long, two pump rooms are provided along with a ballast pump in some cases.

R

Reefer Containers — Cargo containers having their own refrigeration units. *Also see* Containers and Container Terminal.

Refrigerated Vessel — Vessel specially designed and equipped for the transportation of food products (meat, fruit, fish, butter, and eggs) under cold storage; cargo space is insulated for this purpose.

Refrigerating Plant — Installation of machinery for the purposes of cooling designated spaces aboard a vessel and for manufacturing ice.

Righting Arm — Moment that tends to return a vessel to the upright position after any small rotational displacement. Also called *righting moment* or *restoring moment.*

Riser — Pipe leading from the fire main to the fire station (hydrants) on upper deck levels of a vessel.

Ro/Ro — Abbreviation for Roll on/Roll off.

Roll on/Roll off (Ro/Ro) — Form of cargo handling using a vessel designed to carry vehicles that are loaded and unloaded by driving them onto/off the vessel by means of ramps. A vehicle ferry is a ro/ro vessel.

S

Sag — To curve downward in the middle as a result of improper loading. *Also see* Sagging, Hog, and Hogging.

Sagging — Straining of a vessel that tends to make the middle portion lower than the bow and stern. *Also see* Sag, Hog, and Hogging.

Sail Area — Area of a vessel (viewed from the side) that is above the waterline and is subject to the force of the wind.

Scantlings — Dimensions of the various parts of a vessel (frames, girders, plating, etc.).

Scupper — Opening in the side of a vessel to allow water falling on deck to drain overboard.

Sea Chest — (1) Enclosure attached to the inside of a vessel's underwater shell open to the sea and fitted with a portable strainer plate; passes seawater into the vessel for cooling, fire fighting, or sanitary purposes. (2) Storage chest for mariner's personal property.

Seaworthy — In fit condition to go safely to sea.

Secure — (1) To make fast such as secure a line to a cleat. (2) Close in a manner to avoid accidental opening or operation.

Self-Closing Door — Installation in which watertight doors are remotely operated by a hydraulic pressure system, allowing them to be closed simultaneously from the bridge or separately at the doors from either side of the bulkhead.

Shaft Alley — Narrow, watertight compartment between the engine room and the stern of a vessel that houses the propeller shaft; also called *shaft tunnel.*

Shaftway — Tunnel or alleyway through which the drive shaft or rudder shaft passes.

Shoring (Shoring Timbers) — Heavy timbers used to support bulkheads damaged by collision; also used to secure cargo; also prop or support placed against or beneath anything to prevent sinking or sagging.

SOLAS — Acronym for International Convention for the Safety of Life at Sea.

Sounding — Name of the measurement of the depth of water in which a vessel is floating. *Also see* Sound, To.

Sound, To — Operation of measuring the depth of water in which a vessel is floating. *Also see* Sounding.

SS — Prefix to the name of a vessel with a steam propulsion plant.

Stability — Tendency of a floating vessel to return to an upright position when inclined from the vertical by an external force (winds, waves, etc.). When a vessel returns to or remains at rest after being acted upon, it is either in stable or *neutral equilibrium.* If it continues to move unchecked in reaction to the external force, it is in *unstable equilibrium.* If an unstable vessel does not find a point of stable or neutral stability, it continues to incline until it capsizes. *Also see* Free Surface Effect, Static (Initial) Stability and Longitudinal Stability.

Stack — Ducting through which exhaust gases and often supply gas are routed; chimney. *Also see* Economizer and Fiddley.

Starboard Side — Right-hand side of a vessel as a person faces forward.

Static (Initial) Stability — Ability of a vessel to initially resist heeling from the upright position. Initial stability characteristics hold true only for relatively small angles of inclination. At larger angles (over 10 degrees), the ability of a vessel to resist inclining moments is determined by its overall stability characteristics. *Also see* Stability and Longitudinal Stability.

Station Bill — List of all crew members showing where they should be for the various operations involved in operating a vessel; shows the duty stations and duties of the crew by rank.

Steering Gear — All the apparatus by which a vessel is steered; includes wheel, rudder, and ropes or chains connected to them.

Stern — Back end or rear of a vessel.

Stevedore — One who works at or is responsible for the loading and unloading of cargo of a vessel in port; sometimes called *longshoreman*.

Superstructure — Enclosed structure built above the main deck that extends from one side of a vessel to the other. *Also see* House.

Supply/Exhaust Ventilation — Combined supply and exhaust system of mechanical ventilation that is generally used in the ventilation of passenger quarters. *Also see* Ventilation.

SV — Prefix to the name of a vessel propelled by sail.

Swash Plates — Metal plates in the lower part of tanks that prevent the surging of liquids with the motion of a vessel.

T

Tanker — *See* Oil Tanker.

Tankerman — Person qualified and certified to perform all duties included in the handling of bulk liquid cargoes (petroleum products). *Also see* Oil Tanker.

Tank Top — Lowest deck; top plate of the bottom tanks. *Also see* Deck.

Terminal — *See* Break Bulk Terminal, Bulk Terminal, Dry Bulk Terminal, Car Terminal, and Container Terminal.

Tonnage — Amount of internal volume or carrying capacity of a vessel in units of 100 cubic feet (2.8 m³); used for determining port and canal charges.

Topside — General term referring to the weather decks as opposed to belowdeck.

Towboat — Powerful, small vessel designed for pushing larger vessels such as barges on inland waterways. *Also see* Tugboat.

Transshipment — Transfer of cargo from a vessel to another vessel before the place of destination has been reached.

Transverse — Athwartship (side to side) dimensions of a vessel.

Trim — Longitudinal angle of a vessel; relation of a vessel's floating attitude to the water considered from front to back; difference between forward and aft draft readings; to cause a vessel to assume a desirable position in the water by arrangement of ballast, cargo, or passengers. *Also see* Trimming Tank and Ballast.

Trimming Tank — Tank located near the ends of a vessel used for changing the trim of a vessel by admitting or discharging water ballast. *Also see* Trim and Ballast Tank.

Tugboat — Strongly built, powerful boat used for towing and pushing in harbors and inland waterways. *Also see* Towboat.

Tween Deck — Intermediate deck between the main deck and the bottom of a cargo hold; designed to support cargo so that the cargo at the bottom of the hold is not crushed by the weight of cargo above it. *Also see* Deck.

U

Ullage — Amount that a partially filled tank lacks being full; measure of the empty part of a tank. *Also see* Ullage Hole.

Ullage Hole — Opening, usually located in the hatch cover, leading to a liquid cargo tank that allows measuring of liquid cargo. *Also see* Ullage.

United States Coast Guard (USCG) — Federal marine law enforcement and rescue agency in the United States; responsible for the safety, order, and operation of maritime traffic.

Upper Deck — Topmost continuous deck running the whole length and width of a vessel. *Also see* Deck.

USCG — Abbreviation for United States Coast Guard.

V

Vaportight Fixture — Fixture sealed to prevent an explosive atmosphere from entering the device's electrical contacts where an ignition spark could be generated.

Ventilation — Process of replacing foul air in any of a vessel's compartments with pure air. *Also See* Positive-Pressure Ventilation and Supply/Exhaust Ventilation.

Vertical Zone — Area of a vessel between adjacent bulkheads.

Vessel — General term for all craft capable of floating on water and larger than a rowboat.

W

Watch — (1) Division of a day that constitutes a period of duty for a crew member on a vessel; a crew member's assigned duty period. (2) One who watches; lookout assigned to patrol. *Also see* Watch Officer.

Watch Officer — Officer in charge of a watch; has the responsibility of the safe and proper navigation of the vessel during this time period. Also known as *officer of the watch. Also see* Watch.

Waterline — Level at which a vessel floats; line to which water raises on hull.

Watertight Bulkhead — Bulkhead (wall) strengthened and sealed to form a barrier against flooding. *Also see* Bulkhead (1).

Watertight Door — Door designed to keep out water; fitted to ensure integrity of the bulkheads (walls).

Watertight Transverse Bulkhead — Bulkhead (wall) that has no openings through it and extends from tank top to the main deck; built to control flooding. *Also see* Bulkhead (1) and Watertight Bulkhead.

Weather Deck — All parts of main deck and decks above that are exposed to the weather. *Also see* Deck.

Wharf — Place for berthing ships along or at an angle from a shore; constructed by extending bulkheads out from the shore and back-filling the enclosed area to create a flat surface for loading and unloading vessels. *Also see* Pier.

Winch — Stationary, motor-driven hoisting machine having a vertical drum around which a rope or chain winds as a load is lifted. A special form of winch using a horizontal drum is a *windlass.*

Wing Tank — Tank located well outboard next to the side shell plating of a vessel; often a continuation of the double bottom up the sides to a deck.

Index

barge-carrying vessels, CO_2 suppression systems on, 205
beam-application photoelectric smoke detection systems, 188
bearings, maintenance/repair covered in fire prevention programs, 49–50
bilge area inspections, 37
bilge pumps and dewatering, 309
bimetallic fixed-temperature heat detection systems, 190
BLEVE (boiling liquid expanding vapor explosion), defined, 5
blowers in inert gas systems, 238
boiler rooms as no-smoking areas, 27
boilers, maintenance/repair covered in fire prevention programs, 49
boiling liquid expanding vapor explosion (BLEVE), defined, 5
boiling points, defined, 4
branches. *See* hoses, nozzles
break bulk vessels, 62, 297–298
breathing apparatus. *See* self-contained breathing apparatus (SCBA)
British thermal unit (Btu), defined, 5
broken stream nozzles, 168, 170, 173
bulk vessels, 62–63, 297–298
bulkheads
    classification, 64–65
    materials used in construction, 59
    as visual factor in size-up process, 277
bunker gear. *See* personal protective equipment (PPE)
buoyancy and stability, 306
Bureau Veritas (BV) vessel classification, 59
burning process and fire behavior, 12–13
butterfly valves for automatic sprinkler suppression systems, 220

C

cabins. *See* accommodation spaces
calibration, defined, 95
calorie (Cal), defined, 5
captains. *See* masters
carbon dioxide extinguishers, 134, 138–139, 141, 142
carbon dioxide suppression systems
    advantages and disadvantages of using, 202
    alarms, 208
    bulk storage systems, 208
    cargo-hold systems, 205
    components of, 201
    evacuating before discharging, 204, 361–362
    independent systems, 205–206
    inspections and maintenance, 209–210
    local application systems, 206–208
    overview, 201–202
    total flooding systems, 203–204
cargo
    See also names of specific materials
    access and hazard potential, 61
    causing fires, 27, 29, 43
    classifying as regulated or nonregulated, 32–33
    loading/unloading, 33, 34, 38, 42–44
    unknown, attacking fires, 299
    volatile, cleaning tanks after containing, 44
cargo holds
    arrangement (layout) and fire control, 72
    as no-smoking areas, 27
    refrigerated, 62
    repairs as part of damage control, 310
    cargo manifests, identifying hazardous materials using, 28
cargo vessels, 61–63
Carnival Ecstasy (1998) incident, 39, 355–356
casing, 228, 229
causes of fires
    See also hazardous conditions
    chemical reactions, 28–29
    collisions, 44–45
    determining, 257–258

electrical equipment, 30–32
fuel oil systems, 35–37
galley operations, 34–35
hot-work operations, 37–38, 39–40
incendiary/arson fires, 45
overview, 25–26
shipyard operations, 41
shoreside personnel, 38–40
smoking, 26–28
stack maintenance, 295–296
stowage, 32–34
tanker operations, 41–44
ceilings (deckheads/overheads), 10
Celsius (C) or centigrade, defined, 5
centrifugal fire pumps, 228–230
chemical agent suppression systems, 212–215
chemical heat energy, 7
chemicals
    chain reactions, 13, 22, 23
    defined, 33
    detecting and measuring, 96
    fires caused, 28–29
chemistry of fire
    See also fire behavior; fire development; *specific terms*
    definitions of important terms, 3–6
    heat energy sources, 6–8
    heat transfer, 8–10
    overview, 3
chief officers
    drill duties, 334
    emergency response responsibilities, 275
    of fire departments/brigades, 342, 343
training duties, 319
chief steward officers, emergency response responsibilities, 276
cigarette smoking. *See* smoking
Class A foam concentrates, 177, 181
classification of fires
    extinguishers and, 134–135, 143–148
    foam concentrates and, 175–177
    overview, 23–24
    tactical guidelines based on, 290, 292, 293, 294, 295, 298, 299, 300, 301–302
classification of fixed-temperature heat detectors, 189
classification of hoses, 161
classification of vessels and vessel features, 59–60, 71, 72
clean atmosphere, defined, 44
clothing (personal protective clothing), 76–79. *See also* personal protective equipment (PPE)
CO_2 extinguishers. *See* carbon dioxide extinguishers
cognitive learning domain, 325
collisions
    causing fires, 44–45
    Sea Witch/Esso Brussels (New York Harbor, 1973) incident, 44, 358–359
    stability after, 306
colorimetric indicator tubes, 96
combustible cargo, hazards of, 33–34
combustible gas detectors, 96
combustible liquids, defined, 4
combustion
    defined, 3
    heat of, 7
    modes of burning process, 12
    suppression of fires affected by products of, 21–22
Command process, 278–279, 282. *See also* Incident Command Systems (ICS) and Incident Management Systems (IMS)
command roles, training for, 317, 322, 336–338
commanders. *See* masters

solubility, 11–12
vapors evolving from, 10
list, 307, 311–312
Lloyd's Register of Shipping (LR), 59
loading/unloading operations
    Ambassador incident, 341, 354–355
    causing fires, 33, 34
    inspecting fuel transfer systems, 36
    shoreside personnel and, 38
    on tankers, 42–44
locating fires, 242–243
logging communications (record keeping), 50, 282, 323
loss control as part of overhaul operations, 257, 258
lounge inspection checklist, 53–54
lower flammable limit (LFL), 12, 96
low-light enhancement devices, 95

M

machinery spaces
    See also engine rooms
    attacking fires in, 293–295
    foam concentrates useful in fighting fires in, 179
    suppression systems in, 203, 224, 235–236
maintenance and repairs
    See also under specific equipment and vessel areas
    as element of fire prevention programs, 49–50
    as part of damage control, 309–311
manifests. *See* cargo manifests
Manulani (1997) incident, 39, 357
marine all-purpose nozzles, 171
masters
    defined, 32
    determining organizational structures and emergency response
        responsibilities, 273
    drill duties, 334
    emergency response responsibilities, 275
    fire prevention program responsibilities, 46
    harbourmasters (captains of the port), 38
    reviewing training, 319
    training exercise, 336–338
material safety data sheets (MSDSs), 280, 319, 320–321
materials. *See* cargo
measurements, definitions of important terms, 5–6
mechanical heat energy, 8
medical emergencies, 51
medium-expansion foam concentrates, 179–180, 183
merchant vessel masters. *See* masters
mess room inspection checklist, 53–54
metacentric height (GM), 306–307, 308, 313, 316
metals
    chemical reactions causing fires, 29
    extinguishers used in fires involving, 138, 144, 147
    heat conductivity of, 9
    reactive, 138
Model Procedures Guides, 350
monitoring. *See* inspections
monitoring equipment and environmental analysis instruments, 94–96, 238
monitors (foam fixed appliances), 236
mushrooming, defined, 10
muster lists, 264, 266, 343
mustering, 242, 258

N

negative heat balance, defined, 13
newton (N), defined, 6
NFPA (National Fire Protection Association) Standards
    750, 224
    1962, 163

1971, 76
1982, 91
nonregulated cargo, 32–33
nozzles. *See* hoses, nozzles
nuclear heat energy, 8

O

officer of the watch (OOW), 195, 197, 242, 336–338
officers, drill and training duties, 319, 322, 334. *See also specific titles of officers*
oil burners causing fires, 36
oils, static electricity produced by, 43
open areas on passenger vessels, 60–61
open-circuit SCBA, 79–80
organic materials, defined, 5
organizational structures and Incident Command Systems, 273–276, 350–351
orienting new crew members, 50–51, 319
outside screw and yoke (OS&Y) valves, 219
overhaul, 256–258, 288, 348
overheads (deckheads/ceilings), 10
oxidation, 3, 5, 7
oxygen
    as component of fire tetrahedron, 13
    excluding as extinguishment method, 22, 23–24, 138–139
    measuring, 95
oxygen-deficient atmospheres, SCBA used in, 82

P

PASS (personal alert safety systems), 77, 91–93
P-A-S-S procedure, 155
passageways on container vessels, 61
passenger vessels
    attacking fires on, 303–304
    crowd control, 279, 303
    overview and hazard potential, 60–61
personal alert safety systems (PASS), 77, 91–93
personal floatation devices (PFDs), 324
personal protective equipment (PPE)
    See also self-contained breathing apparatus (SCBA)
    clothing, 76–79
    defined, 75
    donning and doffing, 97–99
    insulation from static electricity, 43
    performance test, 337
    skill evolutions practice, 324
    wearing during emergency response, 244
    wearing during overhaul operations, 257
personnel accountability systems, 93–94, 245–246, 258, 343. *See also* search and rescue procedures
photoelectric smoke detection systems, 188, 189
piercing nozzles, 173, 299
piloted ignitions, defined, 14
piping
    in fire main systems, 233–234
    maintenance/repair covered in fire prevention programs, 49
    misuse, 159
    repairs as part of damage control, 310–311
    in suppression systems
        automatic sprinklers, 219, 220
        deck foam, 236
        halogenated agent, 211
        high-pressure water mist, 224
        steam smothering, 211
piston pumps (positive displacement fire pumps), 230–231
plume formation, defined, 15
pneumatic line and spot detectors, 193
positive displacement fire pumps, 230–232
positive heat balance, defined, 13

training

See also drills

as element of fire prevention programs, 47–48

evaluation, 327, 328, 333–334, 335–336, 338, 339

instructors

planning and preparation, 326, 330–331

presentation techniques, 331–333

requirements, 319, 322

self-evaluation, 333

joint exercises with shoreside firefighters, 351–352

lesson plans, 325–330

overview, 317

process, 318

records of, 323

requirements, 319–325

skill practice evolutions, 322, 323–324, 333, 337

using SCBA, 88–89

transferring cargo. *See* loading/unloading operations

Transport Dangerous Goods, Canada (TDG), 32

turnout gear, 76–79. *See also* personal protective equipment (PPE)

turret nozzles, 214

U

Underwriters Laboratories Inc., 144

Unified Command process, 349–350, 351

unified responses with shoreside firefighters, 345–346, 348–350, 351

Universe Explorer (1996) incident, 360–361

upper flammable limit (UFL), defined, 12

U.S. Code of Federal Regulations (CFR), 32, 42

utility hose, defined, 161

V

valve devices in hoses, 164–165

vapor density, 5, 11

vapor pressure, defined, 5

vaporization, 10, 303

vapors

See also gases

defined, 4

evolving from liquid fuels, 10

extinguishers used on fires involving, 136

during fuel transfer, 36

hot-work near, 39

ignited during cargo transfers, 43–44

measuring presence of, 96

suppressing using fluoroprotein foams, 178

vaportight fixtures, 31, 71

vapor-to-air mixture, 12

vehicle fires on ro/ro vessels and ferries, 63, 299–301

ventilation

backdraft potential affected by, 20–21

during cargo transfers, 44

compartment fires and, 18

CO_2 suppression systems and, 201, 204

as damage control, 310

fire development and, 18

during fire fighting procedures, 252–256

fire main pumps and, 232

fire protection implications, 68–70

fire spread and, 272

hot-work and, 40

importance of, 252

positive-pressure ventilation (PPV) used for extinguishment, 296–297

as priority in emergency response process, 278

as priority in *RECEO*, 287–288

shoreside fire fighting and, 341–342

spontaneous heating and, 7

techniques, 324

thermal layering of gases and, 19, 20

types of, 252–256

ventilation controlled fires, defined, 14

ventilation systems, 69, 70, 257, 263

Venturi effect, 253

vertical ventilation, 252, 253

vessel materials as causes of fires, 28–29

vessels

See also names of specific vessels and components

arrangement (layout) and fire control, 71–72

classification of, 59–60

construction, 59, 64–71, 263, 264, 272

types of, 60–64

visible warning devices, 197. *See also* alarms

volatility, defined, 12

volatility of bulk cargo, defined, 33

volute, 229

W

warning devices. *See* alarms

water

as coolant, 23, 225

dewatering, 252, 278, 308, 309

effects on fires, 225–226, 303

estimating quantities during prefire planning, 312–316

pressure, 226–227

properties, 225–226

water curtain nozzles, 173

water extinguishers, 134

water pressure, 226–227, 233

water seals on inert gas systems, 238

water-based suppression systems, 215–224

water-curtain protection, 173, 245

watertight doors, 65–68

weather decks as no-smoking areas, 27

weight, definitions of important terms, 5

welding hazards, 37, 39

wet chemical suppression systems, 215

wet water, 23

wheeled extinguishers, 141–142

wiring. *See* electrical fires

wood, extinguishers used on fires involving, 144

work spaces as no-smoking areas, 28

Y

yacht fires, 303

Z

zero visibility, thermal imaging devices used in, 96

zoning charts for automatic sprinkler suppression systems, 220

Indexed by Kari Kells.

COMMENT SHEET

DATE _____ NAME _____

ADDRESS _____

ORGANIZATION REPRESENTED _____

CHAPTER TITLE _____ NUMBER _____

SECTION/PARAGRAPH/FIGURE _____ PAGE _____

1. Proposal (include proposed wording or identification of wording to be deleted),
   OR PROPOSED FIGURE:

2. Statement of Problem and Substantiation for Proposal:

RETURN TO: IFSTA Editor   SIGNATURE _____
           Fire Protection Publications
           Oklahoma State University
           930 N. Willis
           Stillwater, OK 74078-8045

Use this sheet to make any suggestions, recommendations, or comments. We need your input to make the manuals as up to date as possible. Your help is appreciated. Use additional pages if necessary.

Your Training Connection.....

The International Fire Service Training Association

We have a free catalog describing hundreds of fire and emergency service training materials available from a convenient single source: the International Fire Service Training Association (IFSTA).

Choose from products including IFSTA manuals, IFSTA study guides, IFSTA curriculum packages, Fire Protection Publications manuals, books from other publishers, software, videos, and NFPA standards.

Contact us by phone, fax, U.S. mail, e-mail, internet web page, or personal visit.

Phone
1-800-654-4055

Fax
405-744-8204

U.S. mail
IFSTA, Fire Protection Publications
Oklahoma State University
930 North Willis
Stillwater, OK 74078-8045

E-mail
editors@ifstafpp.okstate.edu

Internet web page
www.ifsta.org

Personal visit
Call if you need directions!

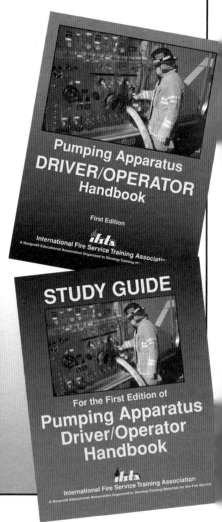